T0212770

Undergraduate Texts in Mathematics

Undergraduate Texts in Mathematics

Undergraduate Texts in Mathematics are generally aimed at third- and fourth-year undergraduate mathematics students at North American universities. These texts strive to provide students and teachers with new perspectives and novel approaches. The books include motivation that guides the reader to an appreciation of interrelations among different aspects of the subject. They feature examples that illustrate key concepts as well as exercises that strengthen understanding.

More information about this series at http://www.springer.com/series/666

Joseph H. Silverman • John T. Tate

Rational Points on Elliptic Curves

Second Edition

 Springer

Joseph H. Silverman
Department of Mathematics
Brown University
Providence, RI, USA

John T. Tate
Department of Mathematics
Harvard University
Cambridge, MA, USA

ISSN 0172-6056   ISSN 2197-5604   (electronic)
Undergraduate Texts in Mathematics
ISBN 978-3-319-30757-2   ISBN 978-3-319-18588-0   (eBook)
DOI 10.1007/978-3-319-18588-0

Springer Cham Heidelberg New York Dordrecht London

Printed on acid-free paper

Springer International Publishing AG Switzerland is part of Springer Science+Business Media (www.springer.com)

Preface

Preface to the Original 1992 Edition

In 1961 the second author delivered a series of lectures at Haverford College on the subject of "Rational Points on Cubic Curves." These lectures, intended for junior and senior mathematics majors, were recorded, transcribed, and printed in mimeograph form. Since that time, they have been widely distributed as photocopies of ever-decreasing legibility, and portions have appeared in various textbooks (Husemöller [25], Chahal [9]), but they have never appeared in their entirety. In view of the recent interest in the theory of elliptic curves for subjects ranging from cryptography (Lenstra [30], Koblitz [27]) to physics (Luck–Moussa–Waldschmidt [31]), as well as the tremendous amount of purely mathematical activity in this area, it seems a propitious time to publish an expanded version of those original notes suitable for presentation to an advanced undergraduate audience.

We have attempted to maintain much of the informality of the original Haverford lecturers. Our main goal in doing this has been to write a textbook in a technically difficult field that is "readable" by the average undergraduate mathematics major. We hope that we have succeeded in this goal. The most obvious drawback to such an approach is that we have not been entirely rigorous in all of our proofs. In particular, much of the foundational material on elliptic curves presented in Chapter 1 is meant to explain and convince, rather than to rigorously prove. Of course, the necessary algebraic geometry can mostly be developed in one moderately long chapter, as we have done in Appendix A. But the emphasis of this book is on number theoretic aspects of elliptic curves, so we feel that an informal approach to the underlying geometry is permissible, since it allows us more rapid access to the number theory. For those who wish to delve more deeply into the geometry, there are several good books on the theory of algebraic curves suitable for an undergraduate course, such as Reid [37], Walker [57], and Brieskorn–Knörrer [8]. In the later chapters we have generally provided all of the details for the proofs of the main theorems.

The original Haverford lectures make up Chapters 1, 2, 3, and the first two sections of Chapter 4. In a few places we have added a small amount of explanatory material, references have been updated to include some discoveries made since 1961, and a large number of exercises have been added. But those who have seen the original mimeographed notes will recognize that the changes have been kept to a minimum. In particular, the emphasis is still on proving (special cases of) the fundamental theorems in the subject: (1) the Nagell–Lutz theorem, which gives a precise procedure for finding all of the rational points of finite order on an elliptic curve; (2) Mordell's theorem, which says that the group of rational points on an elliptic curve is finitely generated; (3) a special case of Hasse's theorem, due to Gauss, which describes the number of points on an elliptic curve defined over a finite field.

In Section 4.4 we have described Lenstra's elliptic curve algorithm for factoring large integers. This is one of the recent applications of elliptic curves to the "real world," to wit, the attempt to break certain widely used public key ciphers. We have restricted ourselves to describing the factorization algorithm itself, since there have been many popular descriptions of the corresponding ciphers.[1]

Chapters 5 and 6 are new. Chapter 5 deals with integer points on elliptic curves. Section 5.2 is loosely based on an IAP undergraduate lecture given by the first author at MIT in 1983. The remaining sections of Chapter 5 contain a proof of a special case of Siegel's theorem, which asserts that an elliptic curve has only finitely many integral points. The proof, based on Thue's method of Diophantine approximation, is elementary, but intricate. However, in view of Vojta's [56] and Faltings' [15] recent spectacular applications of Diophantine approximation techniques, it seems appropriate to introduce this subject at an undergraduate level. Chapter 6 gives an introduction to the theory of complex multiplication. Elliptic curves with complex multiplication arise in many different contexts in number theory and in other areas of mathematics. The goal of Chapter 6 is to explain how points of finite order on elliptic curves with complex multiplication can be used to generate extension fields with Abelian Galois groups, much as roots of unity generate Abelian extensions of the rational numbers. For Chapter 6 only, we have assumed that the reader is familiar with the rudiments of field theory and Galois theory.

---

[1]That was what we said in the first edition, but in this second edition, we have included a discussion of elliptic curve cryptography; see Section 4.5.

Finally, we have included an appendix giving an introduction to projective geometry, with an especial emphasis on curves in the projective plane. The first three sections of Appendix A provide the background needed for reading the rest of the book. In Section A.4 of the appendix we give an elementary proof of Bezout's theorem, and in Section A.5, we provide a rigorous discussion of the reduction modulo p map and explain why it induces a homomorphism on the rational points of an elliptic curve.

The contents of this book should form a leisurely semester course, with some time left over for additional topics in either algebraic geometry or number theory. The first author has also used this material as a supplementary special topic at the end of an undergraduate course in modern algebra, covering Chapters 1, 2, and 4 (excluding Section 4.3) in about four weeks of class. We note that the last five chapters are essentially independent of one another (except Section 4.3 depends on the Nagell–Lutz theorem, proven in Chapter 2). This gives the instructor maximum freedom in choosing topics if time is short. It also allows students to read portions of the book on their own, e.g., as a suitable project for a reading course or honors thesis. We have included many exercises, ranging from easy calculations to published theorems. An exercise marked with a $(*)$ is likely to be somewhat challenging. An exercise marked with $(**)$ is either extremely difficult to solve with the material that we cover or is a currently unsolved problem.

It has been said that "it is possible to write endlessly on elliptic curves."[2] We heartily agree with this sentiment, but have attempted to resist succumbing to its blandishments. This is especially evident in our frequent decision to prove special cases of general theorems, even when only a few additional pages would be required to prove a more general result. Our goal throughout has been to illuminate the coherence and the beauty of the arithmetic theory of elliptic curves; we happily leave the task of being encyclopedic to the authors of more advanced monographs.

Preface to the 2015 Edition

The most important change to the new edition is the addition of two new sections. In Section 4.5 we briefly discuss how and why elliptic curves are used in modern cryptography, and in Section 6.6, we give an overview of how elliptic

---

[2]From the introduction to *Elliptic Curves: Diophantine Analysis*, Serge Lang, Springer-Verlag, New York, 1978. Professor Lang follows his assertion with the statement that "This is not a threat," indicating that he, too, has avoided the temptation to write a book of indefinite length.

curves play a key role in Wiles' proof of Fermat's Last Theorem. We have also taken the opportunity to make numerous corrections, both typographical and mathematical, to add a few new problems, and to update historical material to reflect some of the exciting advances of the past 25 years.

Electronic Resources

The interested reader will find additional material and a list of errata on the Rational Points on Elliptic Curves home page:

$$\texttt{www.math.brown.edu/~jhs/RPECHome.html}$$

This web page includes some of the numerical exercises in the book, allowing the reader to cut and paste them into other programs, rather than having to retype them.

There are now many commercial and free computer packages that perform calculations of varying levels of sophistication on elliptic curves,[3] including, for example,

Sage: `http://www.sagemath.org`
Pari/GP: `http://pari.math.u-bordeaux.fr`

No book is ever free from error or incapable of being improved. We would be delighted to receive comments, good or bad, and corrections from our readers. You can send mail to us at

$$\texttt{jhs@math.brown.edu}$$

Acknowledgments

First Edition, First Printing: The authors would like to thank Rob Gross, Emma Previato, Michael Rosen, Seth Padowitz, Chris Towse, Paul van Mulbregt, Eileen O'Sullivan, and the students of Math 153 (especially Jeff Achter and Jeff Humphrey) for reading and providing corrections to the original draft. They would also like to thank Davide Cervone for producing beautiful illustrations from their original jagged diagrams.

The first author owes a tremendous debt of gratitude to Susan for her patience and understanding, to Debby for her fluorescent attire brightening up

---

[3]This was not the case when the first edition of this book appeared in 1992, at which time the first author had created a small stand-alone application for Macintosh computers and a somewhat more highly featured set of routines for *Mathematica*. These antique packages are no longer available.

the days, to Danny for his unfailing good humor, and to Jonathan for taking timely naps during critical stages in the preparation of this manuscript.

The second author would like to thank Louis Solomon for the invitation to deliver the Philips Lectures at Haverford College in the Spring of 1961.

Providence, USA   Joseph H. Silverman
Cambridge, USA   John T. Tate
March 27, 1992

First Edition (Second Printing) and Second Edition: We, the authors, would like the thank the following individuals for sending comments and corrections: G. Allison, T. Anderson, P. Berman, D. Appleby, K. Bender, G. Bender, A. Berkovich, J. Blumenstein, P. de Boor, J. Brillhart, D. Clausen, S. Datta, Z. Fang, D. Freeman, L. Goldberg, F. Goldstein, A. Guth, D. Gupta, A. Granville, R. Hoibakk, I. Igusic, M. Kida, P. Kahn, J. Kraft, C. Levesque, B. Levin, J. Lipman, R. Lipes, A. Mazel-Gee, M. Mossinghoff, K. Nolish, B. Pelz, R. Pennington, R. Pries, A. Rajan, K. Ribet, M. Reid, H. Rose, L. Gómez-Sánchez, R. Schwartz, D. Schwein J.-P. Serre, M. Szydlo, L. Tartar, J. Tobey, R. Urian, C.R. Videla, J. Wendel, A. Ziv.

Providence, USA   Joseph H. Silverman
Cambridge, USA   John T. Tate
March 27, 2015

Contents

Introduction

The theory of Diophantine equations is that branch of number theory that deals with the solution of polynomial equations in either integers or rational numbers. The subject itself is named after one of the greatest of the ancient Greek algebraists, Diophantus of Alexandria,[4] who formulated and solved many such problems.

Most readers will undoubtedly be familiar with Fermat's Last Theorem. This theorem, which Fermat stated in the seventeenth century, says that if $n \geq 3$ is an integer, then the equation

$$X^n + Y^n = Z^n$$

has no solutions in nonzero integers X, Y, and Z. Equivalently, it asserts that the only solutions in rational numbers to the equation

$$x^n + y^n = 1$$

are those with either $x = 0$ or $y = 0$.[5]

---

[4]Diophantus lived sometime before the third century AD. He wrote the *Arithmetica*, a treatise on algebra and number theory in 13 volumes, of which 6 volumes have survived.

[5]In the first edition of this book in 1992, we noted that Fermat's Last Theorem was a conjecture, not a theorem. Fermat wrote his "theorem" as a marginal note in his copy of Diophantus' *Arithmetica*, but also wrote that the margin was unfortunately too small for him to write down the proof. And for 350 years, no one managed to find a proof. However, this all changed in 1995, when Andrew Wiles, with assistance from Richard Taylor on one point, proved Fermat's assertion [53, 60]. We will have more to say about Wiles' proof, which is intimately connected with the theory of elliptic curves, in Section 6.6.

As another example of a Diophantine equation, we consider the problem of writing an integer as the difference of a square and a cube. In other words, we fix an integer $c \in \mathbb{Z}$ and look for solutions to the Diophantine equation[6]

$$y^2 - x^3 = c.$$

Suppose that we are interested in solution in rational numbers $x, y \in \mathbb{Q}$. An amazing property of this equation is the existence of a *duplication formula*, discovered by Bachet in 1621. If (x, y) is a solution with x and y rational and $y \neq 0$, then it is not hard to check that the pair

$$\left(\frac{x^4 - 8cx}{4y^2}, \frac{-x^6 - 20cx^3 + 8c^2}{8y^3} \right)$$

is a solution in rational numbers to the same equation. Further, it is possible to prove, although Bachet was unable to do so, that if $c \notin \{1, -432\}$ and if the original solution satisfies $xy \neq 0$, then repeating this process leads to infinitely many distinct solutions. So except for 1 and -432, if an integer can be expressed as the difference of a square and a cube using nonzero rational numbers, then it can be so expressed in infinitely many ways. For example, if we start with the solution $(3, 5)$ to the equation

$$y^2 - x^3 = -2$$

and apply Bachet's duplication formula, we find a sequence of solutions that starts

$$(3, 5), \quad \left(\frac{129}{10^2}, -\frac{383}{10^3} \right), \quad \left(\frac{2340922881}{7660^2}, \frac{113259286337279}{7660^3} \right), \ldots$$

As you can see, the numerators and denominators rapidly become extremely large.

Next we'll take the same equation,

$$y^2 - x^3 = c,$$

and ask for solutions in integers $x, y, \in \mathbb{Z}$. In the 1650s Fermat posed as a challenge to the English mathematical community the problem of showing that the equation $y^2 - x^3 = -2$ has only two solutions in integers,

---

[6]This equation is sometimes called Bachet's equation, after the seventeenth-century mathematician who originally discovered the duplication formula. It is also known as Mordell's equation, in honor of the twentieth-century mathematician L.J. Mordell, who made fundamental contributions to the solution of this and many similar Diophantine equations. We will prove a special case of Mordell's theorem in Chapter 3.

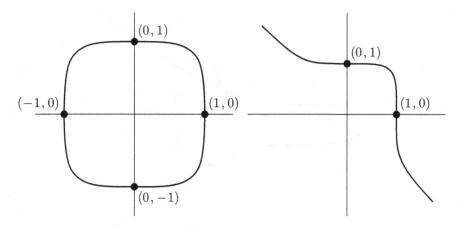

Figure 1: The Fermat curves $x^4 + y^4 = 1$ and $x^5 + y^5 = 1$

namely, $(3, \pm 5)$. This is in marked contrast to the question of solutions in rational numbers, since we have just seen that there are infinitely many of those. None of Fermat's contemporaries appears to have solved the problem, which was given an incomplete solution by Euler in the 1730s and a correct proof 150 years later! Then in 1908, Axel Thue[7] made a tremendous breakthrough; he showed that for any nonzero integer c, the equation $y^2 - x^3 = c$ has only finitely many solutions in integers x and y. This is a tremendous (qualitative) generalization of Fermat's challenge problem, since it says that among the potentially infinitely many solutions in rational numbers, only finitely many of them can be in integers.

The seventeenth century witnessed Descartes' introduction of coordinates into geometry, a revolutionary development that allowed geometric problems to be solved algebraically and algebraic problems to be studied geometrically. For example, if n is even, then the real solutions to Fermat's equation $x^n + y^n = 1$ in the xy-plane form a geometric object that looks like a squashed circle. Fermat's theorem is then equivalent to the assertion that the only points on that squashed circle having rational coordinates are the four points $(\pm 1, 0)$ and $(0, \pm 1)$. The Fermat equations with odd exponents look a bit different. We have illustrated the Fermat curves with exponents 4 and 5 in Figure 1.

---

[7]Axel Thue made important contributions to the theory of Diophantine equations, especially to the problem of showing that certain equations have only finitely many solutions in integers. These theorems about integer solutions were generalized by C.L. Siegel during the 1920s and 1930s. We will prove a version of the Thue–Siegel theorem, actually a special case of Thue's original result, in Chapter 5.

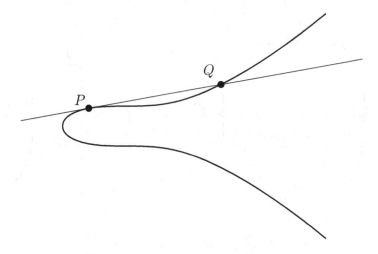

Figure 2: Bachet's equation $y^2 - x^3 = c$

Similarly, we can look at Bachet's equation $y^2 - x^3 = c$, which we have graphed in Figure 2. Recall that Bachet discovered a duplication formula which he used to take a given rational solution and produce a new rational solution. Bachet's formula is rather complicated, and one might wonder from whence it comes. The answer is that it comes from geometry! Thus suppose that we let $P = (x, y)$ be our original solution, so P is a point on the curve as illustrated in Figure 2. Next we draw the tangent line to the curve at the point P, an easy exercise for a first semester calculus course.[8] This tangent line will intersect the curve in one further point, which we have labeled Q. Then, if you work out the algebra to calculate the coordinates of Q, you will find Bachet's duplication formula. So Bachet's complicated algebraic formula has a simple geometric interpretation in terms of the intersection of a tangent line with a curve. This is our first intimation of the fruitful interplay that is possible among algebra, number theory, and geometry.

The simplest sort of Diophantine equation is a polynomial equation in one variable,

$$a_n x^n + a_{n-1} x^{n-1} + \cdots + a_1 x + a_0 = 0.$$

Assuming that $a_0, \ldots, a_n$ are integers, how can we find all integer and all rational solutions? Gauss' lemma provides a simple answer. If p/q is a rational solution written in lowest terms, then Gauss' lemma tells us that q divides a_n and p divides a_0. This gives us a small list of possible rational solutions, and

---

[8]Of course, Bachet had neither calculus nor analytic geometry, so he probably discovered his formula by clever algebraic manipulation.

we can substitute each of them into the equation to determine the actual solutions. So Diophantine equations in one variable are easy.[9]

When we move to Diophantine equations in two variables, the situation changes dramatically. Suppose we take a polynomial $f(x, y)$ with integer coefficients and look at the equation

$$f(x, y) = 0.$$

For example, Fermat's and Bachet's equations have this form. Here are some natural questions that we might ask:

(a) Are there any solutions in integers?
(b) Are there any solutions in rational numbers?
(c) Are there infinitely many solutions in integers?
(d) Are there infinitely many solutions in rational numbers?

In this generality, only question (c) has been fully answered, although much progress has recently been made on (d).[10]

The set of real solutions to an equation $f(x, y) = 0$ forms a curve in the xy-plane. Such curves are called *algebraic curves* to indicate that they are the set of solutions of a polynomial equation. In trying to answer questions (a)–(d), we might begin by looking at simple polynomials, such as polynomials of degree 1 (also called *linear polynomials*, because their graphs are straight lines). For a linear equation

$$ax + by = c$$

with integer coefficients, it is easy to answer our questions.[11] There are always infinitely many rational solutions, there are no integer solutions if $\gcd(a, b)$ does not divide c, and there are infinitely many integer solutions if $\gcd(a, b)$ does divide c. So linear equations in two variables are even easier to analyze than higher-degree equations in one variable.

---

[9]In practice, it may be easier to approximate the real roots to high accuracy and then check which, if any, of these roots can be written in the form b/a_n for some integer b. This avoids having to find the prime factorization of a_0 and a_n.

[10]For polynomials $f(x_1, \ldots, x_n)$ with more than two variables, our four questions have only been answered for some very special sorts of questions. Even worse, work of Davis, Matijasevič, and Robinson has shown that in general it is not possible to find a solution to question (a). That is, there does not exist an algorithm which takes as input the polynomial f and produces as output either YES or NO as an answer to question (a).

[11]We assume that a and b are not both zero, since if $a = b = 0$, there are either no solutions if $c \neq 0$, while every (x, y) is a solution if $c = 0$.

Next we turn to polynomials of degree 2, also called *quadratic polynomials*. Their graphs are conic sections. It turns out that if such an equation has one rational solution, then it has infinitely many. The complete set of solutions can be described very easily using geometry. We will briefly explain how this is done in Section 1.1. We will also briefly indicate how to answer question (b) for quadratic polynomials. So although it would be untrue to say that quadratic polynomials are easy, it is fair to say that their solutions are completely understood.

This brings us to the main topic of this book, namely, the solution of degree 3 polynomial equations in rational numbers and in integers. One example of such an equation is Bachet's equation $y^2 - x^3 = c$ that we looked at earlier. Some other examples that will appear during our studies are

$$y^2 = x^3 + ax^2 + bx + c \quad \text{and} \quad ax^3 + by^3 = c.$$

The solutions to these equations using real numbers are called *cubic curves* or *elliptic curves*.[12] In contrast to linear and quadratic equations, the rational and integer solutions to cubic equations are still not completely understood, and even in those cases where the complete answers are known, the proofs involve a subtle blend of techniques from algebra, number theory, and geometry. Our primary goal in this book is to introduce you to the beautiful subject of Diophantine equations by studying in depth the first case of such equations that is still imperfectly understood, namely, cubic equations in two variables. To give you an idea of the sorts of results that we will be studying, we briefly indicate what is known about questions (a)–(d) for cubic curves.

First, Siegel proved in the 1920s that a cubic equation has only finitely many integer solutions,[13] and in 1970 Baker and Coates gave an explicit upper bound for the largest solution in terms of the coefficients of the polynomials. This provides a satisfactory answer to (a) and (c), although the Baker–Coates bounds for the largest solution are generally too large to be practical.[14] In Chapter 5 we will prove a special case of Siegel's theorem for equations of the form $ax^3 + by^3 = c$.

---

[12]Despite its name, an elliptic curve is not an ellipse, since ellipses are conic sections, and conic sections are given by quadratic equations! The curious chain of events that led to elliptic curves being so named is recounted in Section 1.3.

[13]Actually, Siegel's theorem applies only to "nonsingular" cubic equations. However, most cubic equations are nonsingular, and in practice, it is generally quite easy to check whether a given equation is nonsingular.

[14]Techniques developed since 1970 are practical enough to find all integer solutions on many cubic equations, as long as the coefficients are not too large.

Second, all of the possibly infinitely many rational solutions to a cubic equation may be found by starting with a finite set of solutions and repeatedly applying a geometric procedure similar to Bachet's duplication formula. The fact that there always exists a finite generating set was suggested by Poincaré in 1901 and proven by L.J. Mordell in 1923. We will prove a special case of Mordell's theorem in Chapter 3. However, we must in truth point out that Mordell's theorem does not really answer questions (b) and (d). As we shall see, the proof of Mordell's theorem gives a procedure that *often* allows one to find a finite generating set for the set of rational solutions. But it is only conjectured, and not yet proven, that Mordell's method always yields a generating set. So even for special sorts of cubic equations such as $y^2 - x^3 = c$ and $ax^3 + by^3 = c$, there is no general method (algorithm) currently known that is guaranteed to answer question (b) or (d).

We have mentioned several times the idea that the study of Diophantine equations involves an interplay among algebra, number theory, and geometry. The geometric component is clear, since the equation itself defines (in the case of two variables) a curve in the plane, and we have already seen how it may be useful to consider the intersection of that curve with various lines. The number theory is also clearly present, since we are searching for solutions in either integers or rational numbers, and what is the heart of number theory other than the study of relations between integers and/or rational numbers. But what of the algebra? We could point out that polynomials are essentially algebraic objects. However, algebra plays a far more important role.

Recall that Bachet's duplication formula may be described as follows: start with a point P on a cubic curve, draw the tangent line at P, and take the third point of intersection of the line with the curve. Similarly, if we start with two points P_1 and P_2 on the curve, we can draw the line through P_1 and P_2 and look at the third intersection point P_3. This will work for most choices of P_1 and P_2, since most lines intersect a cubic curve in exactly three points. We might describe this procedure, which is illustrated in Figure 3, as a way to "add" two points on the curve and get a third point. Amazingly, it turns out that with a slight modification, this geometric operation turns the set of rational solutions to a cubic equation into an Abelian group! And Mordell's theorem, alluded to earlier, may be rephrased as saying that this group has a finite number of generators. So here is algebra, number theory, and geometry all packaged together in one of the greatest theorems of the twentieth century.

We hope that the preceding introduction has convinced you of some of the beauty and elegance to be found in the theory of Diophantine equations. But the study of Diophantine equations, in particular the theory of elliptic curves,

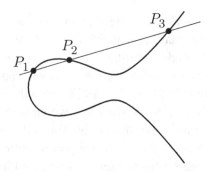

Figure 3: "Adding" two points on a cubic curve

also has its practical applications. We will study two such applications in this book.

Everyone is familiar with the Fundamental Theorem of Arithmetic, which asserts that every positive integer factors uniquely into a product of primes. However, if the integer is fairly large, say on the order of 10^{300} to 10^{600}, it may be virtually impossible in practice to perform that factorization. This is true even though there are quick ways to check if an integer of that size is not prime. In other words, if someone hands you a composite integer N having, say, 450 digits, then you can easily prove that N is not prime, even though you probably won't be able to find any prime factors of N. This curious state of affairs was used by Rivest, Shamir, and Adleman to construct the first practical and secure public key cryptosystem, called RSA. It then becomes of practical importance to find the best possible algorithms to factor large numbers. One such algorithm, which is particularly effective when N has factors of somewhat different magnitudes, is due to Hendrik Lenstra and uses elliptic curves defined over finite fields. We describe Lenstra's algorithm in Section 4.4.

Just as factoring large numbers is hard, it turns out that expressing a given point on an elliptic curve as a multiple of some other given point on the curve is hard, and indeed, based on current algorithms, it appears to be significantly harder than factoring. This is called the *elliptic curve discrete logarithm problem*, and it has been used as the basis for a public key cryptosystem that is, in some ways, more efficient than RSA due to the added difficulty of the underlying hard mathematical problem. We give a brief introduction to elliptic curve cryptography in Section 4.5.

Chapter 1

Geometry and Arithmetic

1.1 Rational Points on Conics

Everyone knows what a rational number is, a quotient of two integers. We call a point (x, y) in the plane a *rational point* if both of its coordinates are rational numbers. We call a line a *rational line* if the equation of the line can be written with rational numbers, that is, if it has an equation

$$ax + by + c = 0$$

with a, b, and c rational. Now it is pretty obvious that if you have two rational points, then the line through them is a rational line. And it is neither hard to guess nor hard to prove that if you have two rational lines, then the point where they intersect is a rational point. Equivalently, if you have two linear equations with rational numbers as coefficients and you solve them, you get rational numbers as answers.

The general subject of this book is rational points on curves, especially cubic curves. But as an introduction, we will start with conics. Let

$$ax^2 + bxy + cy^2 + dx + ey + f = 0$$

be a conic. We will say that the conic is *rational* if the coefficients of its equation are rational numbers.

Now what about the intersection of a rational line with a rational conic? Will it be true that the points of intersection are rational? By writing down some example, it is easy to see that the answer is, in general, no. If you use

© Springer International Publishing Switzerland 2015
J.H. Silverman, J.T. Tate, *Rational Points on Elliptic Curves*,
Undergraduate Texts in Mathematics, DOI 10.1007/978-3-319-18588-0_1

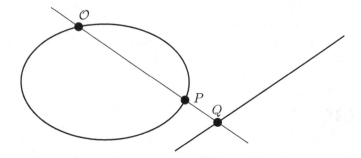

Figure 1.1: Projecting a conic onto a line

analytic geometry to find the coordinates of these points, you will come out with a quadratic equation for the x-coordinates of the intersection points. And if the conic is rational and the line is rational, the quadratic equation will have rational coefficients. So the two points of intersection will be rational if and only if the roots of that quadratic equation are rational. But in general, they might be conjugate quadratic irrationalities.

However, if one of the intersection points is rational, then so is the other. This is true because if a quadratic polynomial $ax^2 + bx + c$ with rational coefficients has one rational root, then the other root is rational, because the sum of the roots is $-b/a$. This very simple idea enables one to completely describe the rational points on a conic. Given a rational conic, the first question is whether or not there are any rational points on it. We will return to this question later, and we suppose for now that we know of one rational point $\mathcal{O}$ on our rational conic. Then we can get all of the rational points very simply. We just draw some rational line and project the conic onto the line from the point $\mathcal{O}$. (To project $\mathcal{O}$ itself onto the line, we use the tangent line to the conic at $\mathcal{O}$.)

A line meets a conic in two points, so for every point P on the conic we get a point Q on the line. Conversely, for every point Q on the line, by joining Q to the point $\mathcal{O}$, we get a point P on the conic. (See Figure 1.1.) In this way we get a one-to-one correspondence between the points on the conic and the points on the line.[1] But now you see by the remarks that we have made that if the point P on the conic has rational coordinates, then the

---

[1]More precisely, the is a one-to-one correspondence between the points of the line and all but one of the points of the conic. The missing point on the conic is the unique point $\mathcal{O}'$ on the conic such that the line connecting $\mathcal{O}$ and $\mathcal{O}'$ is parallel to the line onto which we are projecting. However, if we work in projective space and use homogeneous coordinates, then this problem disappears and we get a perfect one-to-one correspondence. See Appendix A for details.

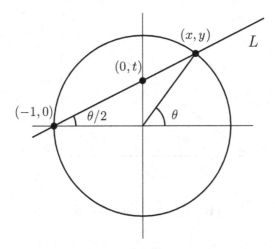

Figure 1.2: A rational parametrization of the circle

points Q on the line will have rational coordinates. And conversely, if Q is rational, then because $\mathcal{O}$ is assumed to be rational, the line through $\mathcal{O}$ and Q is rational and meets the conic in two points, one of which is rational. So the other point is rational, too. Thus the rational points on the conic are in one-to-one correspondence with the rational points on the line. Of course, the rational points on the line are easily described in terms of rational values of some parameter.

Let's carry out this procedure for the circle

$$x^2 + y^2 = 1.$$

We will project from the point $(-1, 0)$ onto the y-axis. Let's call the intersection point $(0, t)$; see Figure 1.2. If we know x and y, then we can easily find t. The equation of the line L connecting $(-1, 0)$ to $(0, t)$ is

$$y = t(1 + x).$$

The point (x, y) is assumed to be on the line L and also on the circle, so we get the relation

$$1 - x^2 = y^2 = t^2(1 + x)^2.$$

For a fixed value of t, this is a quadratic equation whose roots are the x-coordinates of the two intersections of the line L and the circle. Clearly $x = -1$ is a root, because the point $(-1, 0)$ is on both L and the circle. To find the other root, we cancel a factor of $1 + x$ from both sides of the equation. This gives the linear equation

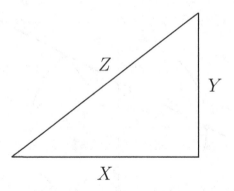

Figure 1.3: A right triangle

$$1 - x = t^2(1 + x).$$

Solving this for x in terms of t, and then using the relation $y = t(1 + x)$ to find y, we obtain

$$x = \frac{1 - t^2}{1 + t^2}, \qquad y = \frac{2t}{1 + t^2}. \qquad\qquad (*)$$

This is the familiar rational parametrization of the circle. And now the assertion made above is clear from these formulas. That is, if x and y are rational numbers, then $t = y/(1 + x)$ will be a rational number. And conversely, if t is a rational number, then it is obvious from the formulas $(*)$ that the coordinates x and y are rational numbers. So this is the way that you get rational points on a circle, simply plug in an arbitrary rational number for t. That will give you all points except $(-1, 0)$. (If you want to get $(-1, 0)$, then you must "substitute" infinity for t!)

These formula may be used to solve the elementary problem of describing all right triangles with integer sides. Let us consider the problem of finding some other triangles, besides 3, 4, 5, which have whole number sides. Let us call the lengths of the sides X, Y, Z; see Figure 1.3. That means we want to find all integers such that

$$X^2 + Y^2 = Z^2.$$

We first observe that if we have such integers where X, Y, and Z have a common factor, then we can take the common factor out. So we may as well assume that the three of them do not have any common factors. Right triangles whose integer sides have no common factor are called *primitive*. But then it follows that no two of the sides have a common factor, either. For example, if there is some prime dividing both Y and Z, the it would divide $X^2 = Z^2 - Y^2$, hence it would divide X, contrary to our assumption that X, Y, Z have no common factor. So if we make the trivial reduction to the case of primitive triangles, then no two of the sides have a common factor.

In particular, the point (x, y) defined by

$$x = \frac{X}{Z}, \qquad y = \frac{Y}{Z},$$

is a rational point on the circle $x^2 + y^2 = 1$. Further, the rational numbers are in lowest terms.

Since X and Y have no common factor, they cannot both be even. We claim that neither can they both be odd. The point is that the square of an odd number is congruent to 1 modulo 4. If X and Y were both odd, then $X^2 + Y^2$ would be congruent to 2 modulo 4. But $X^2 + Y^2 = Z^2$, and Z^2 is congruent to either 0 or 1 modulo 4. Therefore X and Y are not both odd, say X is odd and Y is even.

The point (x, y) is a rational point on the circle, so there is some rational number t so that x and y are given by the formulas (*) that we derived earlier. Write $t = m/n$ in lowest terms. Then

$$\frac{X}{Z} = x = \frac{n^2 - m^2}{n^2 + m^2}, \quad \frac{Y}{Z} = y = \frac{2mn}{n^2 + m^2}.$$

Since X/Z and Y/Z are in lowest terms, this means that there is some integer λ satisfying

$$\lambda Z = n^2 + m^2, \qquad \lambda Y = 2mn, \qquad \lambda X = n^2 - m^2.$$

We want to show that $\lambda = 1$. Because λ divides both $n^2 + m^2$ and $n^2 - m^2$, it divides their sum $2n^2$ and their difference $2m^2$. But m and n have no common divisors. Hence λ divides 2, so either $\lambda = 1$ or $\lambda = 2$. If $\lambda = 2$, then $n^2 - m^2 = \lambda X$ is divisible by 2, but not by 4, because we are assuming that X is odd. In other words, $n^2 - m^2$ is congruent to 2 modulo 4. But n^2 and m^2 are each congruent to either 0 or 1 modulo 4, so this is not possible. Hence $\lambda = 1$.

This proves that to get all primitive triangles, you take two relatively prime integers m and n, one odd and one even, and let

$$X = n^2 - m^2, \qquad Y = 2mn, \qquad Z = n^2 + m^2,$$

be the sides of the triangle. These are the ones with X odd and Y even. The others are obtained by interchanging X and Y.

The formulas have other uses. You may have met them in calculus. In Figure 1.2, we have

$$x = \cos\theta \quad \text{and} \quad y = \sin\theta, \quad \text{and so} \quad t = \tan\frac{1}{2}\theta = \frac{\sin\theta}{1 + \cos\theta}.$$

So the formulas (∗) given earlier allow us to express sine and cosine rationally in terms of the tangent of the half-angle:

$$x = \cos\theta = \frac{1 - t^2}{1 + t^2}, \qquad y = \sin\theta = \frac{2t}{1 + t^2}.$$

If you have some complicated identity in sine and cosine that you want to test, all that you have to do is substitute these formulas, collect powers of t, and see if you get zero.[2]

Another use comes from the observation that these formulas let us express all trigonometric functions of an angle θ as rational expressions in $t = \tan(\theta/2)$. We also note that

$$\theta = 2\arctan(t), \qquad d\theta = \frac{2\,dt}{1 + t^2}.$$

So if you have an integral that involves $\cos\theta$ and $\sin\theta$ and $d\theta$ and if you make the appropriate substitutions, then you can transform your integral into an integral in t and dt. If the integral is a rational function of $\sin\theta$ and $\cos\theta$, you come out with the integral of a rational function of t. Since rational functions can be integrated in terms of elementary functions, it follows that any rational function of $\sin\theta$ and $\cos\theta$ can be integrated in terms of elementary functions.

What if we take the circle

$$x^2 + y^2 = 3$$

and are asked to find the rational points on it? This is the easiest problem of all, because the answer is that there are none. It is impossible for the sum of two squares of rational numbers to equal 3. How can we see that it is impossible?

Suppose that there is a rational point and write it as

$$x = \frac{X}{Z} \quad \text{and} \quad y = \frac{Y}{Z}$$

---

[2]If they had told you this in high school, the whole business of trigonometric identities would have become a trivial exercise in algebra!

for some integers X, Y, and Z. Then

$$X^2 + Y^2 = 3Z^2.$$

If X, Y, Z have a common factor, then we may remove it, so we may assume that they have no common factor. It follows that neither X nor Y is divisible by 3. This is true because if 3 were to divide X, then 3 divides $Y^2 = 3Z^2 - X^2$, so 3 divides Y. But then 9 divides $X^2 + Y^2 = 3Z^2$, so 3 divides Z, contradicting the fact that X, Y, Z have no common factors. Hence 3 does not divide X, and a similar argument shows that 3 does not divide Y.

Since X and Y are not divisible by 3, we have

$$X \equiv \pm 1 \pmod 3 \quad \text{and} \quad Y \equiv \pm 1 \pmod 3,$$

and hence

$$X^2 + Y^2 \equiv 1 + 1 \equiv 2 \pmod 3.$$

However, we also have

$$X^2 + Y^2 = 3Z^2 \equiv 0 \pmod 3.$$

This contradiction shows that no two rational numbers have squares whose sum is 3.

We have seen by the projection argument that if you have one rational point on a rational conic, then all of the rational points on the conic may be described in terms of a rational parameter t. But how can we check whether there are any rational points? The argument that we gave for $x^2 + y^2 = 3$ provides a clue. We showed that this conic has no rational points by checking that a certain related equation has no solutions modulo 3.

There is a general method to test, in a finite number of steps, whether a given rational conic has a rational point. The method consists in checking whether a certain congruence has a solution. The theorem goes back to Legendre. Let us take a simple, but not trivial, case, and consider whether the equation

$$aX^2 + bY^2 = cZ^2$$

has a solution in integers. Legendre's theorem states that there is an integer m, depending in a simple fashion on a, b, and c, so that the above equation has a solution in integers, not all zero, if and only if it has a real solution with X, Y, Z not all zero and also the congruence

$$aX^2 + bY^2 \equiv cZ^2 \mod m$$

has a solution in integers that are relatively prime to m.

There is a more elegant way to state this theorem, due to Hasse:

> A homogeneous quadratic equation in several variables is solvable by integers, not all zero, if and only if it is solvable in real numbers and in p-adic numbers for each prime p.

Once one has Hasse's result, then one gets Legendre's theorem in a fairly elementary way. Legendre's theorem, combined with the work that we did earlier, provides a very satisfactory answer to the question of rational points on rational conics. So now we move on to cubics.

1.2 The Geometry of Cubic Curves

Now we are ready to begin our study of cubics. Let

$$ax^3 + bx^2y + cxy^2 + dy^3 + ex^2 + fxy + gy^2 + hx + iy + j = 0 \quad (**)$$

be the equation for a general cubic. We will say that a cubic is *rational* if the coefficients of its equation are rational numbers. A famous example is

$$x^3 + y^3 = 1,$$

or in homogeneous form,

$$X^3 + Y^3 = Z^3.$$

To find a rational solution of $x^3 + y^3 = 1$ amounts to finding integer solution of $X^3 + Y^3 = Z^3$, the first non-trivial case of Fermat's last theorem.

We cannot directly use the geometric principle that worked so well for conics because a line generally meets a cubic in three points. And if we have one rational point, we cannot project the cubic onto a line, because each point on the line would then correspond to two points on the curve.

But there is a geometric principle that we can use. If we can find two rational points on the curve, then we can generally find a third one. Namely, draw the line connecting the two points that you know. This will be a rational line, and it meets the cubic in one more point. If we look and see what happens when we try to find the three intersections of a rational line with a rational cubic, we find that we come out with a cubic equation with rational coefficients. If two of the roots of this equation are rational, then the third must be, too. So this gives a kind of composition law: Starting with two points P and Q, we draw the line through P and Q and let $P * Q$ denote the third point of intersection of the line with the cubic; see Figure 1.4

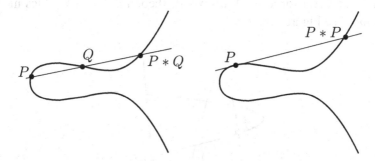

Figure 1.4: The composition of points on a cubic

Even if we only have one rational point P, we can still generally get another. Drawing the tangent line to the cubic at P, we are essentially drawing the line through P and P. The tangent line meets the cubic twice at P, and the same argument shows that the third intersection point is rational. Then we can draw lines through these new points and get more points. So if we start with a few rational points, then drawing lines and taking intersections will generally get us lots of others.

One of the main theorems that we want to prove in this book is the theorem of Mordell (1922) which states that if C is a non-singular rational cubic curve, then there is a *finite* set of rational points such that all other rational points can be obtained by repeatedly drawing lines and taking intersections. We will prove Mordell's theorem for a wide class of cubic curves, using only elementary number theory of the ordinary integers. The principle of the proof in the general case is the same, but requires some tools and facts from the theory of algebraic numbers.[3]

Mordell's theorem may be reformulated to be more enlightening. To do this, we first describe an elementary geometric property of cubics. We will not give a complete proof, but we will make it very plausible, which should suffice. (Further details are given in Appendix A.) In general, two cubic curves meet in nine points. To make this statement correct, one should first of all use

---

[3]For those who have studied some algebraic number theory, the required facts are the finiteness of the class group and the finite generation of the unit group in number fields.

the projective plane, which has extra points at infinity. Secondly, one should introduce multiplicities of intersections, counting points of tangency for example as intersections of multiplicity great than one. And finally, one must allow complex numbers for coordinates. We will ignore these technicalities. Then a curve of degree m and a curve of degree n meet in mn points. This is Bezout's theorem, one of the basic theorems in the theory of plane curves. (See Appendix A.4 for a proof a Bezout's theorem.) So two cubics meet in nine points; see Figure 1.5.

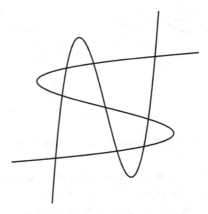

Figure 1.5: The intersection of two cubic curves

The theorem that we want to use is the following:

> Let C, C_1, and C_2 be cubic curves. Suppose that C goes through eight of the nine intersection points of C_1 and C_2. Then C goes through the ninth intersection point.

Why should this be true, at least in general? The trick is to consider the problem of constructing a cubic curve that goes through a certain number of points. To define a cubic curve (**), we have to give ten coefficients $a, b, c, d, e, f, g, h, i, j$. If we multiply all of the coefficients by a non-zero constant, then we get the same curve. So really the set of all possible cubics is, so to speak, nine dimensional. And if we want the cubic to go through a point whose coordinates are given, that imposes one linear condition on the coefficients of the cubic polynomial. The set of cubics that go through one given point is, so to speak, eight dimensional. Each time that we impose the condition that the cubic should contain another specified point, we impose another linear condition on the coefficients, which reduces by one the dimension of the set of all such cubics.[4] In particular, the family of all cubics that go through eight given intersection points $P_1, \ldots, P_8$ of C_1 and C_2 is a one-dimensional family.

Let $F_1(x, y) = 0$ and $F_2(x, y) = 0$ be the cubic equations giving C_1 and C_2. Then for every choice of numbers λ_1 and λ_2, the linear combination $\lambda_1 F_1 + \lambda_2 F_2$ is a cubic going through $P_1, \ldots, P_8$. Since there is only a one-dimensional family of such cubics, the set of cubics $\lambda_1 F_1 + \lambda_2 F_2$ must be that family. In particular, the cubic C is given by an equation $\lambda_1 F_1 + \lambda_2 F_2 = 0$ for a suitable choice of λ_1 and λ_2.

Now what about the ninth point P_9 in the intersection of C_1 and C_2? Since P_9 is on both C_1 and C_2, we know that $F_1(x, y)$ and $F_2(x, y)$ both vanish at P_9. It follows that $\lambda_1 F_1 + \lambda_2 F_2$ also vanishes at P_9, so C contains P_9.

In passing we mention that there is no known method that is guaranteed to determine, in a finite number of steps, whether a given rational cubic has a rational point. There is no analogue of Hasse's theorem for cubics. That question is still open, and it is a very important question. Even looking modulo m for all integers m is not sufficient. Selmer gave the example

$$3X^3 + 4Y^3 + 5Z^3 = 0.$$

This is a cubic, and Selmer showed by an ingenious argument that it has no integer solutions other than $(0, 0, 0)$. On the other hand, one can check that for every positive integer m, the congruence

$$3X^3 + 4Y^3 + 5Z^3 \equiv 0 \pmod{m}$$

has a solution in integers with no common factor. So for general cubics, the existence of a non-trivial solution modulo m for all m does not ensure that a rational solution exists. We put this difficult problem aside and assume henceforth that our cubic has a rational point, which we denote by $\mathcal{O}$.

We want to reformulate Mordell's theorem in a way that has great aesthetic and technical advantages. We have seen that if we have any two rational points on a rational cubic, say P and Q, then we can draw the line joining P to Q and obtain a third point that we denote $P * Q$. This has the flavor of many of the constructions that you have studied in modern algebra. If we consider the set of all rational points on the cubic, we can say that there is a law of composition that sends the pair (P, Q) to the point $P * Q$. What sort of algebraic structure does this composition law put on the set of rational points?

---

[4]Note that this is really just a plausibility argument; in order to make it rigorous, we would need to prove that each new linear condition is independent of the previous ones.

For example, is it a group law? Unfortunately, we do not get a group, since to start with, it is fairly clear that there is no identity element.

However, by playing around a bit, we can make the set of rational points into a group in such a way that the given rational point $\mathcal{O}$ becomes the identity element. We will denote the group law by $+$ because it is going to be a commutative group, but we stress that this new "cubic curve addition" has nothing to do with ordinary addition. The rule is as follows:

> To add P and Q, take the third intersection point $P * Q$, join it to $\mathcal{O}$ by a line, and then take the third intersection point to be $P + Q$. In other words, set $P + Q = \mathcal{O} * (P * Q)$.

The group law is illustrated in Figure 1.6, and the fact that $\mathcal{O}$ acts as the identity element is shown in Figure 1.7.

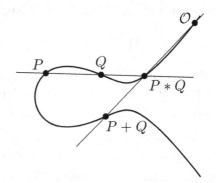

Figure 1.6: The group law on a cubic

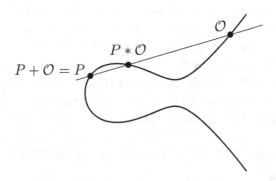

Figure 1.7: Verifying that $\mathcal{O}$ is the identity element

It is clear that this operation is commutative, that is,

$$P + Q = Q + P,$$

since the line through P and Q is the same as the line through Q and P, so $P * Q = Q * P$. We claim that also $P + \mathcal{O} = P$, so $\mathcal{O}$ acts as the identity element. Why is that? Well, if we join P to $\mathcal{O}$, then we get the point $P * \mathcal{O}$ as the third intersection point. Next we join $P * \mathcal{O}$ to $\mathcal{O}$ and take the third intersection point. That third intersection point is clearly P. So

$$P + \mathcal{O} = P.$$

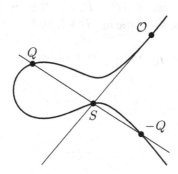

Figure 1.8: The negative of a point

It is a little harder to get inverses, but not very hard. Draw the tangent line to the cubic at $\mathcal{O}$, and let the tangent meet the cubic at the additional point S, i.e., $S = \mathcal{O} * \mathcal{O}$. (We are assuming that the cubic is non-singular, so there is a tangent line at every point.) Then given a point Q, we join Q to S, and the third intersection point $Q * S$ will be $-Q$; see Figure 1.8. To check that this is so, we add Q to $-Q$. To do this, we take the third intersection of the line through Q and $-Q$, which is S. Then we join S to $\mathcal{O}$ and take the third intersection point $S * \mathcal{O}$. But the line through S and $\mathcal{O}$ meets the cubic once at S and twice at $\mathcal{O}$, because it is tangent to the cubic at $\mathcal{O}$. (You must interpret things properly.) So the third intersection is the second time it meets the cubic at $\mathcal{O}$. Therefore

$$Q + (-Q) = \mathcal{O}.$$

If we only knew that $+$ was associative, then we would have a group. Let us try to prove the associative law. Let P, Q, and R be three points on the curve. We want to prove that

$$(P + Q) + R = P + (Q + R).$$

To get $P+Q$, we form $P*Q$ and take the third point of intersection of the line connecting $P*Q$ to $\mathcal{O}$. To add $P+Q$ to R, we draw the line through $P+Q$. That meets the curve at $(P + Q) * R$, so to get $(P + Q) + R$, we have to join $(P+Q)*R$ to $\mathcal{O}$ and take the third intersection. Now that does not show up too well in the picture, but to show $(P + Q) + R = P + (Q + R)$, it will be enough to show that $(P + Q) * R = P * (Q + R)$. To form $P * (Q + R)$, we have to find $Q * R$, join that to $\mathcal{O}$, and take the third intersection, which is $Q + R$. Then we must join $Q + R$ to P, which gives the point $P * (Q + R)$, and that is supposed to be the same as $(P + Q) * R$. In Figure 1.9, each of the points

$$\mathcal{O}, \ P, \ Q, \ R, \ P*Q, \ P+Q, \ Q*R, \ Q+R \tag{$\dagger$}$$

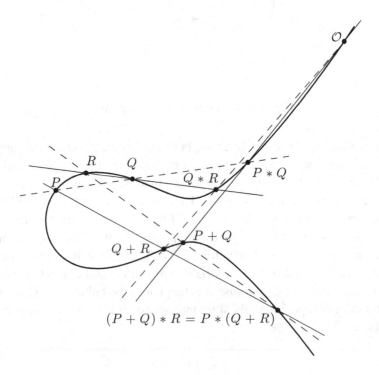

Figure 1.9: Verifying the associative law

lies on one of the dashed lines and one of the solid lines. Let us consider the dashed line through $P + Q$ and R and the solid line through P and $Q + R$. Does their intersection lie on the cubic? If so, then we will have proven that $P * (Q + R) = (P + Q) * R$.

We have nine points, namely the eight points listed in (†) and the intersection of the solid and dashed lines. So we have two (degenerate) cubics that go through the nine points, since a line has a linear equation, so if we have three linear equations and multiply them together, we get a cubic equation. The set of solutions to that cubic equation is just the union of the three lines. Now we apply our theorem, taking for C_1 the union of the three dashed lines and for C_2 the union of the three solid lines. By construction, the two cubics go through the nine points. But the original cubic curve C goes through the eight points given by (†), and therefore it also goes through the ninth. Thus the intersection of the two lines lies on C, which proves that $(P + Q) * R = P * (Q + R)$.

We will not do any more toward proving that the operation $+$ makes the points of C into a group. Later, when we have a normal form, we will have explicit formulas for adding points. So if our use of unproven assertions bothers you, then you can spend some time computing with those explicit formulas and verify directly that associativity holds.

We also want to mention that there is nothing special about our choice of $\mathcal{O}$. If we choose a different point $\mathcal{O}'$ to be the identity element of our group, then we get a group with exactly the same structure. In fact, the map

$$P \longmapsto P + \mathcal{O}'$$

is an isomorphism from the group $(C, \mathcal{O}, +)$ to the group $(C, \mathcal{O}', +')$, where the new addition law is defined by

$$P +' Q = P + Q - \mathcal{O}'.$$

Maybe we should explain that we have dodged some subtleties. If the line through P and Q is tangent to the curve at P, then the third point of intersection must be interpreted as P. And if you think of that tangent line as the line through P and P, then the third intersection is Q. Further, if P is a point of inflection on C, then the tangent line at P meets the curve three times at P. So in this case the third point of intersection for the line through P and P is again P. In other words, if P is an inflection point, then $P * P = P$. You just have to count intersections in the correct way, and it is clear why if you think of the points as varying a little bit. But to put everything on solid ground is a big task. If you are going into this business, it is important to start with better foundations and from a more general point of view. Then all these questions will be taken care of.

How does what we've done allow us to reformulate Mordell's theorem? Mordell's theorem says that we get all of the rational point by starting with a finite set of points, drawing lines through those points to get new points, then drawing lines through the new points to get yet more points, and so on. In terms of the group law, this says that the group of rational points is finitely generated. So we have the following statement of Mordell's theorem.

> **Mordell's Theorem**. If a non-singular rational plane cubic curve has a rational point, then the group of rational points is finitely generated.

This version is obviously technically a much better form because we can use a little elementary group theory, nothing very deep, but a convenient device in the proof.

1.3   Weierstrass Normal Form

We are going to prove Mordell's theorem as Mordell did, using explicit formulas for the addition law. To make these formulas as simple as possible, it is important to know that any cubic with a rational point can be transformed into a certain special form called *Weierstrass normal form*. We will not completely prove this, but we will give enough of an indication of the proof so that anyone who is familiar with projective geometry can carry out the details. (See Appendix A for an introduction to projective geometry.) Also, we will work out a specific example to illustrate the general theory. After that, we will restrict attention to cubics that are given in Weierstrass form, which classically consists of equations that look like

$$y^2 = 4x^3 - g_2x - g_3.$$

We will also use the slightly modified and more general equation

$$y^2 = x^3 + ax^2 + bx + c,$$

and we will call either of them Weierstrass form. What we need to show is that any cubic is, as one says, birationally equivalent to a cubic of this type. We now explain what this means, assuming that the reader knows a (very) little bit of projective geometry.

We start with a cubic curve, which we view as being in the projective plane. The idea is to choose axes in the projective plane so that the equation for the curve has a simple form. We assume that we are given a rational point $\mathcal{O}$ on C, so we begin by taking $Z = 0$ to be the tangent line to C at $\mathcal{O}$. This tangent line intersects C at one other point, and we take the $X = 0$ axis to be tangent to C at this new point. Finally, we choose $Y = 0$ to be any line (other than $Z = 0$) that goes through $\mathcal{O}$. See Figure 1.10.[5]

If we choose axes in this fashion and let $x = X/Z$ and $y = Y/Z$, then we get some linear conditions on the form that the equation will take in these coordinates. This is called a projective transformation. We will not work out the algebra, but will just tell you that at the end the equation for C takes the form

$$xy^2 + (ax + b)y = cx^2 + dx + e.$$

Next we multiply through by x,

$$(xy)^2 + (ax + b)xy = cx^3 + dx^2 + ex.$$

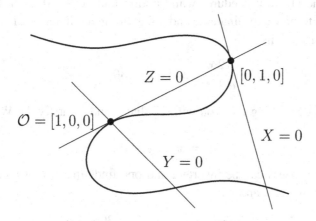

Figure 1.10: Choosing axes to put C into Weierstrass form

Now if we give a new name to xy, we will just call it y again, then we obtain

$$y^2 + (ax + b)y = \text{cubic in } x.$$

---

[5]We are assuming the $\mathcal{O}$ is not a point of inflection. Otherwise we can take $X = 0$ to be any line not containing $\mathcal{O}$.

Replacing y by $y - \frac{1}{2}(ax + b)$, which is another linear transformation, amounts to completing the square on the left-hand side of the equation, and we obtain

$$y^2 = \text{cubic in } x.$$

The cubic in x might not have leading coefficient 1, but we can adjust that by replacing x and y by λx and $\lambda^2 y$, where λ is the leading coefficient of the cubic. So we do finally get an equation in Weierstrass form. And if we want to get rid of the x^2 term in the cubic, we can replace x by $x - \alpha$ for an appropriate choice of α.

An example should make all of this clear.[6] Suppose that we start with a cubic of the form

$$u^3 + v^3 = \alpha,$$

where α is a given rational number. The homogeneous form of this equation is

$$U^3 + V^3 = \alpha W^3,$$

so in the projective plane this curve contains the rational point $[1, -1, 0]$. Applying the above procedure (while noting that $[1, -1, 0]$ is an inflection point) leads to new coordinates x and y that are given in terms of u and v by the rational functions

$$x = \frac{12\alpha}{u + v} \quad \text{and} \quad y = 36\alpha \frac{u - v}{u + v}.$$

If you work everything out, you will see that x and y satisfy the Weierstrass equation

$$y^2 = x^3 - 432\alpha^2.$$

Further, the process can be inverted, and one finds that u and v can be expressed in terms of x and y by

$$u = \frac{36\alpha + y}{6x} \quad \text{and} \quad v = \frac{36\alpha - y}{6x}.$$

Thus if we have a rational solution to $u^3 + v^3 = \alpha$, then we get rational x and y that satisfy the equation $y^2 = x^3 - 432\alpha^2$. And conversely, if we have a rational solution of $y^2 = x^3 - 432\alpha^2$, then we get rational numbers u and v satisfying $u^3 + v^3 = \alpha$. Of course, if $u = -v$, then the denominators in the expressions for x and y are zero, but there are only a finite number

---

[6]This example is somewhat special. For a more typical example with messier computations and larger numbers, see Appendix B.

of such exceptions, and they are easy to find. So the problem of finding rational points on $u^3 + v^3 = \alpha$ is the same as the problem of finding rational points on $y^2 = x^3 - 432\alpha^2$. And the general argument sketched above indicates that the same is true for any cubic. Of course, the normal form has an entirely different shape from the original equation. But there is a one-to-one correspondence between the rational points on one curve and the rational points on the other (up to a few easily catalogued exceptional points). So the problem of rational points on general cubic curves having one rational point is reduced to studying rational points on cubic curves in Weierstrass normal form.

The transformations that we used to put the curve in normalized form do not map straight lines to straight line. Since we defined the group law on our curve using lines connecting points, it is not at all clear that our transformation preserves the structure of the group. In other words, is our transformation a group homomorphism? It is, but that is not at all obvious. The point is that our description of addition of points on the curve is not a good one, because it seems to depend on the way that the curve is embedded in the plane. But in fact the addition law is an intrinsic operation that may be described on the curve and is invariant under birational transformations. This follows from basic facts about algebraic curves, but is not so easy (virtually impossible?) to prove simply by manipulating the explicit equations.

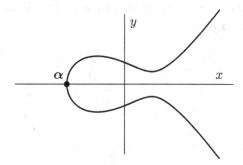

Figure 1.11: A cubic curve with one real component

A cubic equation in normal form looks like

$$y^2 = f(x) = x^3 + ax^2 + bx + c.$$

Assuming that the (complex) roots of $f(x)$ are distinct, such a curve is called an *elliptic curve*. (More generally, any curve that is birationally equation to such a curve is called an elliptic curve.) Where does this name come from, since these curves are certainly not ellipses? The answer is that these curves arose in studying the problem of how to compute the arc length of an ellipse. If one writes down the integral that gives the arc length of an elliptic and makes an elementary substitution, the integrand will involve the square root of a cubic or quartic polynomial. So to compute the arc-length of an ellipse, one integrates a function involving $y = \sqrt{f(x)}$, and the answer is given in terms of certain functions on the "elliptic" curve $y^2 = f(x)$.

Now we take the coefficients a, b, c of $f(x)$ to be rational, so in particular they are real. Hence the cubic polynomial $f(x)$ has at least one real root. In real numbers, we can factor it as

$$f(x) = (x - \alpha)(x^2 + \beta x + \gamma) \quad \text{with } \alpha, \beta, \gamma \text{ real.}$$

Of course, it might have three real roots. If it has one real root, the curve looks something like Figure 1.11, because $y = 0$ when $x = \alpha$. If $f(x)$ has three real roots, then the curve looks like Figure 1.12. In this case the real points form two connected components.

All of this is valid provided that the roots of $f(x)$ are distinct. What is the significance of that condition? We have been assuming all along that our cubic curve is non-singular. If we write the equation as $F(x, y) = y^2 - f(x) = 0$ and take partial derivatives,

$$\frac{\partial F}{\partial x} = -f'(x), \qquad \frac{\partial F}{\partial y} = 2y,$$

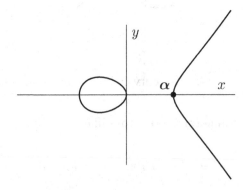

Figure 1.12: A cubic curve with two real components

then by definition the curve is non-singular provided that there is no point on the curve at which both partial derivatives simultaneously vanish. This will mean that every point on the curve has a well-defined tangent line. Now suppose that the partial derivatives were to vanish simultaneously at a point (x_0, y_0) on the curve. Then $y_0 = 0$, and hence $f(x_0) = y_0^2 = 0$, and also $f'(x_0) = 0$, so $f(x)$ and $f'(x)$ have the common root x_0. Thus x_0 is a double root of f. Conversely, if f has a double root x_0, then $(x_0, 0)$ is a singular point on the curve.

There are three possible pictures for the singularity. Which one occurs depends on whether f has a double root or triple root, and if a double root, whether the tangent directions are real or complex. In the case that f has a double root, typical equations are

$$y^2 = x^2(x + 1) \quad \text{and} \quad y^2 = x^2(x - 1).$$

The former curve has a singularity with distinct tangent directions as illustrated in Figure 1.13, while the latter has an isolated singular point at $(0, 0)$ as shown in Figure 1.14.[7]

If $f(x)$ has a triple root, then after translating x to put the triple root at 0, we obtain an equation

$$y^2 = x^3,$$

which is a semicubical parabola with a cusp at the origin as illustrated in Figure 1.15. These are examples of singular cubics in Weierstrass form, and the general case looks the same after a change of coordinates.

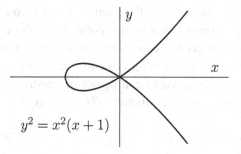

Figure 1.13: A singular cubic with distinct tangent directions

---

[7]To understand the curve $y^2 = x^2(x - 1)$, we should really draw its complex solutions in $\mathbb{C}^2$, in which case we would see that it has distinct complex tangent directions at $(0, 0)$.

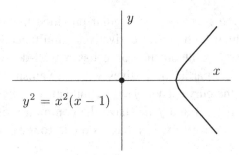

Figure 1.14: A singular cubic with an isolated singular point

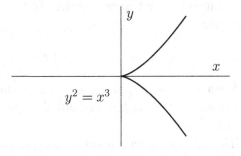

Figure 1.15: A singular cubic with a cusp

Why have we concentrated attention only on non-singular cubics? It is not just to be fussy. Singular cubics and non-singular cubics have completely different sorts of behavior. For instance, singular cubics are just as easy to treat as conics. If we project from the singular point onto some line, we see that the line going through that singular point meets the cubic twice at the singular point, so it meets the cubic only once more. The projection of a singular cubic curve onto a line is thus one-to-one. So just as for a conic, the rational points on a singular cubic can be put in one-to-one correspondence with the rational points on a line. In fact, it is very easy do so explicitly with formulas.

We illustrate with the singular cubic $y^2 = x^2(x+1)$. If we let $r = y/x$, then the equation becomes

$$r^2 = x + 1,$$

and hence

$$x = r^2 - 1 \quad \text{and} \quad y = rx = r^3 - r.$$

So if we take any rational number r and use these equations to define x and y, then we obtain a rational point on the cubic; and if we start with a rational point $(x, y) \neq (0, 0)$ on the cubic, we obtain a corresponding rational number $r = x/y$. These operations are inverses of each other and are defined at all rational points except the singular point $(0, 0)$. So in this way we get all rational points on the curve.

The curve $y^2 = x^3$ is even simpler. We just take

$$x = t^2 \quad \text{and} \quad y = t^3.$$

So the rational points on singular cubics are trivial to analyze, and Mordell's theorem does not hold for them. Actually, we have not yet explained how to get a group law for these singular curves, but if one avoids the singularity and uses the procedure that we described earlier, then one does get a group. We will study these singular groups in more detail at the end of Chapter 3, and in particular we will see that they are not finitely generated.

1.4 Explicit Formulas for the Group Law

We are going to look at the group of points on a non-singular cubic a little more closely. If you are familiar with projective geometry, then you will not have any trouble; and if not, then you will have to accept a point at infinity, but only one. (If you have never studied any projective geometry, you might also want to look at the first two sections of Appendix A.)

We start with the equation

$$y^2 = x^3 + ax^2 + bx + c$$

and make it homogeneous by setting $x = X/Z$ and $y = Y/Z$, yielding

$$Y^2 Z = X^3 + aX^2 Z + bXZ^2 + cZ^3.$$

What is the intersection of this cubic with the line at infinity $Z = 0$? Substituting $Z = 0$ into the equation gives $X^3 = 0$, which has a triple root $X = 0$. This means that the cubic meets the line at infinity in three points, but the three points are all the same! So a cubic has exactly one point at infinity, namely the point at infinity where vertical lines (that is, lines $x = $ constant) meet. The point at infinity is an inflection point of the cubic, the tangent line at that point is the line at infinity, and that tangent line meets the curve with multiplicity three. And one easily checks that the point at infinity is a non-singular point by looking at the partial derivatives there. So for a cubic in Weierstrass form, there is one point at infinity, and it is non-singular. We will call that point $\mathcal{O}$.

The point $\mathcal{O}$ is counted as a rational point, and we take it as the identity element when we make the set of points into a group. So to make the game work, we have to make the convention that the points on our cubic consist of the ordinary points in the ordinary affine xy-plane together with one other point $\mathcal{O}$ that you cannot see. And now we find that it is really true that every line meets the cubic in three points. Thus the line at infinity meets the cubic at the point $\mathcal{O}$ three times, vertical lines meet the cubic at two points in the xy-plane and also at the point $\mathcal{O}$, and non-vertical lines meet the cubic in three points in the xy-plane. (Of course, we may have to allow x and y to be complex numbers.)

Now we are going to discuss the group structure a little more closely. How do we add two points P and Q on a cubic equation in Weierstrass form? First we draw the line through P and Q and find the third intersection point $P * Q$. Then we draw the line through $P * Q$ and $\mathcal{O}$, which is just the vertical line through $P * Q$. A cubic curve in Weierstrass form is symmetric about the x-axis, so to find $P + Q$, we just take $P * Q$ and reflect it about the x-axis. This procedure is illustrated in Figure 1.16.

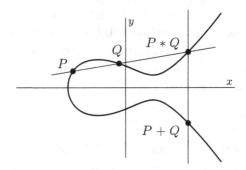

Figure 1.16: Adding points on a Weierstrass cubic

What is the negative of a point Q? The negative of Q is the reflected point, i.e., if $Q = (x, y)$, then $-Q = (x, -y)$; see Figure 1.17. To check this, suppose that we add Q to the point that we claim is $-Q$. The line through Q and $-Q$ is vertical, so the third point of intersection is $\mathcal{O}$. Now connect $\mathcal{O}$ to $\mathcal{O}$ and take the third intersection. Connecting $\mathcal{O}$ to $\mathcal{O}$ gives the line at infinity, and the third intersection is again $\mathcal{O}$. This shows that $Q + (-Q) = \mathcal{O}$, so $-Q$ is the negative of Q. Of course, this reasoning does not apply to the case $Q = \mathcal{O}$, but it is easy to see that $-\mathcal{O} = \mathcal{O}$. We also mention that if P, Q, R are distinct points, then $P + Q + R = \mathcal{O}$ if and only if P, Q, R are colinear.

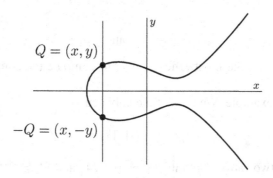

Figure 1.17: The negative of a point on a Weierstrass cubic

Now we develop some formulas to allow us to compute $P+Q$ efficiently. Let us change notation. We set

$$P_1 = (x_1, y_1), \quad P_2 = (x_2, y_2), \quad P_1 * P_2 = (x_3, y_3), \quad P_1 + P_2 = (x_3, -y_3);$$

see Figure 1.18. We assume that (x_1, y_1) and (x_2, y_2) are given, and we want to compute (x_3, y_3).

We first look at the equation of the line joining (x_1, y_1) to (x_2, y_2). This line has the equation

$$y = \lambda x + \nu, \quad \text{where} \quad \lambda = \frac{y_2 - y_1}{x_2 - x_1} \quad \text{and} \quad \nu = y_1 - \lambda x_1 = y_2 - \lambda x_2.$$

By construction, this line intersects the cubic in the two points (x_1, y_1) to (x_2, y_2). How do we get the third point of intersection? We substitute $y = \lambda x + \nu$ into the equation of the curve to obtain

$$y^2 = (\lambda x + \nu)^2 = x^3 + ax^2 + bx + c.$$

Putting everything to one side yields

$$0 = x^3 + (a - \lambda^2)x^2 + (b - 2\lambda\nu)x + (c - \nu^2).$$

This is a cubic equation in x, and its three roots x_1, x_2, x_3 give us the x-coordinates of the three intersection points. Thus

$$x^3 + (a - \lambda^2)x^2 + (b - 2\lambda\nu)x + (c - \nu^2) = (x - x_1)(x - x_2)(x - x_3).$$

Equating the coefficients of the x^2 term on either side, we find that

$$a - \lambda^2 = -x_1 - x_2 - x_3,$$

and so

$$x_3 = \lambda^2 - a - x_1 - x_2 \quad \text{and} \quad y_3 = \lambda x_3 + \nu.$$

These formulas are the most efficient way to compute the sum of two (distinct) points.

Let's do an example. We look at the cubic curve

$$y^2 = x^3 + 17,$$

which has the two rational points $P_1 = (-1, 4)$ and $P_2 = (2, 5)$. To compute $P_1 + P_2$, we find the line through P_1 and P_2. This is the line

$$y = \frac{1}{3}x + \frac{13}{3}, \quad \text{so} \quad \lambda = \frac{1}{3} \quad \text{and} \quad \nu = \frac{13}{3}.$$

Next

$$x_3 = \lambda^2 - x_1 - x_2 = -\frac{8}{9} \quad \text{and} \quad y_3 = \lambda x_3 + \nu = \frac{109}{27}.$$

Finally, we find that

$$P_1 + P_2 = (x_3, -y_3) = \left(-\frac{8}{9}, -\frac{109}{27} \right).$$

So doing computations really is not that bad.

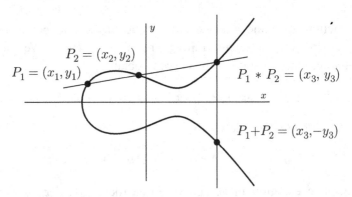

Figure 1.18: Deriving a formula for the addition law

The formulas that we have given for $P_1 + P_2$ involve the slope of the line connecting P_1 to P_2. What if the two points are the same? So suppose that we have $P_0 = (x_0, y_0)$ and we want to find $P_0 + P_0 = 2P_0$. We need to find the line joining P_0 to P_0. Because $x_1 = x_2$ and $y_1 = y_2$, we cannot use the slope formula $\lambda = \frac{y_2 - y_1}{x_2 - x_1}$. But the recipe that we described for adding a point to itself says that the line joining P_0 to P_0 is the tangent line to the cubic at P_0. From the relation $y^2 = f(x)$, we find by implicit differentiation that

$$\lambda = \frac{dy}{dx}\bigg|_{P_0} = \frac{f'(x_0)}{2y_0},$$

so that is what we use when we want to double the point $P_0 = (x_0, y_0)$.

Continuing with our example curve $y^2 = x^3 + 17$ and point $P_1 = (-1, 4)$, we compute $2P_1$ as follows. First, the slope of the tangent line is

$$\lambda = \frac{f'(x_1)}{2y_1} = \frac{f'(1)}{8} = \frac{3}{8}.$$

Then using the fact that the tangent line goes through P_1, we find that the tangent line is $y = \frac{3}{8}x + \frac{35}{8}$, so $\nu = \frac{35}{8}$. Finally using these values for λ and ν, we apply the formulas for x_3 and y_3 to eventually find that $2P_1 = \left(\frac{137}{64}, -\frac{2651}{512}\right)$.

Sometimes it is convenient to have an explicit expression for $2P$ in terms of the coordinates of P. If we substitute $\lambda = f'(x)/2y$ into our formulas, put everything over a common denominator, and replace y^2 by $f(x)$, then we find that

$$x\text{-coordinate of } 2(x, y) = \frac{x^4 - 2bx^2 - 8cx + b^2 - 4ac}{4x^3 + 4ax^2 + 4bx + 4c}.$$

This formula for $x(2P)$ is called the *duplication formula*. It will come in very handy later for both theoretical and computational purposes. We will leave it to you to verify the duplication formula, as well as to derive a companion formula for the y-coordinate of $2P$.

These are the basic formulas for the addition of points on a cubic when the cubic is in Weierstrass form. We will use these formulas extensively to prove many facts about rational points on cubic curves, including Mordell's theorem. Further, if you were not satisfied with our incomplete proof that the addition law is associative, you can just take three points at random and compute. Of course, there are a lot of special cases to consider, such as when one of the points is the negative of another or when two of the points coincide. But in a few days[8] you will be able to check associativity using these formulas. So we need say nothing more about the proof of the associative law!

---

[8] This tongue-in-cheek estimate of "a few days" was made back in the paper-and-pencil era of the 1960s. Although still tedious, the verification takes much less time now using a good computer algebra system.

Exercises

1.1. (a) If P and Q are distinct rational points in the xy-plane, prove that the line connecting them is a rational line.

(b) If L_1 and L_2 are distinct non-parallel rational lines in the xy-plane, prove that their intersection is a rational point.

1.2. Let C be the conic given by the equation

$$F(x, y) = ax^2 + bxy + cy^2 + dx + ey + f = 0,$$

and let δ be the determinant

$$\delta = \det \begin{pmatrix} 2a & b & d \\ b & 2c & e \\ d & e & 2f \end{pmatrix}.$$

(a) Show that if $\delta \neq 0$, then C has no singular points, i.e., show that there are no points (x, y) satisfying

$$F(x, y) = \frac{\partial F}{\partial x}(x, y) = \frac{\partial F}{\partial y}(x, y) = 0.$$

(b) Conversely, show that if $\delta = 0$ and $b^2 - 4ac \neq 0$, then there is a unique singular point on C.

(c) Let L be the line $y = \alpha x + \beta$ with $\alpha \neq 0$. Show that the intersection of L and C consists of either zero, one, or two points.

(d) Determine the conditions on the coefficients which ensure that the intersection $L \cap C$ consists of exactly one point. What is the geometric significance of these conditions. (Note that there will be more than one case to consider.)

1.3. Let C be the conic given by the equation

$$x^2 - 3xy + 2y^2 - x + 1 = 0.$$

(a) Check that C is non-singular. (Use Exercise 1.2.)

(b) Let L be the line $y = \alpha x + \beta$. Suppose that the intersection $L \cap C$ contains the point (x_0, y_0). Assuming that the intersection consists of two distinct points, find the second point of $L \cap C$ in terms of α, β, x_0, y_0.

(c) If L is a rational line and P_0 is a rational point, i.e., if $\alpha, \beta, x_0, y_0 \in \mathbb{Q}$, prove that the second point of $L \cap C$ is also a rational point.

1.4. Find all primitive integral right triangles whose hypotenuse has length less than 30.

1.5. Describe all rational points on the circle

$$x^2 + y^2 = 2$$

by projecting from the point $(1, 1)$ onto an appropriate rational line. (Your formulas will be simpler if you are clever in your choice of the line.)

1.6. (a) Let a, b, c, d, e, f be non-zero real numbers. Use the substitution $t = \tan(\theta/2)$ to transform the integral

$$\int \frac{a + b\cos\theta + c\sin\theta}{d + e\cos\theta + f\sin\theta}\, d\theta$$

into the integral of a rational function of t.

(b) Evaluate the integral

$$\int \frac{a + b\cos\theta + c\sin\theta}{1 + \cos\theta + \sin\theta}\, d\theta.$$

1.7. For each of the following conics, either find a rational point or prove that there are no rational points.

(a) $x^2 + y^2 = 6$

(b) $3x^2 + 5y^2 = 4$

(c) $3x^2 + 6y^2 = 4$

1.8. (a) Prove that for every exponent $k \geq 1$, the congruence

$$x^2 + 1 \equiv 0 \pmod{5^k}$$

has a solution $x_k \in \mathbb{Z}/5^k\mathbb{Z}$.

(b) Prove that the solutions in (a) can be chosen to satisfy

$$x_{k+1} \equiv x_k \pmod{5^k} \quad \text{for every } k \geq 1.$$

(c) Prove that if we require the list of solutions $x_1, x_2, x_3, \ldots$ to satisfy (b), then there are exactly two lists of solutions, the first being characterized by $x_1 \equiv 2 \pmod 5$ and the second by $x_1 \equiv 3 \pmod 5$.

Hint. Use induction on k. (This problem says that the equation $x^2 + 1 = 0$ has exactly two solutions in the 5-adic numbers. It is a special case of Hensel's lemma.)

1.9. Let C_1 and C_2 be the cubics given by the following equations:

$$C_1 : x^3 + 2y^3 - x - 2y = 0, \qquad C_2 = 2x^3 - y^3 - 2x + y = 0.$$

(a) Find the nine points of intersection of C_1 and C_2.

(b) Let $\{(0,0), P_1, \ldots, P_8\}$ be the nine points from (a). Prove directly that if a cubic curve goes through $P_1, \ldots, P_8$, then it must go through the ninth point $(0,0)$. (Do not simply quote the theorem in Section 1.2. This exercise is asking you to prove that theorem for particular curves C_1 and C_2.)

1.10. Define a composition law on the points of a cubic C by the following rules as described in the text: Given $P, Q \in C$, then $P * Q$ is the point on C so that P, Q, and $P * Q$ are colinear.

(a) Explain why this law is commutative, $P * Q = Q * P$.

(b) Prove that there is no identity element for this composition law, that is, prove that there is no point $P_0 \in C$ such that $P_0 * P = P$ for all $P \in C$.

(c) Prove that this composition law is not associative, that is, prove that in general
$P * (Q * R) \neq (P * Q) * R$.

(d) Explain why $P * (P * Q) = Q$.

(e) Suppose that the line through $\mathcal{O}$ and S is tangent to C at $\mathcal{O}$. Explain why

$$\mathcal{O} * \big(Q * (Q * S)\big) = \mathcal{O}.$$

This is an algebraic verification that the point that we called $-Q$ is the additive inverse of Q.

1.11. Let S be a set with a composition law $*$ having the following two properties:

(i) $P * Q = Q * P$ for all $P, Q \in S$.

(ii) $P * (P * Q) = Q$ for all $P, Q \in S$.

Fix an element $\mathcal{O} \in S$ and define a new composition law $+$ on S by the rule

$$P + Q = \mathcal{O} * (P * Q).$$

(a) Prove that $P + Q = Q + P$ and $P + \mathcal{O} = P$, i.e., prove that $+$ is commutative and that $\mathcal{O}$ serves as the identity element.

(b) Prove for any given $P, Q \in S$, the equation $X + P = Q$ has a unique solution in S, namely $X = P * (Q * \mathcal{O})$. In particular, if we define $-P$ to be $P * (\mathcal{O} * \mathcal{O})$, then $-P$ is the unique solution in S to the equation $X + P = \mathcal{O}$.

(c) Prove that $+$ is associative, and thus that $(S, +)$ is a group, if and only if

(iii) $R * (\mathcal{O} * (P * Q)) = P * (\mathcal{O} * (Q * R))$ for all $P, Q, R \in S$.

(d) Let $\mathcal{O}' \in S$ be another point, and define a composition law $+'$ by

$$P +' Q = \mathcal{O}' * (P * Q).$$

Assume that $+$ is associative. Prove that $+'$ is associative, so we obtain two group structures $(S, +)$ and $(S, +')$, and then prove that the map

$$P \longmapsto \mathcal{O} * (\mathcal{O}' * P)$$

is a group isomorphism from $(S, +)$ to $(S, +')$.

(e) * Find a set S with a composition law $*$ satisfying (i) and (ii) such that $(S, +)$ is not a group.

1.12. The cubic curve $u^3 + v^3 = \alpha$ (with $\alpha \neq 0$) has a rational point $[1, -1, 0]$ at infinity, i.e., this is the point on the homogenized equation $U^3 + V^3 = \alpha W^3$. Taking $[1, -1, 0]$ to be $\mathcal{O}$, we can make the points on the curve into a group.

(a) Derive a formula for the sum $P_1 + P_2$ of two distinct points $P_1 = (u_1, v_1)$ and $P_2 = (u_2, v_2)$.

(b) Derive a duplication formula for $2P$ in terms of $P = (u, v)$.

1.13. (a) Verify that if u and v satisfy the relation $u^3 + v^3 = \alpha$, then the quantities

$$x = \frac{12\alpha}{u+v} \quad \text{and} \quad y = 36\alpha \frac{u-v}{u+v}$$

satisfy the relation $y^2 = x^3 - 432\alpha^2$.

(b) Conversely, if x and y satisfy the relation $y^2 = x^3 - 432\alpha^2$, prove that the quantities

$$u = \frac{36\alpha + y}{6x} \quad \text{and} \quad v = \frac{36\alpha - y}{6x}$$

satisfy the relation $u^3 + v^3 = \alpha$.

(c) Prove that the maps in (a) and (b) are inverses, and hence give a birational transformation between the curves $u^3 + v^3 = \alpha$ and $y^2 = x^3 - 432\alpha^2$.

(d) Prove that this birational transformation is an isomorphism of groups, using the group law formulas for $u^3 + v^3 = \alpha$ that you derived in Exercise 1.12.

1.14. Let C be the cubic curve $u^3 + v^2 = u + v + 1$. In the projective plane, this curve has a point $[1, -1, 0]$ at infinity. Find rational functions $x = x(u, v)$ and $y(u, v)$ so that x and y satisfy a cubic equation C' in Weierstrass normal form and that define a birational transformation from C to C' sending $[1, -1, 0]$ to the point at infinity on C'.

1.15. Let $g(t)$ be a quartic polynomial, and let α be a root of $g(t)$. Let $\beta \neq 0$ be any number.

(a) Prove that the equations

$$x = \frac{\beta}{t-\alpha}, \qquad y = x^2 u = \frac{\beta^2 u}{(t-\alpha)^2}$$

give a birational transformation between the curve $u^2 = g(t)$ and the curve $y^2 = f(x)$, where $f(x)$ is the cubic polynomial

$$f(x) = g'(\alpha)\beta x^3 + \frac{1}{2}g''(\alpha)\beta^2 x^2 + \frac{1}{6}g'''(\alpha)\beta^3 x + \frac{1}{24}g''''(\alpha)\beta^4.$$

(b) Prove that if g has distinct (complex) roots, then f also has distinct roots, and so $u^2 = g(t)$ is an elliptic curve.

1.16. Let $0 < \beta \leq \alpha$, and let C be the ellipse

$$\frac{x^2}{\alpha^2} + \frac{y^2}{\beta^2} = 1.$$

(a) Prove that the arc length of C is given by the integral

$$4\alpha \int_0^{\pi/2} \sqrt{1 - k^2 \sin^2 \theta} \, d\theta$$

for an appropriate choice of the constant k depending on α and β.

(b) Check your value for k in (a) by verifying that when $\alpha = \beta$, the integral yields the correct value for the arc length of a circle.

(c) Prove that the integral in (a) is also equal to

$$4\alpha \int_0^1 \sqrt{\frac{1 - k^2 t^2}{1 - t^2}} \, dt = 4\alpha \int_0^1 \frac{1 - k^2 t^2}{\sqrt{(1 - t^2)(1 - k^2 t^2)}} \, dt.$$

(d) Prove that if the ellipse C is not a circle, then the equation

$$u^2 = (1 - t^2)(1 - k^2 t^2)$$

defines an elliptic curve, cf. Exercise 1.15. Hence the problem of determining the arc length of an ellipse comes down to evaluating the integral

$$\int_0^1 \frac{1 - k^2 t^2}{u} \, dt \quad \text{on the elliptic curve } u^2 = (1 - t^2)(1 - k^2 t^2).$$

And this is how elliptic curves received their unfortunate moniker!

1.17. Let C be a cubic curve in the projective plane given by the homogeneous equation
$$Y^2 Z = X^3 + aX^2 Z + bXZ^2 + cZ^3.$$

Verify that the point $[0, 1, 0]$ at infinity is a non-singular point of C.

1.18. The cubic curve
$$y^2 = x^3 + 17$$

has the following five rational points:

$$Q_1 = (-2, 3), \ Q_2 = (-1, 4), \ Q_3 = (2, 5), \ Q_4 = (4, 9), \ Q_5 = (8, 23).$$

(a) Show that Q_2, Q_4, and Q_5 can be expressed as $mQ_1 + nQ_3$ for appropriate choices of integers m and n.

(b) Compute the points

$$Q_6 = -Q_1 + 2Q_3 \quad \text{and} \quad Q_7 = 3Q_1 - Q_3.$$

(c) Notice that the points $Q_1, Q_2, Q_3, Q_4, Q_5, Q_6, Q_7$ and their inverses all have integer coordinates. There is exactly one more rational point on this curve that has integer coordinates and $y > 0$. Find that point.

(d) ** Prove the assertion in (c) that there are exactly eight rational points (x, y) on this curve with $y > 0$ and x and y both integers. (This is an extremely difficult problem, and you will almost certainly not be able to do it with the tools that we have developed. But it is also a very interesting problem that is well worth thinking about.)

1.19. Suppose that $P = (x, y)$ is a point on the cubic curve

$$y^2 = x^3 + ax^2 + bx + c.$$

(a) Verify that the x-coordinate of the point $2P$ is given by the *duplication formula*

$$x(2P) = \frac{x^4 - 2bx^2 - 8cx + b^2 - 4ac}{4x^3 + 4ax^2 + 4bx + 4c}.$$

(b) Derive a similar formula for the y-coordinate of $2P$ in terms of x and y.

(c) Find a polynomial in x whose roots are the x-coordinates of the point $P = (x, y)$ satisfying $3P = \mathcal{O}$. (*Hint.* The relation $3P = \mathcal{O}$ can also be written as $2P = -P$.)

(d) For the particular curve $y^2 = x^3 + 1$, solve the equation in (c) to find all points satisfying $3P = \mathcal{O}$. Note that you will need to use complex numbers.

1.20. Consider the point $P = (3, 8)$ on the cubic curve

$$y^2 = x^3 - 43x + 166.$$

Compute P, $2P$, $4P$, and $8P$. Comparing P to $8P$, what can you conclude?

1.21. Let $y^2 = f(x) = x^3 + ax^2 + bx + c$ be an elliptic curve in Weierstrass form.

(a) Prove that an alternative form for the duplication formula is

$$x(2P) = \frac{f'(x)^2 - (a + 2x)f(x)}{4f(x)}.$$

(b) Using (a), or some other method, prove that if $f(x)$ has distinct (complex) roots, then the numerator and the denominator of the formula

$$x(2P) = \frac{x^4 - 2bx^2 - 8cx + b^2 - 4ac}{4x^3 + 4ax^2 + 4bx + 4c}$$

have no common (complex) roots.

1.22. Let C and W be the projective curves

$$C : XY^2 + (aX + bZ)YZ = cX^2Z + dXZ^2 + eZ^3,$$
$$W : Y^2Z + (aX + bZ)YZ = cX^3 + dX^2Z + eXZ^2,$$

and let $\mathcal{O}$, P, and Q be the points on C given by

$$\mathcal{O} = [1, 0, 0], \quad P = [0, 1, 0], \quad Q = [0, e, b].$$

(We assume that e and b are not both zero, since otherwise C decomposes as the line $X = 0$ and the conic $Y^2 + aYZ = cXZ + dZ^2$.)

(a) In Section 1.3 we defined a map from C to W. Prove that under this map, the points on W corresponding to $\mathcal{O}, P, Q$ are the points

$$\mathcal{O}' = [0, 1, 0], \quad P' = [0, -b, 1], \quad Q' = [0, 0, 1].$$

(b) Write down conditions on the coefficients of C for it to be nonsingular at $\mathcal{O}$, P, and Q, and similarly write down conditions on the coefficients of W for it to be nonsingular at $\mathcal{O}'$, P', and Q'.

(c) Use (b) to show that $\mathcal{O}$, P, and Q are nonsingular points of C if and only if $\mathcal{O}'$, P', and Q', respectively, are nonsingular points of W.

(d) Let $R = [x, y, 1] \in C$ and $R' = [x, xy, 1] \in W$ with $x \neq 0$. Prove that R is a nonsingular point on C if and only if R' is a nonsingular point on W.

Chapter 2

Points of Finite Order

2.1 Points of Order Two and Three

An element P of any group is said to have *order* m if

$$mP = \underbrace{P + P + \cdots + P}_{m \text{ summands}} = \mathcal{O},$$

but $m'P \neq \mathcal{O}$ for all integers $1 \leq m' < m$. If such an m exists, then P has *finite order*, otherwise it has *infinite order*. We begin our study of points of finite order on cubic curves by looking at points of order two and order three. As usual, we will assume that our non-singular cubic curve is given by a Weierstrass equation

$$y^2 = f(x) = x^3 + ax^2 + bx + c,$$

and that the point at infinity $\mathcal{O}$ is taken to be the zero element for the group law.

Which points in our group satisfy $2P = \mathcal{O}$, but $P \neq \mathcal{O}$? Instead of $2P = \mathcal{O}$, it is easier to look at the equivalent condition $P = -P$. Since $-(x, y) = (x, -y)$, these are just the points with $y = 0$, i.e., the points

$$P_1 = (\alpha_1, 0), \quad P_2 = (\alpha_2, 0), \quad P_3 = (\alpha_3, 0),$$

where $\alpha_1, \alpha_2, \alpha_3$ are the (complex) roots of the cubic polynomial $f(x)$. So if we allow complex coordinates, there are exactly three points of order two, because the non-singularity of the curve ensures that $f(x)$ has distinct roots. If all three roots of $f(x)$ are real, then the picture looks like Figure 2.1.

© Springer International Publishing Switzerland 2015
J.H. Silverman, J.T. Tate, *Rational Points on Elliptic Curves*,
Undergraduate Texts in Mathematics, DOI 10.1007/978-3-319-18588-0_2

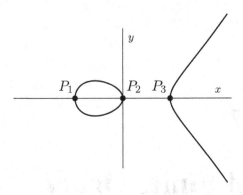

Figure 2.1: Points of order two

If we take all of the points satisfying $2P = \mathcal{O}$, including $P = \mathcal{O}$, then we get the set $\{\mathcal{O}, P_1, P_2, P_3\}$. It is easily seen that in any abelian group, the set of solutions to the equation $2P = \mathcal{O}$ forms a subgroup. (More generally, for any m, the set of solutions to $mP = \mathcal{O}$ forms a subgroup.) So we have an abelian group of order four, and since every element has order one or two, it is clear that this group is the *Four Group*, i.e., a direct product of two groups of order two. This means that the sum of any two of the points P_1, P_2, P_3 should equal the third, which is obvious from the fact that the three points are colinear. So now we know exactly what the group of points P such that $2P = \mathcal{O}$ looks like. If we allow complex coordinates, it is the Four Group. If we allow only real coordinates, it is either the Four Group or a cyclic group of order two, depending on whether $f(x)$ has three real roots or one real root. And if we restrict attention to points with rational coordinates, there are three possibilities, it is either the Four Group, cyclic of order two, or trivial, depending on whether $f(x)$ has three, one, or zero rational roots.

Next we look at the points of order three. Instead of $3P = \mathcal{O}$, we write $2P = -P$, so a point of order three will satisfy $x(2P) = x(-P) = x(P)$. Conversely, if $P \neq \mathcal{O}$ satisfies $x(2P) = x(P)$, then $2P = \pm P$, so either $P = \mathcal{O}$ (excluded by assumption) or $3P = \mathcal{O}$. In other words, the points of order three are exactly the points satisfying $x(2P) = x(P)$.

To find the points satisfying this condition, we use the duplication formula and set the x-coordinate of $2P$ equal to the x-coordinate of P. If we write $P = (x, y)$, then we have shown in Section 1.4 that the x-coordinate of $2P$ equals

$$\frac{x^4 - 2bx^2 - 8cx + b^2 - 4ac}{4x^3 + 4ax^2 + 4bx + 4c}.$$

Setting this expression equal to x, cross-multiplying, and doing a little algebra, we have completed a proof of part (c) of the following proposition.

Theorem 2.1 (Points of Order Two and Three). *Let C be a non-singular cubic curve*

$$C : y^2 = f(x) = x^3 + ax^2 + bx + c.$$

(Recall that C is non-singular provided $f(x)$ and $f'(x)$ have no common complex roots, or equivalently, if $f(x)$ does not have a double root.)

(a) *A point $P = (x, y) \neq \mathcal{O}$ on C has order two if and only if $y = 0$.*

(b) *The curve C has exactly four points of order dividing two. These four points form a group that is a product of two cyclic groups of order two.*

(c) *A point $P = (x, y) \neq \mathcal{O}$ on C has order three if and only if x is a root of the polynomial*

$$\psi_3(x) = 3x^4 + 4ax^3 + 6bx^2 + 12cx + 4ac - b^2.$$

(d) *The curve C has exactly nine points of order dividing three. These nine points form a group that is a product of two cyclic groups of order three.*

Proof. We proved (a) and (b) above, and we also proved (c) except for a little bit of algebra, which we will leave to you. We now give the proof of (d). Since the x-coordinate of $2P$ is equal to

$$x(2P) = \frac{f'(x)^2}{4f(x)} - a - 2x,$$

we see that an alternative expression for $\psi_3(x)$ is

$$\psi_3(x) = 2f(x)f''(x) - f'(x)^2.$$

We claim that $\psi_3(x)$ has four distinct (complex) roots. To verify this, we need to check that $\psi_3(x)$ and $\psi_3'(x)$ have no common roots. But

$$\psi_3'(x) = 2f(x)f'''(x) = 12f(x),$$

so a common root of $\psi_3(x)$ and $\psi_3'(x)$ would be a common root of

$$2f(x)f''(x) - f'(x)^2 \quad \text{and} \quad 12f(x),$$

and thus would be a common root of $f(x)$ and $f'(x)$. This contradicts our assumption that C is non-singular. We conclude that $\psi_3(x)$ indeed has four distinct complex roots.

Let $\beta_1, \beta_2, \beta_3, \beta_4$ be the four complex root of $\psi_3(x)$, and for each β_i, let δ_i be one of the square roots $\delta_i = \sqrt{f(\beta_i)}$. Then from (c), the set

$$\{(\beta_1, \pm\delta_1),\ (\beta_2, \pm\delta_2),\ (\beta_3, \pm\delta_3),\ (\beta_4, \pm\delta_4)\}$$

is the complete set of points of order three on C. Further, we observe that no δ_i can equal zero, because otherwise the point $(\beta_i, \delta_i) = (\beta_i, 0)$ would have order two, contradicting the fact that it has order three. Therefore the set contains eight distinct points, so C contains eight points of order three. The only other point on C with order dividing three is the point of order one, namely $\mathcal{O}$, which completes the proof that C has exactly nine points of order dividing three.

Finally, we note that there is only one (abelian) group with nine elements such that every element has order dividing three, namely the product of two cyclic groups of order three.   □

So we now know that if we allow complex numbers, then the points of order dividing three form a group of order nine that is the direct product of two cyclic groups of order three. It turns out that the real points of order three always form either a cyclic group of order three or the trivial group. We discuss this further in the next section.

There is also a nice geometric way to describe the points of order three. They are the inflection points on C, that is, the points where the tangent line to the cubic has a triple order contact. We can see this geometrically. The condition $2P = -P$ means that when we draw the tangent at the point P, then take the third intersection point and connect it with $\mathcal{O}$, we get $-P$. Now that is the case only if the third intersection of the tangent at P is the same point P. So $2P = -P$ if and only if P is a point of inflection. Of course, this can also be shown analytically. We leave the analytic proof as an exercise.

2.2   Real and Complex Points on Cubic Curves

The real points on our cubic curve

$$y^2 = f(x) = x^3 + ax^2 + bx + c \qquad\qquad (*)$$

form either one or two components, depending on whether $f(x)$ has one or three real roots. We illustrated the case of three real roots in Figure 2.1, and, of course, the case of one real root looks like Figure 2.2.

This picture shows the real points, that is, the points with real coordinates. Actually, the equation for our cubic curve defines several sets of points.

We write $C(\mathbb{Q})$ for the set of points on the curve whose coordinates happen to be rational,

$$C(\mathbb{Q}) = \{(x,y) \in C : x, y \in \mathbb{Q}\} \cup \{\mathcal{O}\}.$$

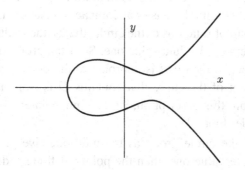

Figure 2.2: A cubic curve with one real component

Similarly, we write

$$C(\mathbb{R}) = \{(x,y) \in C : x, y \in \mathbb{R}\} \cup \{\mathcal{O}\}$$

for the set of points pictured in Figures 2.1 and 2.2 whose coordinates are arbitrary real numbers, and

$$C(\mathbb{C}) = \{(x,y) \in C : x, y \in \mathbb{C}\} \cup \{\mathcal{O}\}$$

for the set of pairs of complex numbers that satisfy the Weierstrass equation $(*)$. Note that we include the point $\mathcal{O}$ at infinity in all of these sets. In Section 1.4 we explained how to make the points on the curve C into a group. This construction was purely algebraic, so it works in any of these three cases.

Thus the points on the curve with complex coordinates form a group. The points with real coordinates form a subgroup because if two points have real coordinates, then so do their sum and difference. And since we are assuming that the coefficients a, b, c are rational numbers, it is even true that the rational points form a subgroup of the group of real points. So we have a big group and some subgroups,

$$\{\mathcal{O}\} \subset C(\mathbb{Q}) \subset C(\mathbb{R}) \subset C(\mathbb{C}).$$

One can use the methods of analysis to study the group of real points or complex points, and that is what we do in the rest of this section.

It is intuitively clear that the addition of real points on the curve is continuous, since if we move two points a little bit, the line connecting them and the third intersection point with C also move just a little bit. So the group of real points is a one-dimensional Lie group, and it is in fact compact, although it does not look it, because it has the point at infinity. There is only one such connected group. Any one-dimensional compact connected Lie group is isomorphic to the group of rotations of the circle, that is, the multiplicative group of complex numbers of absolute value one. So if the group of real points on the curve is connected, then it is isomorphic to the circle group, and in any case, the component of the curve that contains $\mathcal{O}$ is isomorphic to the circle group. And from this description, we can immediately see what the real points of finite order look like.

If we think of the circle group as the multiplicative group of complex numbers of absolute value one, then the points of finite order in that group are the roots of unity. And for each positive integer m, the points of order dividing m form a cyclic group of order m. Explicitly, this set of complex m'th roots of unity is

$$\left\{ 1, e^{2\pi i/m}, e^{4\pi i/m}, \ldots, e^{2(m-1)\pi i/m} \right\}.$$

So if $C(\mathbb{R})$ has one component, then the points of order dividing m in $C(\mathbb{R})$ form a cyclic group of order m.

If there are two connected components, then the group $C(\mathbb{R})$ is the direct product of the circle group with a group of order two. In this case, there are two possibilities for the points of order dividing m. If m is odd, we again get a cyclic group of order m, whereas if m is even, then we find the direct product of a cyclic group of order two and a cyclic group of order m.

In particular, we see that the real points of order dividing three always form a cyclic group of order three. Since we saw in Section 2.1 that there are eight complex points of order three, it is never possible for all of the complex points of order three to be real, and certainly they cannot all be rational. Notice that the x-coordinates of the points of order three are the roots of the quartic polynomial $\psi_3(x)$ described in Section 2.1. This quartic has real coefficients, so it has either zero, two, or four real roots. Since each x gives two possible values for y, this shows that the curve has either zero, four, or eight points of order three with real x-coordinate. However, our discussion shows that there must be exactly one real value of x for which the two corresponding y's are real. This can also be proven directly from the equations, a task that we leave for the exercises.

Before continuing with our discussion of rational points, we briefly digress to describe the structure of $C(\mathbb{C})$. Substituting $x - \frac{1}{3}a$ for x, we can eliminate the ax^2 term, and then replacing x and y by $4x$ and $4y$, we end up with the classical form of the Weierstrass equation,

$$y^2 = 4x^3 - g_2 x - g_3. \qquad (**)$$

As always, the cubic polynomial on the right is assumed to have distinct roots.

In the Weierstrass theory of elliptic functions, it is shown that whenever you have two complex numbers g_2 and g_3 so that the polynomial $4x^3 - g_2 x - g_3$ has distinct roots, i.e., such that $g_2^3 - 27 g_3^2 \neq 0$, then you can find complex numbers ω_1 and ω_2 called *periods* in the complex u-plane by evaluating certain definite integrals. These periods are $\mathbb{R}$-linearly independent, and one looks at the group formed by taking all of their $\mathbb{Z}$-linear combinations,

$$L = \mathbb{Z}\omega_1 + \mathbb{Z}\omega_2 = \{n_1\omega_1 + n_2\omega_2 : n_1, n_2 \in \mathbb{Z}\}.$$

Such subgroups of the complex plane are called *lattices*. Although there are many choices for the generators ω_1, ω_2 of L, it turns out that the coefficients g_2 and g_3 uniquely determine the group L itself. Conversely, the group L uniquely determines g_2 and g_3 via the formulas

$$g_2 = 60 \sum_{\substack{\omega \in L \\ \omega \neq 0}} \frac{1}{\omega^4} \quad \text{and} \quad g_3 = 140 \sum_{\substack{\omega \in L \\ \omega \neq 0}} \frac{1}{\omega^6}.$$

One also uses the periods to define a function $\wp(u)$ by the series

$$\wp(u) = \frac{1}{u^2} + \sum_{\substack{\omega \in L \\ \omega \neq 0}} \left(\frac{1}{(u-\omega)^2} - \frac{1}{\omega^2} \right).$$

This meromorphic function is called the *Weierstrass $\wp$-function*. It visibly has poles at the points of L, and no other poles in the complex u-plane. Less obvious is the fact that $\wp$ is doubly periodic, that is,

$$\wp(u + \omega_1) = \wp(u) \quad \text{and} \quad \wp(u + \omega_2) = \wp(u) \quad \text{for all complex numbers } u.$$

From this it follows that

$$\wp(u + \omega) = \wp(u) \quad \text{for all } u \in \mathbb{C} \text{ and all } \omega \in L.$$

Notice the similarity to trigonometric and exponential functions, which have single periods. For example, the function $f(u) = \sin(u)$ has period 2π, and the function $f(u) = e^u$ has period $2\pi i$.

One can show that this doubly periodic function $\wp(u)$ satisfies the differential equation

$$\wp'(u)^2 = 4\wp(u) - g_2\wp(u) - g_3.$$

Thus for every complex number u we get a point

$$P(u) = \big(\wp(u), \wp'(u)\big)$$

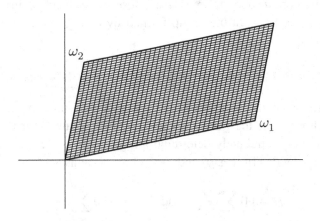

Figure 2.3: The period parallelogram

on the cubic curve $(**)$, albeit in general a point with complex coordinates. So we obtain a map from the complex u-plane to $C(\mathbb{C})$. (Of course, we send the points in L, which are poles of $\wp$ and $\wp'$, to the point $\mathcal{O}$ at infinity.)

The facts about this map are as follows. The map is onto the curve, i.e., every pair (x, y) of complex numbers satisfying $y^2 = 4x^3 - g_2x - g_3$ comes from some u. Because $\wp$ is doubly periodic, the map cannot be one-to-one. If u and v have the property that their difference $u - v$ equals $m_1\omega_1 + m_2\omega_2$ for some integers m_1 and m_2, i.e., if $u-v \in L$, then $P(u) = P(v)$. So instead we just look at values of u that lie in the *period parallelogram,* which is the parallelogram whose sides are the periods ω_1 and ω_2; see Figure 2.3. Then it is true that to a given point (x, y) on the curve there is exactly one u in the period parallelogram that is mapped to (x, y), provided that one makes suitable conventions about the boundary of the parallelogram.

Thus the period parallelogram is mapped one-to-one onto the complex points of the curve. The mapping $u \mapsto P(u)$ has the property

$$P(u_1 + u_2) = P(u_1) + P(u_2).$$

Note that the sum $u_1 + u_2$ is just ordinary addition of complex numbers, whereas $P(u_1) + P(u_2)$ is the addition law on the cubic curve. This equation amounts to the famous addition formula for $\wp(u)$. It says that the functions $\wp$ and $\wp'$, evaluated at $u_1 + u_2$, can be expressed rationally in terms of their values at u_1 and u_2. The formulas are the ones that we gave earlier in Section 1.4 expressing $(x_3, -y_3) = P_1 + P_2$ in terms of (x_1, y_1) and (x_2, y_2).

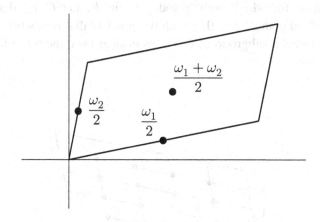

Figure 2.4: Points of order two on a complex torus

The mapping $u \mapsto P(u)$ is thus a homomorphism from the additive group of complex numbers onto the group of complex points of our cubic, and the kernel of that homomorphism is the lattice L that we considered earlier. The quotient group of the complex u-plane modulo the lattice L is isomorphic to the group of complex points on our curve, the isomorphism being given by convergent complex power series. Thus the group of complex points on our cubic is a torus, the direct product of two circle groups.

Using this description, we can describe the complex points of finite order. Suppose that we want a point of order two. This means we need a complex number $u \notin L$ such that $2u$ is in L. Looking modulo L, there are three such points,

$$\frac{\omega_1}{2}, \quad \frac{\omega_2}{2}, \quad \frac{\omega_1 + \omega_2}{2},$$

as illustrated in Figure 2.4

Similarly, to find the points of order dividing m, we look for complex numbers u in the period parallelogram such that $mu \in L$. The case $m = 5$ is illustrated in Figure 2.5. There are 25 such points in all, and it is clear that they form the direct product of two cyclic groups of order five. In general, the complex points of order dividing m form a group of order m^2 that is the direct product of two cyclic groups of order m. So over the complex numbers and over the real numbers, we have a very good description of the points of finite order on our cubic curve.

Before returning to the rational numbers, we briefly comment on other fields. If F is any subfield of the complex number and if the coefficients a, b, c of the cubic equation lie in F, then we can look at the set of solutions (x, y) of the equation for which both x and y lie in F. Let $C(F)$ denote this set of "F-valued points," together with the point $\mathcal{O}$ that is always included. Then $C(F)$ forms a subgroup of $C(\mathbb{C})$, as is clear from the formulas giving the addition law.

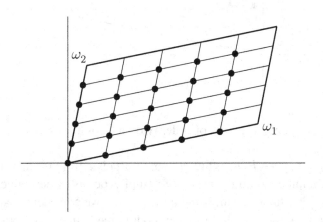

Figure 2.5: Points of order dividing five on a complex torus

More generally, there is no need in all of this to start with the field of complex numbers. All of our operations, such as the addition law, are purely algebraic. If, for instance, we take F to be the field of integers modulo p and take a, b, c to be elements of that field, then we can look for solutions of the equation in the finite field. Of course, there are only a finite number of solutions, since there are only finitely many possible values for x and y.

But again those solution, together with the point at infinity, form a group. Just use the formulas giving the addition law. You can't visualize it, but the formulas work perfectly well for any field.[1]

Because in this case the group is finite, we see that every point has finite order, but one can ask about points of various orders. It turns out that the points of order p form either a cyclic group of order p or the trivial group, but if q is some prime different from p, then the points of order q form either a trivial group, a cyclic group of order q, or the direct product of two cyclic groups of order q.

2.3 The Discriminant

After our digression into real and complex analysis, we return to the field of rational numbers. As always, we take our curve in its normal form

$$y^2 = f(x) = x^3 + ax^2 + bx + c,$$

where a, b, c are rational numbers. If we let $X = d^2 x$ and $Y = d^3 y$, then our equation becomes

$$Y^2 = X^3 + d^2 a X^2 + d^4 b X + d^6 c.$$

By choosing a large integer d, we can clear any denominators in a, b, and c. So from now on we will assume that our cubic curve is given by an equation having integer coefficients.

Our goal in this chapter is to prove a theorem, first proven (independently) by Nagell and Lutz in the 1930s, which will tell us how to find all of the rational points of finite order. Their theorem has two parts. The first part says that if (x, y) is a rational point of finite order, then its coordinates are integers. The second part says that either $y = 0$, in which case it is a point of order two, or else $y \mid D$, where D is the discriminant of the polynomial $f(x)$. In particular, a cubic curve has only a finite number of rational points of finite order.

---

[1]However, there are two caveats. First, as with the case of rational, real, or complex numbers, we must assume that the cubic polynomial $x^3 + ax^2 + bx + c$ does not have a double root in the algebraic closure of the finite field. Second, the formulas do not work for fields of characteristic 2. The problem occurs when we try to go from a general cubic equation to an equation of the form $y^2 = f(x)$. This transformation requires dividing by 2 and completing the square; see Section 1.3. To work with cubic equations in characteristic 2, one uses more general Weierstrass equations of the form $y^2 + a_1 xy + a_3 y = x^3 + a_2 x^2 + a_4 x + a_6$.

The *discriminant of $f(x)$* is the quantity

$$D = -4a^3c + a^2b^2 + 18abc - 4b^3 - 27c^2.$$

You may be familiar with this when $a = 0$, in which case $D = -4b^3 - 27c^2$. If we factor f over the complex numbers,

$$f(x) = (x - \alpha_1)(x - \alpha_2)(x - \alpha_3),$$

then one can check that

$$D = (\alpha_1 - \alpha_2)^2(\alpha_1 - \alpha_3)^2(\alpha_2 - \alpha_3)^2.$$

It follows that $D \neq 0$ if and only if the roots of $f(x)$ are distinct.

Thus using the Nagell–Lutz theorem, the question of finding the rational points of finite order can be settled in a finite number of steps. You take the integer D and consider each of the finitely many integers y that divide D. You take these y values and substitute them into the equation $y^2 = f(x)$. The polynomial $f(x)$ has integer coefficients and leading coefficient 1, so if it has an integer root, that root will divide the constant term. Thus there are only a finite number of things to check, and in this way we will be sure to find all the points of finite order in a finite number of steps.

Warning. We are not asserting that every point (x, y) with integer coordinates and $y \mid D$ must have finite order. The Nagell–Lutz theorem is not an "if and only if" statement.

If $f(x)$ is any polynomial with leading coefficient 1 in the ring $\mathbb{Z}[x]$ of polynomials with integer coefficients, then the discriminant of $f(x)$ will always be in the ideal of $\mathbb{Z}[x]$ generated by $f(x)$ and $f'(x)$. This follows from the general theory of discriminants, but for our particular polynomial $f(x) = x^3 + ax^2 + bx + c$, the quickest proof is just to write out an explicit formula:

$$\begin{aligned} D = {} & \Big((18b - 6a^2)x - (4a^3 - 15ab + 27c)\Big)f(x) \\ & + \Big((2a^2 - 6b)x^2 + (2a^3 - 7ab + 9c)x + (a^2b + 3ac - 4b^2)\Big)f'(x). \end{aligned}$$

We leave it to you to multiply this out and verify that it is correct. The important thing to remember is that there are polynomials $r(x)$ and $s(x)$ with *integer* coefficients so that D can be written as

$$D = r(x)f(x) + s(x)f'(x).$$

Why do we want this formula for D? If we assume the first part of the Nagell–Lutz theorem, namely that points of finite order have integer coordinates, then we can use the formula to prove the second part, i.e., that either $y = 0$ or $y \mid D$. More precisely, if P has finite order, then clearly $2P$ also has finite order, so the first part of the Nagell–Lutz theorem implies that both P and $2P$ have integer coordinates. Hence the second part of the Nagell–Lutz theorem follows from the next result.

Lemma 2.2. *Let $P = (x, y)$ be a point on our cubic curve such that both P and $2P$ have integer coordinates. Then either $y = 0$ or $y \mid D$.*

Proof. We assume that $y \neq 0$ and prove that $y \mid D$. Because $y \neq 0$, we know that $2P \neq \mathcal{O}$, so we may write $2P = (X, Y)$. By assumption, x, y, X, Y are all integers. The duplication formula says that

$$2x + X = \lambda^2 - a, \quad \text{where} \quad \lambda = \frac{f'(x)}{2y}.$$

Since x, X, and a are all integers and λ is a rational number, it follows that λ is also an integer. Since $2y$ and $f'(x)$ are integers, we see that $2y \mid f'(x)$, and in particular $y \mid f'(x)$. But $y^2 = f(x)$, so also $y \mid f(x)$. Now we use the relation

$$D = r(x)f(x) + s(x)f'(x).$$

The coefficients of r and s are integers, so $r(x)$ and $s(x)$ take on integer values when evaluated at the integer x. It follows that y divides D. $\square$

2.4 Points of Finite Order Have Integer Coordinates

Now we come to the most interesting part of the Nagell–Lutz theorem, the proof that a rational point (x, y) of finite order must have integer coordinates. We will show that x and y are integers in a rather indirect way. We observe that one way to show that a positive integer equals 1 is to show that it is not divisible by any primes. Thus we can break the problem up into an infinite number of subproblems, namely we show that when the rational numbers x and y are written in lowest terms, their denominators are not divisible by 2, and they are not divisible by 3, and they are not divisible by 5, and so on.

So we let p be some prime, and we try to show that p does not divide the denominator of x and does not divide the denominator of y. That leads us to consider the set of rational points (x, y) where p does divide the denominator of x or y.

It will be helpful to set some notation. Every non-zero rational number may be written uniquely in the form $\dfrac{m}{n}p^\nu$, where m and n are integers that are prime to p and with $n \geq 1$ and where the fraction m/n is in lowest terms. We define the *order* of such a rational number to be the exponent ν, and we write

$$\operatorname{ord}\left(\frac{m}{n}p^\nu\right) = \nu.$$

To say that p divides the denominator of a rational number is the same as saying that its order is negative, and similarly to say that p divides the numerator of a rational number is the same as saying that its order is positive. The order of a rational number is zero if and only if p divides neither its numerator nor its denominator.

Let us look at a rational point (x, y) on our cubic curve, where we assume that p divides the denominator of x. Thus

$$x = \frac{m}{np^\mu} \quad \text{and} \quad y = \frac{u}{wp^\sigma},$$

where $\mu > 0$ and where p does not divide m, n, u, or w. We plug this point into the equation for our cubic. Putting things over a common denominator, we find that

$$\frac{u^2}{w^2 p^{2\sigma}} = \frac{m^3 + am^2 np^\mu + bmn^2 p^{2\mu} + cn^3 p^{3\mu}}{n^3 p^{3\mu}}.$$

We know that $p \nmid u^2$ and $p \nmid w^2$, so

$$\operatorname{ord}\left(\frac{u^2}{w^2 p^{2\sigma}}\right) = -2\sigma.$$

Further, since $\mu > 0$ and $p \nmid m$, it follows that

$$p \nmid m^3 + am^2 np^\mu + bmn^2 p^{2\mu} + cn^3 p^{3\mu},$$

and hence

$$\operatorname{ord}\left(\frac{m^3 + am^2 np^\mu + bmn^2 p^{2\mu} + cn^3 p^{3\mu}}{n^3 p^{3\mu}}\right) = -3\mu.$$

Thus $2\sigma = 3\mu$. In particular, $\sigma > 0$, and so p divides the denominator of y. Further, the relation $2\sigma = 3\mu$ means that $2 \mid \mu$ and $3 \mid \sigma$, so we have $\mu = 2\nu$ and $\sigma = 3\nu$ for some integer $\nu > 0$.

Similarly if we assume that p divides the denominator of y, we find by the same calculation that the exact same result holds, namely that $\mu = 2\nu$

and $\sigma = 3\nu$ for some integer $\nu > 0$. Thus if p appears in the denominator of either x or y, then it is in the denominator of both of them, and in that case, the exact power is $p^{2\nu}$ for x and $p^{3\nu}$ for y for some positive integer ν.

This suggests that we make the following definition. We will let $C(p^\nu)$ be the set of rational points (x, y) of the cubic curve such that $p^{2\nu}$ divides the denominator of x and $p^{3\nu}$ divides the denominator of y. In other words,

$$C(p^\nu) = \{(x, y) \in C(\mathbb{Q}) : \text{ord}(x) \leq -2\nu \text{ and } \text{ord}(y) \leq -3\nu\}.$$

For instance, $C(p)$ is the set where p appears in the denominator of x and y, and then there is at least a p^2 in x and a p^3 in y. Obviously, we have inclusions

$$C(\mathbb{Q}) \supset C(p) \supset C(p^2) \supset C(p^3) \supset \cdots.$$

By convention, we also include the zero element $\mathcal{O}$ in every $C(p^\nu)$.

Recall that our objective is to show that if (x, y) is a point of finite order, then x and y are integers, and our strategy is to show that for every prime p, the denominators of x and y are not divisible by p. With our new notation, this means that we want to show that a point of finite order cannot lie in $C(p)$. The first step in showing this is to prove that each of the sets $C(p^\nu)$ is a subgroup of $C(\mathbb{Q})$.

Those of you who know about p-adic numbers will see that it makes good sense to consider this descending chain of subgroups. A high power of p in the denominator means, in the p-adic sense, that the number is very big. As we go down the chain of subgroups $C(p^\nu)$, we find points (x, y) with bigger and bigger coordinates in the p-adic sense. These are points that are getting closer and closer to infinity, and hence closer and closer to the zero element of our group. The $C(p^\nu)$'s are neighborhoods of $\mathcal{O}$ in the p-adic topology. But this is all by way of motivation, we will not actually need to know anything about p-adic numbers for the proof.

First we are going to change coordinates and move the point at infinity to a finite place. We will let

$$t = \frac{x}{y} \quad \text{and} \quad s = \frac{1}{y}.$$

Then $y^2 = x^3 + ax^2 + bx + c$ becomes

$$s = t^3 + at^2 s + bts^2 + cs^3$$

in the (t, s)-plane. We can always get back the old coordinates, of course, because $y = 1/s$ and $x = t/s$. In the (t, s)-plane, we have all of the points in the old (x, y)-plane except the points where $y = 0$, and the zero element $\mathcal{O}$ of our curve is now at the origin $(0, 0)$ in the (t, s)-plane.

You can visualize the situation this way. We have two views of the curve. The view in the (x, y)-plane shows us everything except $\mathcal{O}$. The view in the (t, s)-plane shows us $\mathcal{O}$ and everything except the points of order two. Other than $\mathcal{O}$ and the points of order two, there is a one-to-one correspondence between points on the curve in the (x, y)-plane and points on the curve in the (t, s)-plane; see Figure 2.6.

Further, a line $y = \lambda x + \nu$ in the (x, y)-plane corresponds to a line in the (t, s)-plane. Namely, if we divide $y = \lambda x + \nu$ by νy, we get

$$\frac{1}{\nu} = \frac{\lambda}{\nu}\frac{x}{y} + \frac{1}{y}, \quad \text{so} \quad s = -\frac{\lambda}{\nu}t + \frac{1}{\nu}.$$

Thus we can add points in the (t, s)-plane by the same procedure as in the (x, y)-plane. We need to find explicit formulas.

It is convenient to work in a certain ring which we denote by R, or by R_p if we want to stress that R depends on p. This ring R is the set of all rational numbers with no p in the denominator. Note that R is a ring, since if α and β are rational numbers with no p in their denominators, then the same is true of $\alpha \pm \beta$ and $\alpha\beta$.

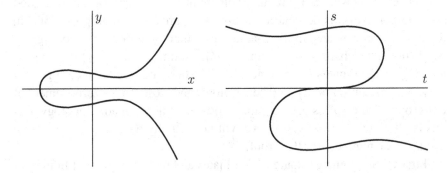

Figure 2.6: Two views of a cubic curve

Another way to describe R is to say that it consists of zero together with all non-zero rational numbers such that $\mathrm{ord}(x) \geq 0$, or if we make the convention that $\mathrm{ord}(0) = \infty$, then

$$R = \{\alpha \in \mathbb{Q} : \mathrm{ord}(\alpha) \geq 0\}.$$

The ring R is a subring of the field of rational numbers. It is a marvelous ring in the sense that it has unique factorization, and it has only one maximal ideal, which is the ideal generated by p. The units in R are just the rational numbers of order zero, that is, rational numbers with numerator and denominator prime to p.

Let's look at the divisibility of our new s and t coordinates by powers of p, in particular for points in $C(p)$. Let (x, y) be a rational point in the (x, y)-plane lying in $C(p^\nu)$, so we can write

$$x = \frac{m}{np^{2(\nu+i)}} \quad \text{and} \quad y = \frac{u}{wp^{3(\nu+i)}}$$

for some $i \geq 0$. Then

$$t = \frac{x}{y} = \frac{mw}{nu}p^{\nu+i} \quad \text{and} \quad s = \frac{1}{y} = \frac{w}{u}p^{3(\nu+i)}.$$

Thus our point (t, s) is in $C(p^\nu)$ if and only if $t \in p^\nu R$ and $s \in p^{3\nu} R$. This says that p^ν divides the numerator of t and $p^{3\nu}$ divides the numerator of s.

To prove that the $C(p^\nu)$'s are subgroups, we have to add points and show that if a certain power of p divides the t-coordinate of two points, then that power of p divides the t-coordinate of their sum. This is just a matter of writing down the formulas.

Let $P_1 = (t_1, s_1)$ and $P_2 = (t_2, s_2)$ be distinct points in $C(p^\nu)$. If $t_1 = t_2$, then the vertical line $t = t_1$ intersects C at P_1, P_2, and a third point $P_3 = (t_1, s_3)$, where P_3 may equal P_1 or P_2. Then $P_1 + P_2 = (-t_1, -s_3)$, so the t-coordinate of $P_1 + P_2$ is in $p^\nu R$, which shows that $P_1 + P_2 \in C(p^\nu)$.

So we are reduced to studying the case that $t_1 \neq t_2$. Let $s = \alpha t + \beta$ be the line through P_1 and P_2. The slope α of this line is given by

$$\alpha = \frac{s_2 - s_1}{t_2 - t_1}.$$

We can rewrite this as follows. The points (t_1, s_1) and (t_2, s_2) satisfy the equation

$$s = t^3 + at^2 s + bts^2 + cs^3.$$

Subtracting the equation for P_1 from the equation for P_2 and factoring gives

$$s_2 - s_1 = (t_2^3 - t_1^3) + a(t_2^2 s_2 - t_1^2 s_1) + b(t_2 s_2^2 - t_1 s_1^2) + c(s_2^3 - s_1^3)$$
$$= (t_2^3 - t_1^3) + a\left((t_2^2 - t_1^2)s_2 + t_1^2(s_2 - s_1)\right)$$
$$+ b\left((t_2 - t_1)s_2^2 + t_1(s_2^2 - s_1^2)\right) + c(s_2^3 - s_1^3).$$

Some of the terms are divisible by $s_2 - s_1$, and some of the terms are divisible by $t_2 - t_1$. Factoring these quantities out, we can express their ratio in terms of what is left, finding (after some calculation)

$$\alpha = \frac{s_2 - s_1}{t_2 - t_1} = \frac{t_2^2 + t_1 t_2 + t_1^2 + a(t_2 + t_1)s_2 + bs_2^2}{1 - at_1^2 - bt_1(s_2 + s_1) - c(s_2^2 + s_1 s_2 + s_1^2)}. \tag{†}$$

The point of all of this, as we will see, was to get the 1 in the denominator of α, so the denominator of α will be a unit in R.

Similarly, if $P_1 = P_2$, then the slope of the tangent line to C at P_1 is

$$\alpha = \frac{ds}{dt}(P_1) = \frac{3t_1^2 + 2at_1 s_1 + bs_1^2}{1 - at_1^2 - 2bt_1 s_1 - 3cs_1^2}.$$

Notice that this is the same slope that we get if we substitute $t_2 = t_1$ and $s_2 = s_1$ into the right-hand side of (†). So we may use (†) in all cases.

Let $P_3 = (t_3, s_3)$ be the third point of intersection of the line $s = \alpha t + \beta$ with the curve; see Figure 2.7. To get the equation whose roots are t_1, t_2, t_3, we substitute $\alpha t + \beta$ for s in the equation of the curve,

$$\alpha t + \beta = t^3 + at^2(\alpha t + \beta) + bt(\alpha t + \beta)^2 + c(\alpha t + \beta)^3.$$

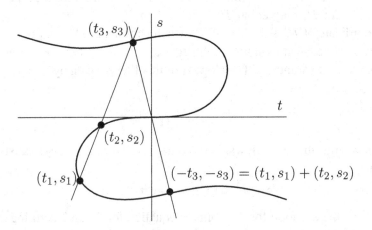

Figure 2.7: Adding points in the (t, s) plane

Multiplying this out and collecting powers of t gives

$$0 = (1 + a\alpha + b\alpha^2 + c\alpha^3)t^3 + (\alpha\beta + 2ba\beta + 3c\alpha^2\beta)t^2 + \cdots.$$

This equation has roots t_1, t_2, t_3, so the right-hand side equals

$$\text{constant} \cdot (t - t_1)(t - t_2)(t - t_3).$$

Comparing coefficients of t^3 and t^2, we find that the sum of the roots is

$$t_1 + t_2 + t_3 = -\frac{\alpha\beta + 2b\alpha\beta + 3c\alpha^2\beta}{1 + a\alpha + b\alpha^2 + c\alpha^3}.$$

These are all the formulas that we will need except for the trivial one

$$\beta = s_1 - \alpha t_1$$

saying that the line goes through P_1.

We now have a formula for t_3, so how do we find $P_1 + P_2$? We draw the line through (t_3, s_3) and the zero element $(0, 0)$ and take the third intersection point with the curve. It is clear at once from the equation of the curve that if (t, s) is on the curve, then so is $(-t, -s)$. So the third intersection point is $(-t_3, -s_3)$.

Let's look more closely at the expression for α. The numerator of α lies in $p^{2\nu}R$, because each of t_1, s_1, t_2, s_2 is in $p^\nu R$. For the same reason, the quantity

$$-at_1^2 - bt_1(s_2 + s_1) - c(s_2^2 + s_1s_2 + s_1^2)$$

is in $p^{2\nu}R$, so the denominator of α is a unit in R. And now you see why we wanted the 1 in the denominator. It follows that $\alpha \in p^{2\nu}R$.

Next, since $s_1 \in p^{3\nu}R$ and $\alpha \in p^{2\nu}R$ and $t_1 \in p^\nu R$, it follows from the formula $\beta = s_1 - \alpha t_1$ that $\beta \in p^{3\nu}R$. Further, we see that the denominator $1 + a\alpha + b\alpha^2 + c\alpha^3$ of $t_1 + t_2 + t_3$ is a unit in R. Looking at the expression for $t_1 + t_2 + t_3$ given above, we have

$$t_1 + t_2 + t_3 \in p^{3\nu}R.$$

Since $t_1, t_2 \in p^\nu R$, it follows that $t_3 \in p^\nu R$, and hence also that $-t_3 \in p^\nu R$.

This proves that if the t-coordinates of P_1 and P_2 lie in $p^\nu R$, i.e., if P_1 and P_2 are in $C(p^\nu)$, then the t-coordinate of $P_1 + P_2$ also lies in $p^\nu R$. Further, if the t-coordinate of $P = (t, s)$ lies in $p^\nu R$, then it is clear that the t-coordinate of $-P = (-t, -s)$ also lies in $p^\nu R$. This shows that $C(p^\nu)$ is closed under addition and taking negatives, hence it is a subgroup of $C(\mathbb{Q})$.

In fact, we have proven something a bit stronger. We have shown that if $P_1, P_2 \in C(p^\nu)$, then

$$t(P_1) + t(P_2) - t(P_1 + P_2) \in p^{3\nu}R.$$

Here we are writing $t(P)$ to denote the t-coordinate of P, so if P is given in (x, y)-coordinates as $(x(P), y(P))$, then $t(P) = x(P)/y(P)$.

This last formula tells us more than the mere fact that $C(p^\nu)$ is a subgroup of $C(\mathbb{Q})$. A more suggestive way to write it is

$$t(P_1 + P_2) \equiv t(P_1) + t(P_2) \pmod{p^{3\nu} R}.$$

Note that the $+$ in $P_1 + P_2$ is the addition on our cubic curve, which is given by quite complicated formulas, whereas the $+$ in $t(P_1) + t(P_2)$ is addition in R, which is just addition of rational numbers. So the map $P \longmapsto t(P)$ is practically a homomorphism from $C(p^\nu)$ into the additive group of rational numbers. It does not quite define a homomorphism, because $t(P_1 + P_2)$ is not actually equal to $t(P_1) + t(P_2)$. However, what we do get is a homomorphism from $C(p^\nu)$ to the quotient group $p^\nu R / p^{3\nu} R$ by sending P to the congruence class of $t(P)$, and the kernel of this homomorphism consists of all points P with $t(P) \in p^{3\nu} R$. Thus the kernel is $C(p^{3\nu})$, so we finally obtain a one-to-one homomorphism

$$\frac{C(p^\nu)}{C(p^{3\nu})} \longrightarrow \frac{p^\nu R}{p^{3\nu} R},$$

$$P = (x, y) \longmapsto t(P) = \frac{x}{y}.$$

It is not hard to see that the quotient group $p^\nu R / p^{3\nu} R$ is a cyclic group of order $p^{2\nu}$. It follows that the quotient group $C(p^\nu)/C(p^{3\nu})$ is a cyclic group of order p^σ for some $0 \le \sigma \le 2\nu$. We summarize our results so far in the following proposition.

Proposition 2.3. *Let p be a prime, let R be the ring of rational numbers with denominator prime to p, and let $C(p^\nu)$ be the set of rational points (x, y) on our curve for which x has denominator divisible by $p^{2\nu}$, together with the point $\mathcal{O}$.*

(a) *$C(p)$ consists of all rational points (x, y) for which the denominator of either x or y is divisible by p.*

(b) *For every $\nu \ge 1$, the set $C(p^\nu)$ is a subgroup of the group of rational points $C(\mathbb{Q})$.*

(c) *The map*

$$\frac{C(p^\nu)}{C(p^{3\nu})} \longrightarrow \frac{p^\nu R}{p^{3\nu} R}, \qquad P = (x, y) \longmapsto t(P) = \frac{x}{y},$$

is a one-to-one homomorphism. (By convention, we send $\mathcal{O} \mapsto 0$.)

Using this proposition, it is not hard to prove our claim that points of finite order have integers coordinates.

Corollary 2.4. (a) *For every prime p, the only point of finite order in the group $C(p)$ is the identity point $\mathcal{O}$.*
(b) *Let $P = (x, y) \in C(\mathbb{Q})$ be a rational point of finite order. Then x and y are integers.*

Proof. Let $P \in C(\mathbb{Q})$ be a point of order m with $m \geq 2$. Take any prime p. We need to show that $P \notin C(p)$. Suppose to the contrary that $P \in C(p)$. We will derive a contradiction.

The point $P = (x, y)$ may be contained in a smaller group $C(p^\nu)$, but it cannot be contained in all of the groups $C(p^\nu)$, because the denominator of x cannot be divisible by arbitrarily high powers of p. So we can find some $\nu > 0$ so that $P \in C(p^\nu)$ and $P \notin C(p^{\nu+1})$, specifically $\nu = -\frac{1}{2} \operatorname{ord}(x)$. We separate the proof into two cases depending on whether m is divisible by p.

Suppose first that $p \nmid m$. Repeated application of the congruence

$$t(P_1 + P_2) \equiv t(P_1) + t(P_2) \pmod{p^{3\nu} R}$$

gives the formula

$$t(mP) \equiv mt(P) \pmod{p^{3\nu} R}.$$

Since $mP = \mathcal{O}$, we have $t(mP) = t(\mathcal{O}) = 0$. On the other hand since m is prime to p, it is a unit in R. Therefore

$$0 \equiv t(P) \pmod{p^{3\nu} R}.$$

This means that $P \in C(p^{3\nu})$, contradicting the fact that $P \notin C(p^{\nu+1})$.

Next we suppose that $p \mid m$. The proof in this case is similar. First, we write $m = pn$ and look at the point $P' = nP$. Since P has order m, it is clear that P' has order p. Further, since $P \in C(p)$ and $C(p)$ is a subgroup of $C(\mathbb{Q})$, we see that $P' \in C(p)$. Writing $P' = (x', y')$, we let $\nu = -\frac{1}{2} \operatorname{ord}(x')$, so $P' \in C(p^\nu)$ and $P' \notin C(p^{\nu+1})$. Then, just as before, we find that

$$0 = t(\mathcal{O}) = t(pP') \equiv pt(P') \pmod{p^{3\nu} R}.$$

This means that

$$t(P') \equiv 0 \pmod{p^{3\nu-1} R}.$$

Since $3\nu - 1 \geq \nu + 1$, this contradicts the fact that $P' \notin C(p^{\nu+1})$, which completes the proof of part (a) of the corollary.

But now part (b) is easy, because if $P = (x, y)$ is a point of finite order, then we know from (a) that $P \notin C(p)$ for all primes p. This means that the denominators of x and y are divisible by no primes, and hence that x and y are integers. $\square$

2.5 The Nagell–Lutz Theorem and Further Developments

We have really finished the proof of the Nagell–Lutz theorem, but to wrap everything up we will state it formally and remind you of the two parts of the proof.

Theorem 2.5 (Nagell–Lutz Theorem). *Let*

$$y^2 = f(x) = x^3 + ax^2 + bx + c$$

be a non-singular cubic curve with integer coefficients a, b, c, and let D be the discriminant of the cubic polynomial,

$$D = -4a^3 c + a^2 b^2 + 18abc - 4b^3 - 27c^2.$$

Let $P = (x, y)$ be a rational point of finite order. Then x and y are integers, and either $y = 0$, in which case P has order two, or else y divides D.

Proof. In Section 2.4 we showed that a point of finite order has integer coordinates. If P has order two, then we know from Section 2.1 that $y = 0$, so we are done. Otherwise $2P \neq \mathcal{O}$. But $2P$ is also a point of finite order, so it also has integer coordinates. In Section 2.3 we showed that if both $P = (x, y)$ and $2P$ have integer coordinates, then y divides D, which completes the proof of the Nagell–Lutz theorem. $\square$

Remark 2.6. For computational purposes, there is a stronger form of the Nagell–Lutz theorem that is often useful. It says that if $P = (x, y)$ is a rational point of finite order with $y \neq 0$, then y^2 divides the discriminant of D. We leave the proof of this stronger statement to the exercises; see Exercise 2.11.

Warning. We want to reiterate that the Nagell–Lutz theorem is not an "if and only if" statement. It is quite possible to have points with integer coordinates and with y dividing D that are not points of finite order. The Nagell–Lutz theorem can be used to compile a list of points that includes all points of finite order, but it can never be used to prove that any particular point actually has finite order. To verify that a point P has finite order, one must find an integer $n \geq 1$ such that $nP = \mathcal{O}$.

On the other hand, the Nagell–Lutz theorem can often be used to prove that a given point has *infinite* order. The idea is to compute $P, 2P, 3P, \ldots$ until one arrives at a multiple nP whose coordinates are not integers. Then one knows that nP, and a fortiori also P, cannot have finite order. This computation can be accelerated by computing instead only the x-coordinates of $2P, 4P, 8P, \ldots$ by repeatedly applying the duplication formula until some x-coordinate is not an integer.[2]

The question naturally arises as to what points of finite order can occur. We have already seen that it is easy to get points of order two by taking the cubic polynomial to have a rational root. Similarly, using our description of the points of order three, it is not hard to find cubic curves such that $C(\mathbb{Q})$ has a point of order three. On the other hand, we have indicated why it is not possible to find two independent points of order three, or indeed of any order greater than two, because it is not even possible to do this in the larger group $C(\mathbb{R})$.

However, it is possible to find individual points of higher order. For example, the point $P = (1, 1)$ on the curve

$$y^2 = x^3 - x^2 + x$$

has order four, since one easily checks that $2P = (0, 0)$, and we know that $(0, 0)$ has order two. Then $3P = -P = (1, -1)$ is also a point of order four. We also note that the other two roots of $x^3 - x^2 + x$ are complex, so the only point of order two is $(0, 0)$.

We can use the Nagell–Lutz theorem to check that there are no other points of finite order on this curve. The discriminant is $D = -3$, so the only possible values for y are ± 1 and ± 3. We already know that $y = \pm 1$ gives points of order four, so we check $y = \pm 3$. This leads to the equation $x^3 - x^2 + x - 9 = 0$. The only possible rational roots are integers dividing 9, and one quickly checks that ± 1, ± 3, and ± 9 are not roots. So the only points of finite order are the ones that we know, and the subgroup of points of finite order is a cyclic group of order four.

In fact, there are infinitely many curves with a rational point of order four. For every rational number t except $t = 0$ and $t = \frac{1}{4}$, the point (t, t) on the non-singular cubic curve

$$y^2 = x^3 - (2t - 1)x^2 + t^2 x$$

---

[2]Iteration of the duplication map $x \to \frac{x^4 - 2bx^2 - 8cx + b^2 - 4ac}{4x^3 + 4ax^2 + 4bx + 4c}$ also plays an important role in the theory of dynamical systems, where Lattès used it in 1918 to give the first example of a rational map whose behavior is everywhere chaotic. See [29, 35].

is a point of order four. (You should check this. Also see Exercise 2.13 for a converse statement.)

In a similar fashion, one can write down infinitely many examples of curves with rational points of order 5, 6, 7, 8, 9, 10, and 12. In essence, these examples were written down during the second half of the nineteenth century. But no one was ever able to find even a single example of a cubic curve with a rational point of order 11. There is a good reason for this, because Billing and Mahler [4] proved in 1940 that no such curve exists.

Many people worked on the problem of determining which orders are possible, culminating in the 1970s with a very beautiful and very difficult theorem of Mazur [32, 33]. We will not even be able to indicate how the proof goes, but the statement, which is easy to understand, is as follows.

Theorem 2.7 (Mazur's Theorem). *Let C be a non-singular rational cubic curve, and suppose that $C(\mathbb{Q})$ contains a point of finite order m. Then either*

$$1 \le m \le 10 \quad or \quad m = 12.$$

More precisely, the set of points of finite order in $C(\mathbb{Q})$ forms a subgroup that has one of the following forms:

(i) *A cyclic group of order N with $1 \le N \le 10$ or $N = 12$.*
(ii) *The product of a cyclic group of order two and a cyclic group of order $2N$ with $1 \le N \le 4$.*

Exercises

2.1. Let A be an abelian group and, for every integer $m \ge 1$, let

$$A_m = \{P \in A : mP = \mathcal{O}\}$$

be the set of elements of order dividing m.

(a) Prove that A_m is a subgroup of A.
(b) Suppose that A has order M^2, and further suppose that for every integer m dividing M, the subgroup A_m has order m^2. Prove that A is the direct product of two cyclic groups of order m.
(c) Find an example of a non-abelian group G and an integer m such that the set $G_m = \{g \in G : g^m = e\}$ is not a subgroup of G.

2.2. Let C be a non-singular cubic curve given by the usual Weierstrass equation

$$y^2 = f(x) = x^3 + ax^2 + bx + c,$$

(a) Prove that
$$\frac{d^2y}{dx^2} = \frac{2f''(x)f(x) - f'(x)^2}{4yf(x)} = \frac{\psi_3(x)}{4yf(x)}.$$

(See Theorem 2.1 for the definition of $\psi_3(x)$.) Use this to deduce that a point $P = (x, y) \in C$ is a point of order three if and only if $P \neq \mathcal{O}$ and P is a point of inflection on the curve C.

(b) Suppose now that a, b, c are in $\mathbb{R}$. Prove that $\psi_3(x)$ has exactly two real roots, say α_1 and α_2 with $\alpha_1 < \alpha_2$. Prove that $f(\alpha_1) < 0$ and $f(\alpha_2) > 0$. Use this to deduce that the points in $C(\mathbb{R})$ of order dividing 3 form a cyclic subgroup of order three.

2.3. Let $\omega_1, \omega_2 \in \mathbb{C}$ be two complex numbers that are $\mathbb{R}$-linearly independent, and let
$$L = \mathbb{Z}\omega_1 + \mathbb{Z}\omega_2 = \{n_1\omega_1 + n_2\omega_2 : n_1, n_2 \in \mathbb{Z}\}$$
be the lattice in $\mathbb{C}$ that they generate.

(a) Show that the series
$$\wp(u) = \frac{1}{u^2} + \sum_{\substack{w \in L \\ w \neq 0}} \left(\frac{1}{(u-w)^2} - \frac{1}{w^2} \right).$$

defining the Weierstrass $\wp$-function is absolutely and uniformly convergent on any compact subset of the complex u-plane that does not contain any of the points of L. Conclude that $\wp$ is a meromorphic function with a double pole at each point of L and no other poles.

(b) Prove that $\wp$ is an even function, i.e., prove that $\wp(-u) = \wp(u)$.

(c) Prove that $\wp$ is a doubly periodic function, that is, show that
$$\wp(u + w) = \wp(u) \quad \text{for every } u \in \mathbb{C} \text{ and every } w \in L.$$

(*Hint.* From (a), you can calculate the derivative $\wp'(u)$ by differentiating each term of the series defining $\wp(u)$. First prove that $\wp'(u + w) = \wp'(u)$, then integrate and use (b) to find the constant of integration.)

2.4. Let C be the cubic curve
$$y^2 = x^3 + 1.$$

(a) For each prime $5 \leq p < 30$, describe the group of points on this curve having coordinates in the finite field with p elements.

(b) For each prime in (a), let M_p be the number of points in the group. (Don't forget to include the point at infinity.) For the set of primes satisfying $p \equiv 2 \pmod{3}$, can you see a pattern for the values of M_p? Make a general conjecture for the value of M_p when $p \equiv 2 \pmod{3}$ and prove that your conjecture is correct.

(c) ** Try to find a pattern for the value of M_p for the set of primes satisfying $p \equiv 1 \pmod{3}$. Compute M_{31} and see if it fits your pattern. If not, make a new conjecture and compute the next few M_p's to test you conjecture.

(d) Answer the same questions as in (a) and (b) for the cubic curve $y^2 = x^3 + x$. Note that in (b) you will have to replace the condition $p \equiv 2 \pmod 3$ with some other congruence condition.

2.5. (a) Let $f(x) = x^2 + ax + b = (x - \alpha_1)(x - \alpha_2)$ be a quadratic polynomial with the indicated factorization. Prove that

$$(\alpha_1 - \alpha_2)^2 = a^2 - 4b.$$

(b) Let

$$f(x) = x^3 + ax^2 + bx + c = (x - \alpha_1)(x - \alpha_2)(x - \alpha_3)$$

be a cubic polynomial with the indicated factorization. Prove that

$$(\alpha_1 - \alpha_2)^2 (\alpha_1 - \alpha_3)^2 (\alpha_2 - \alpha_3)^2 = -4a^3 c + a^2 b^2 + 18abc - 4b^3 - 27c^2.$$

(c) * Let

$$f(x) = x^n + a_1 x^{n-1} + \cdots + a_n = (x - \alpha_1)(x - \alpha_2) \cdots (x - \alpha_n)$$

be a polynomial of degree n with the indicated factorization. The *discriminant of f* is defined to be

$$\mathrm{Disc}(f) = \prod_{i=1}^{n-1} \prod_{j=i+1}^{n} (\alpha_i - \alpha_j)^2,$$

so $\mathrm{Disc}(f) = 0$ if and only if f has a double root. Prove that $\mathrm{Disc}(f)$ can be expressed as a polynomial in the coefficients $a_1, \ldots, a_n$ of f.

2.6. Let p be a prime, and for a rational number $r = \dfrac{m}{n} p^\nu$ with m and n prime to p, let $\mathrm{ord}(r) = \nu$ be as in Section 2.4. Also, by convention, we set $\mathrm{ord}(0) = \infty$.
(a) Prove that

$$\mathrm{ord}(r_1 r_2) = \mathrm{ord}(r_1) + \mathrm{ord}(r_2) \quad \text{for all rational numbers } r_1, r_2.$$

(b) Prove that

$$\mathrm{ord}(r_1 + r_2) \geq \min\{\mathrm{ord}(r_1), \mathrm{ord}(r_2)\} \quad \text{for all rational numbers } r_1, r_2.$$

(c) Prove that if $\mathrm{ord}(r_1) \neq \mathrm{ord}(r_2)$, then the inequality in (b) is an equality.
(d) Define an "absolute value" on the rational numbers by the rule

$$\|r\| = \frac{1}{p^{\mathrm{ord}(r)}},$$

where by convention we set $\|r\| = 0$. Prove that $\| \cdot \|$ has the following properties:

(i) $\|r\| \geq 0$, and $\|r\| = 0$ if and only if $r = 0$.

(ii) $\|r_1 r_2\| = \|r_1\| \cdot \|r_2\|$.

(iii) $\|r_1 + r_2\| \leq \max\{\|r_1\|, \|r_2\|\}$.

Notice that property (iii) is stronger than the usual triangle inequality. The absolute value $\| \cdot \|$ is called the *p-adic absolute value* on the rational numbers. It can be used to define a topology on the rational numbers, the *p-adic topology*.

2.7. Continuing with the notation from the previous exercise, let p be a prime, and let

$$R = \{x \in \mathbb{Q} : \operatorname{ord}(x) \geq 0\} = \{x \in \mathbb{Q} : \|x\| \leq 1\}.$$

So the set R is the p-adic analogue of the interval $[-1, 1]$ on the real line or of the unit disk $\{z \in \mathbb{C} : |z| \leq 1\}$ in the complex plane.

(a) Prove that R is a subring of the rational numbers.

(b) Prove that $p \in R$ and that the ideal generated by p is a maximal ideal. Describe the quotient field R/pR.

(c) Prove that the unit group of R consists of all rational numbers a/b such that p does not divide ab. Deduce that every element of R is either a unit or else is in the ideal generated by p.

(d) Prove that R is a unique factorization domain.

(e) Describe all of the ideals of R and use this description to prove that pR is the only maximal ideal of R. (Rings that have exactly one maximal ideal are called *local rings*.)

2.8. Let p and R be as in the previous exercise. Let $\sigma \geq \nu \geq 0$ be integers. Prove that the quotient group $p^\nu R/p^\sigma R$ is a cyclic group of order $p^{\sigma - \nu}$.

2.9. Let p be a prime and let S be the set of rational numbers whose denominator is a power of p, where $p^0 = 1$ is allowed. Thus S is the set of all rational numbers ap^ν, where a is an integer prime to p and ν is an arbitrary integer.

(a) Prove that S is a subring of the rational numbers.

(b) Prove that the unit group of S consists of all numbers of the form $\pm p^\nu$ with ν any integer.

(c) Let q be a prime other than p. Prove that q generates a maximal ideal of S. Describe the quotient field S/qS, and prove that every maximal ideal of S has this form.

2.10. Let p be a prime, and let C be the cubic curve

$$C : y^2 = x^3 + px.$$

Find all points of finite order in $C(\mathbb{Q})$.

2.11. As usual, let C be a non-singular cubic curve given by an equation

$$y^2 = f(x) = x^3 + ax^2 + bx + c$$

with integer coefficients. We proved in Section 1.4 that if $P = (x, y)$ is a point on C, then the x-coordinate of $2P$ is given by the duplication formula

$$x(2P) = \frac{\phi(x)}{4f(x)} = \frac{x^4 - 2bx^2 - 8cx + b^2 - 4ac}{4(x^3 + ax^2 + bx + c)},$$

where $\phi(x)$ is the indicated quartic polynomial.

(a) Let

$$D = -4a^3c + a^2b^2 + 18abc - 4b^3 - 27c^2$$

be the discriminant of $f(x)$. Find polynomials $F(X)$ and $\Phi(X)$ with integer coefficients so that[3]

$$F(X)f(X) + \Phi(X)\phi(X) = D.$$

(*Hint.* $F(X)$ has degree 3 and $\Phi(X)$ has degree 2.)

(b) Let $P = (x, y)$ be a point of finite order on C. Prove that $2P = \mathcal{O}$ or $y^2 \mid D$. (This is the strong form of the Nagell–Lutz theorem.)

2.12. For each of the following curves, determine the points of finite order. Also determine the structure of the group formed by the points of finite order.

(a) $y^2 = x^3 - 2$

(b) $y^2 = x^3 + 8$

(c) $y^2 = x^3 + 4$

(d) $y^2 = x^3 + 4x$

(e) $y^2 - y = x^3 - x^2$

(f) $y^2 = x^3 + 1$

(g) $y^2 = x^3 - 43x + 166$

(h) $y^2 + 7xy = x^3 + 16x$

(i) $y^2 + xy + y = x^3 - x^2 - 14x + 29$

(j) $y^2 + xy = x^3 - 45x + 81$

(k) $y^2 + 43xy - 210y = x^3 - 210x^2$

(l) $y^2 = x^3 - 4x$

(m) $y^2 + xy - 5y = x^3 - 5x^2$

(n) $y^2 + 5xy - 6y = x^3 - 3x^2$

(o) $y^2 + 17xy - 120y = x^3 - 60x^2$.

Hint. You may need to complete the square on the left before you can use the Nagell–Lutz theorem. Feel free to use the strong form of the Nagell–Lutz theorem described in Exercise 2.11. The results proven in Section 4.3 might also be helpful in limiting the amount of computation that you need to do. After you're done, compare your results to Mazur's theorem (Theorem 2.7).

---

[3] We remark that the resultant of $f(X)$ and $\phi(X)$ is actually D^2, so general theory only predicts an equation of the form $Ff + \Phi\phi = D^2$.

2.13. Let C be the cubic curve

$$C : y^2 = x^3 - (2t - 1)x^2 + t^2x$$

(a) Prove that C is non-singular if and only if $t \notin \{0, \frac{1}{4}\}$.
(b) Assuming C is non-singular, prove that the point (t, t) is a point of order four.
(c) Conversely, let C' be a cubic curve (say given in Weierstrass form), and let P' a point of order four on C'. Prove that there is a change of variables so that C' is equal to C for some value of t and so that P' goes to (t, t).
(d) For a given (C', P') as in (c), how many values of t work?

Chapter 3

The Group of Rational Points

3.1 Heights and Descent

In this chapter we will prove Mordell's theorem that the group of rational points on a non-singular cubic is finitely generated. There is a tool used in the proof called the *height*. In brief, the height of a rational point measures how complicated the point is from the viewpoint of number theory.

We begin by defining the height of a rational number. Let $x = m/n$ be a rational number written in lowest terms. Then we define the *height of* x to be the maximum of the absolute values of the numerator and the denominator of x,

$$H(x) = H\left(\frac{m}{n}\right) = \max\{|m|, |n|\}.$$

The height of a rational number is a positive integer.

Why is the height a good way of measuring the complexity, in a number theoretic sense, of a rational number? For example, why not just take the absolute value $|x|$? Consider the two rational numbers $\frac{1}{2}$ and $\frac{9999}{20000}$. They both have about the same absolute value, but the latter is clearly much more "complicated" than the former, at least if one is interested in doing number theory.[1] If this reason is not convincing enough, then possibly the following property of the height will explain why it is a useful notion.

---

[1] From the perspective of computer science, we might define the complexity of a rational number m/n to be (roughly) the number of bits needed to store m/n on a computer. Including sign, it takes $\lceil \log_2 |m| \rceil + \lceil \log_2 |n| \rceil + 1$ bits to store m/n, so roughly between $\log_2 H(m/n)$ bits and $2 \log_2 H(m/n)$ bits.

© Springer International Publishing Switzerland 2015   65
J.H. Silverman, J.T. Tate, *Rational Points on Elliptic Curves*,
Undergraduate Texts in Mathematics, DOI 10.1007/978-3-319-18588-0_3

Finiteness Property of the Height. The set of all rational numbers whose height is less than some fixed number is a finite set.

The proof of this fact is easy. If the height of $x = m/n$ is less than some fixed constant, then both $|m|$ and $|n|$ are less than that constant, so there are only finitely many possibilities for m and n.

If

$$y^2 = f(x) = x^3 + ax^2 + bx + c$$

is a non-singular cubic curve with integer coefficients a, b, c, and if $P = (x, y)$ is a rational point on the curve, we define the *height of* P to be simply the height of its x-coordinate,

$$H(P) = H(x).$$

(By convention, we set $H(\mathcal{O}) = 1$.) We will see that the height behaves somewhat multiplicatively relative to the addition law on the curve. For example, we will want to compare $H(P + Q)$ to the product $H(P)H(Q)$. For notational reasons it is often convenient to have a function that behaves additively, so we also define the "small h height" by taking the logarithm,

$$h(P) = \log H(P).$$

We observe that $h(P)$ is always a non-negative real number.

Note that the rational points on C also have the finiteness property. If M is any positive number, then

$$\{P \in C(\mathbb{Q}) : H(P) \leq M\}$$

is a finite set, and the same holds if we use $h(P)$ in place of $H(P)$. This is true because points in the set have only finitely many possibilities for their x-coordinates, and for each x-coordinate, there are only two possibilities for the y-coordinate.

Our ultimate goal is to prove that the group of rational points $C(\mathbb{Q})$ is finitely generated. This fact will follow from four lemmas. We are going to state the lemmas now and use them to prove the finite generation of $C(\mathbb{Q})$. After that, we will see about proving the lemmas.

Lemma 3.1. *For every real number M, the set*

$$\{P \in C(\mathbb{Q}) : h(P) \leq M\}$$

is finite.

Lemma 3.2. *Let P_0 be a fixed rational point of C. There is a constant κ_0 that depends on P_0 and on a, b, and c, so that*

$$h(P + P_0) \le 2h(P) + \kappa_0 \quad \text{for all } P \in C(\mathbb{Q}).$$

Lemma 3.3. *There is a constant κ, depending on a, b, and c, so that*

$$h(2P) \ge 4h(P) - \kappa \quad \text{for all } P \in C(\mathbb{Q}).$$

Notice that Lemma 3.3 says that when you double a point, the height goes up quite a bit. So as soon as you get a point with large height, doubling makes a much larger height. Notice also that Lemmas 3.2 and 3.3 relate the group law on C, which is defined geometrically, to the height of points, which is a number theoretic device. So in some sense one can think of the height as a tool to translate geometric information into number theoretic information.

Lemma 3.4. *The index $\big(C(\mathbb{Q}) : 2C(\mathbb{Q})\big)$ is finite.*

We are using the notation $2C(\mathbb{Q})$ to denote the subgroup of $C(\mathbb{Q})$ consisting of points that are twice other points. For any commutative group Γ, the multiplication-by-m map

$$\Gamma \longrightarrow \Gamma, \quad P \longmapsto \underbrace{P + \cdots + P}_{m \text{ terms}} = mP,$$

is a homomorphism, and the image of this homomorphism is the subgroup $m\Gamma$ of Γ. The fourth lemma states that for $\Gamma = C(\mathbb{Q})$, the subgroup 2Γ has finite index in Γ.

These lemmas are in increasing order of difficulty. We have already proven Lemma 3.1. The middle two lemmas are related to the theory of heights of rational numbers, and if you know the formulas for adding and doubling points, then they can be proven without further reference to the curve C. Lemma 3.4 is subtler to prove, and since we want to restrict ourselves to working with rational numbers, we will only be able to prove it for a certain fairly large class of cubic curves.

We now show how these four lemmas imply that $C(\mathbb{Q})$ is a finitely generated group. If you like, you can completely forget about rational points on a curve. Just suppose that we are given a commutative group Γ, written additively, and a (height) function

$$h : \Gamma \longrightarrow [0, \infty)$$

from Γ to the non-negative real numbers. Suppose further that Γ and h satisfy the four lemmas. We restate our hypotheses and prove that Γ is finitely generated.

Theorem 3.5 (Desecent Theorem). *Let Γ be a commutative group, and suppose that there is a function*

$$h : \Gamma \longrightarrow [0, \infty)$$

with the following three properties:
(a) *For every real number M, the set $\{P \in \Gamma : h(P) \leq M\}$ is finite.*
(b) *For every $P_0 \in \Gamma$ there is a constant κ_0 so that*

$$h(P + P_0) \leq 2h(P) + \kappa_0 \quad \text{for all } P \in \Gamma.$$

(c) *There is a constant κ so that*

$$h(2P) \geq 4h(P) - \kappa \quad \text{for all } P \in \Gamma.$$

Suppose further that
(d) *The subgroup 2Γ has finite index in Γ.*
Then Γ is finitely generated.

Proof. The first thing that we do is take a representative for each coset of 2Γ in Γ. We know that there are only finitely many cosets, say n of them, and we let $Q_1, \ldots, Q_n$ be representatives for the cosets. This means that for any element $P \in \Gamma$, there is an index i_1 depending on P such that

$$P - Q_{i_1} \in 2\Gamma.$$

This is true since P has to be in one of the cosets. So we can write

$$P - Q_{i_1} = 2P_1$$

for some $P_1 \in \Gamma$. Now we do the same thing with P_1. Continuing, this proves that we can write

$$P_1 - Q_{i_2} = 2P_2$$
$$P_2 - Q_{i_3} = 2P_3$$
$$\vdots$$
$$P_{m-1} - Q_{i_m} = 2P_m,$$

where $Q_{i_1}, \ldots, Q_{i_m}$ are chosen from the coset representatives $Q_1, \ldots, Q_n$ and where $P_1, \ldots, P_m$ are elements of Γ.

The basic idea is that since P_i is more-or-less equal to $2P_{i+1}$, the height of P_{i+1} is more-or-less one-fourth the height of P_i. So the sequence of points $P, P_1, P_2, \ldots$ should have decreasing height, and eventually we end up in a set of points of bounded height. From property (a), that set is be finite, which will complete the proof. Now we have to turn these vague remarks into a valid proof.

From the first equation we have

$$P = Q_{i_1} + 2P_1.$$

Now substitute the second equation $P_1 = Q_{i_2} + 2P_2$ into this to get

$$P = Q_{i_1} + 2Q_{i_2} + 4P_2.$$

Continuing in this fashion, we obtain

$$P = Q_{i_1} + 2Q_{i_2} + 4Q_{i_3} + \cdots + 2^{m-1}Q_{i_m} + 2^m P_m.$$

In particular, this says that P is in the subgroup of Γ generated by the Q_i's and P_m. We are going to show that by choosing m large enough, we can force P_m to have height less than a certain fixed bound that does not depend on the initial point P. Then the finite set of points with height less than this bound, together with the Q_i's, will generate Γ.

Let's take one of the P_j's in the sequence of points $P, P_1, P_2, \ldots$ and examine the relation between the height of P_{j-1} and the height of P_j. We want to show that the height of P_j is considerably smaller. To do that, we need to specify some constants. If we apply (b) with $-Q_i$ in place of P_0, we get a constant κ_i so that

$$h(P - Q_i) \leq 2h(P) + \kappa_i \quad \text{for all } P \in \Gamma.$$

We do this for each of $Q_1, Q_2, \ldots, Q_n$. Let κ' be the largest of the κ_i's. Then

$$h(P - Q_i) \leq 2h(P) + \kappa' \quad \text{for all } P \in \Gamma \text{ and all } 1 \leq i \leq n.$$

We can do this because there are only finitely many Q_i's. This is one place that we are using property (d), which says that 2Γ has finite index in Γ.

Let κ be the constant from (c). Then we can calculate

$$
\begin{aligned}
4h(P_j) &\leq h(2P_j) + \kappa \\
&= h(P_{j-1} - Q_{i_j}) + \kappa \\
&\leq 2h(P_{j-1}) + \kappa' + \kappa.
\end{aligned}
$$

We rewrite this as

$$h(P_j) \le \frac{1}{2}h(P_{j-1}) + \frac{\kappa' + \kappa}{4}$$
$$= \frac{3}{4}h(P_{j-1}) - \frac{1}{4}\Big(h(P_{j-1}) - (\kappa' + \kappa)\Big).$$

From this we see that if $h(P_{j-1}) \ge \kappa' + \kappa$, then

$$h(P_j) \le \frac{3}{4}h(P_{j-1}).$$

So in the sequence of points $P, P_1, P_2, \ldots$, as long as the point P_j satisfies the condition $h(P_{j-1}) \ge \kappa' + \kappa$, then the next point in the sequence has much smaller height, namely $h(P_j) \le \frac{3}{4}h(P_{j-1})$. But if you start with any number and keep multiplying it by $\frac{3}{4}$, it approaches zero. So eventually we will find an index m so that $h(P_m) \le \kappa' + \kappa$.

We have now shown that every element $P \in \Gamma$ can be written in the form

$$P = a_1Q_1 + a_2Q_2 + \cdots + a_nQ_n + 2^m R$$

for certain integers $a_1, \ldots, a_n$ and some point $R \in \Gamma$ satisfying the inequality $h(R) \le \kappa' + \kappa$. Hence the set

$$\{Q_1, Q_2, \ldots, Q_n\} \cup \{R \in \Gamma : h(R) \le \kappa' + \kappa\}$$

generates Γ. From (a) and (d), this set is finite, which completes the proof that Γ is finitely generated. $\qquad\qquad\square$

We have called this a Descent Theorem because the proof is very much in the style of Fermat's method of infinite descent. One starts with an arbitrary point, in our case a point $P \in C(\mathbb{Q})$, and by clever manipulations one produces (descends to) a smaller point. Of course, one needs to have a way to measure the size of a point. We have used the height for that purpose. If one is lucky, repeated application of this idea leads to one of two possible conclusions. In our case we were led to a finite set of generating points, and then all of the points arise from this finite generating set by reversing the descent procedure. In other cases, one is led to a contradiction, usually the existence of an integer strictly between zero and one. Then one can conclude that there are no solutions. This is the method that Fermat used to show that $x^4 + y^4 = 1$ has no rational solutions with $xy \ne 0$, and it is undoubtedly the idea he had in mind to prove the same thing for $x^n + y^n = 1$. Unfortunately, additional complications arise as n increases, so no one has been able to verify Fermat's claim using these ideas. Wiles's proof of Fermat's last theorem follows a very different path, although, as we will see in Section 6.6, it is a path that uses the theory of elliptic curves in crucial ways.

In view of the Descent Theorem and the proof of Lemma 3.1 that we already gave, it remains to prove Lemmas 3.2–3.4. This will occupy us for the next several sections.

3.2   The Height of $P + P_0$

In this section we will prove Lemma 3.2, which gives a relationship between the heights of P, P_0, and $P + P_0$. Before beginning, we make a couple of remarks.

The first remark is that if $P = (x, y)$ is a rational point on our curve, then x and y have the form

$$x = \frac{m}{e^2} \quad \text{and} \quad y = \frac{n}{e^3}$$

for integers m, n, and e with $e > 0$ and $\gcd(m, e) = \gcd(n, e) = 1$. In other words, when you write the coordinates of a rational point in lowest terms, then the denominator of x is the square of a number whose cube is the denominator of y. We essentially proved this in Section 2.4, where we showed that if p^ν divides the denominator of x, then ν is even and $p^{3\nu/2}$ divides the denominator of y. However, since what we want to know is easy to prove, we will prove it again without resorting to studying one prime at a time.

Thus suppose that we write

$$x = \frac{m}{M} \quad \text{and} \quad y = \frac{n}{N}$$

in lowest terms with $M > 0$ and $N > 0$. Substituting into the equation of the curve gives

$$\frac{n^2}{N^2} = \frac{m^3}{M^3} + a\frac{m^2}{M^2} + b\frac{m}{M} + c,$$

and clearing denominators yields

$$M^3 n^2 = N^2 m^3 + aN^2 M m^2 + bN^2 M^2 m + cN^2 M^3. \tag{$*$}$$

Since N^2 is a factor of all of the terms on the right, we see that $N^2 \mid M^3 n^2$. But $\gcd(n, N) = 1$, so $N^2 \mid M^3$.

Now we want to prove the converse, that is, $M^3 \mid N^2$. This is done in three steps. First, from $(*)$ we immediately see that $M \mid N^2 m^3$, and since

$\gcd(m, M) = 1$, we find that $M \mid N^2$. Using this fact back in $(*)$, we find that $M^2 \mid N^2 m^3$, so $M \mid N$. Finally, using $(*)$ once again, we see that this implies that $M^3 \mid N^2 m^3$, so $M^3 \mid N^2$.

We have now shown that $N^2 \mid M^3$ and $M^3 \mid N^2$, so $M^3 = N^2$. Further, during the proof we showed that $M \mid N$. So if we let $e = N/M$, then we find that

$$e^2 = \frac{N^2}{M^2} = \frac{M^3}{M^2} = M \quad \text{and} \quad e^3 = \frac{N^3}{M^3} = \frac{N^3}{N^2} = N.$$

Therefore $x = m/e^2$ and $y = n/e^3$ have the desired form.

Our second remark concerns how we defined the height of the rational points on our curve. We just took the height of the x-coordinate. If the point P is given in lowest terms as

$$P = \left(\frac{m}{e^2}, \frac{n}{e^3} \right),$$

then the height of P is the maximum of $|m|$ and e^2. In particular,

$$|m| \leq H(P) \quad \text{and} \quad e^2 \leq H(P).$$

We claim that we can also bound the numerator of the y-coordinate in terms of $H(P)$. Precisely, we claim that there is a constant $K > 0$, depending on a, b, c, such that

$$|n| \leq KH(P)^{3/2} \quad \text{for all } P = \left(\frac{m}{e^2}, \frac{n}{e^3} \right) \in C(\mathbb{Q}).$$

To prove this, we just use the fact that the point satisfies the equation. Substituting into the equation and multiplying by e^6 gives

$$n^2 = m^3 + ae^2 m^2 + be^4 m + ce^6.$$

Now take absolute values and use the triangle inequality,

$$|n^2| \leq |m^3| + |ae^2 m^2| + |be^4 m| + |ce^6|$$
$$\leq H(P)^3 + |a|H(P)^3 + |b|H(P)^3 + |c|H(P)^3.$$

So we can take $K = \sqrt{1 + |a| + |b| + |c|}$.

We are now ready to prove Lemma 3.2, which we restate.

Lemma 3.2. *Let P_0 be a fixed rational point of C. There is a constant κ_0 that depends on P_0 and on a, b, and c, so that*

$$h(P + P_0) \leq 2h(P) + \kappa_0 \quad \text{for all } P \in C(\mathbb{Q}).$$

Proof. The proof is really nothing more than writing out the formula for the sum of two points and using the triangle inequality. We first remark that the lemma is trivial if $P_0 = \mathcal{O}$, so we may assume that $P_0 \neq \mathcal{O}$, say $P_0 = (x_0, y_0)$. Next we note that in proving the existence of κ_0, it is enough to prove that the inequality holds for all P except those in some fixed finite set. This is true because, for any finite number of P, we just look at the differences $h(P + P_0) - 2h(P)$ and take κ_0 larger than the finite number of values that occur. Having said this, it suffices to prove Lemma 3.2 for all points $P \notin \{P_0, -P_0, \mathcal{O}\}$.

We write $P = (x, y)$. The reason for avoiding P_0 and $-P_0$ is to have $x \neq x_0$, because then we can avoid using the duplication formula. We write

$$P + P_0 = (\xi, \eta).$$

To get the height of $P + P_0$, we need to calculate the height of ξ, so we need the formula for ξ in terms of (x, y) and (x_0, y_0). The formula that we derived in Section 1.4 looks this way:

$$\xi + x + x_0 = \lambda^2 - a \quad \text{with} \quad \lambda = \frac{y - y_0}{x - x_0}.$$

We need to write this out a little bit.

$$\begin{aligned}
\xi &= \frac{(y - y_0)^2}{(x - x_0)^2} - a - x - x_0 \\
&= \frac{(y - y_0)^2 - (x - x_0)^2(x + x_0 + a)}{(x - x_0)^2}.
\end{aligned}$$

If we multiply this all out, we find that $y^2 - x^3$ appears in the numerator. Since P is on the curve, we may replace $y^2 - x^3$ with the quantity $ax^2 + bx + c$. What we end up with is an expression

$$\xi = \frac{Ay + Bx^2 + Cx + D}{Ex^2 + Fx + G},$$

where A, B, C, D, E, F, G are certain rational numbers that can be expressed in terms of a, b, c and (x_0, y_0). Further, multiplying the numerator and the denominator by the least common denominator of $A, B \ldots, G$, we may assume that $A, B \ldots, G$ are all integers.

In summary, we have integers $A, B \ldots, G$ that depend only on a, b, c and (x_0, y_0) so that for any point $P = (x, y) \notin \{P_0, -P_0, \mathcal{O}\}$, the x-coordinate of $P + P_0$ is equal to

$$\xi = \frac{Ay + Bx^2 + Cx + D}{Ex^2 + Fx + G}.$$

The important fact is that once the curve and the point P_0 are fixed, then this expression is correct for all points P. So it will be all right for our constant κ_0 to depend on $A, B, \ldots, G$, as long as it does not depend on (x, y).

Now substitute $x = m/e^2$ and $y = n/e^3$ and clear denominators by multiplying numerator and denominator by e^4. We find that

$$\xi = \frac{Ane + Bm^2 + Cme^2 + De^4}{Em^2 + Fme^2 + Ge^4},$$

and now the result that we want is almost evident. Notice that we have an expression ξ that is an integer divided by an integer. We do not know that it is in lowest terms, and indeed it might not be, but cancellation will only make the height smaller. Thus

$$H(\xi) \le \max\left\{|Ane + Bm^2 + Cme^2 + De^4|, |Em^2 + Fme^2 + Ge^4|\right\}.$$

Further, we noted earlier that

$$e \le H(P)^{1/2}, \quad n \le KH(P)^{3/2}, \quad \text{and} \quad m \le H(P),$$

where K depends only on a, b, c. Using these and the triangle inequality gives

$$|Ane + Bm^2 + Cme^2 + De^4| \le |Ane| + |Bm^2| + |Cme^2| + |De^4|$$
$$\le (|AK| + |B| + |C| + |D|)H(P)^2$$

and

$$|Em^2 + Fme^2 + Ge^4| \le |Em^2| + |Fme^2| + |Ge^4|$$
$$\le (|E| + |F| + |G|)H(P)^2.$$

Therefore

$$H(P+P_0) = H(\xi) \le \max\left\{|AK|+|B|+|C|+|D|, |E|+|F|+|G|\right\}H(P)^2.$$

Taking logarithms of both sides gives

$$h(P + P_0) \le 2h(P) + \kappa_0,$$

where the constant

$$\kappa_0 = \log\max\Big\{|AK| + |B| + |C| + |D|, |E| + |F| + |G|\Big\}$$

depends only on a, b, c and (x_0, y_0) and does not depend on $P = (x, y)$. This completes the proof of Lemma 3.2.   □

3.3   The Height of $2P$

In the last section we proved that the height of a sum $P + P_0$ is (roughly) less then twice the height of P. In this section we want to prove Lemma 3.3, which says that the height of $2P$ is (roughly) greater than four times the height of P. This is harder, because to get the height to be large, we need to know that there is not too much cancellation in a certain rational number.

We now restate Lemma 3.3 and give the proof.

Lemma 3.3. *There is a constant κ, depending on a, b, and c, so that*

$$h(2P) \geq 4h(P) - \kappa \quad \textit{for all } P \in C(\mathbb{Q}).$$

Proof. Just as in our proof of Lemma 3.2, it is all right to ignore any finite set of points, since we can always take κ larger than $4h(P)$ for all points in that finite set. So we will discard the finitely many points satisfying $2P = \mathcal{O}$.

Let $P = (x, y)$ and write $2P = (\xi, \eta)$. The duplication formula that we derived in Section 1.4 states that

$$\xi + 2x = \lambda^2 - a \quad \text{with} \quad \lambda = \frac{f'(x)}{2y}.$$

Putting everything over a common denominator and using $y^2 = f(x)$, we obtain an explicit formula for ξ in terms of x,

$$\xi = \frac{f'(x)^2 - (8x + 4a)f(x)}{4f(x)} = \frac{x^4 + \cdots}{4x^3 + \cdots}.$$

Note that $f(x) \neq 0$ because $2P \neq \mathcal{O}$.

Thus ξ is the quotient of two polynomials in x with integer coefficients. Since the cubic $y^2 = f(x)$ is non-singular by assumption, we know that $f(x)$ and $f'(x)$ have no common complex roots. It follows that the polynomials in the numerator and denominator of ξ also have no common roots.

Since $h(P) = h(x)$ and $h(2P) = h(\xi)$, we are trying to prove that

$$h(\xi) \geq 4h(x) - \kappa.$$

Thus we are reduced to proving the following general lemma about heights and quotients of polynomials. Notice that this lemma has nothing at all to do with cubic curves.

Lemma 3.6. *Let* $\phi(X)$ *and* $\psi(X)$ *be polynomials with integer coefficients and no common complex roots. Let* d *be the maximum of the degrees of* ϕ *and* ψ.
(a) *There is an integer* $R \geq 1$, *depending on* ϕ *and* ψ, *so that for all rational numbers* m/n,

$$\gcd\left(n^d\phi\left(\frac{m}{n}\right), n^d\psi\left(\frac{m}{n}\right)\right) \quad \textit{divides } R.$$

(b) *There are constants* κ_1 *and* κ_2, *depending on* ϕ *and* ψ, *so that for all rational numbers* m/n *that are not roots of* ψ,

$$dh\left(\frac{m}{n}\right) - \kappa_1 \leq h\left(\frac{\phi(m/n)}{\psi(m/n)}\right) \leq dh\left(\frac{m}{n}\right) + \kappa_2.$$

Proof. (a) First we observe that since ϕ and ψ have degree at most d, the quantities $n^d\phi(m/n)$ and $n^d\psi(m/n)$ are both integers, so it makes sense to talk about their greatest common divisor. The result that we are trying to prove says that there is not too much cancellation when one takes the quotient of these two integers.

Next we note that ϕ and ψ are interchangeable, so for concreteness, we will take $\deg(\phi) = d$ and $\deg(\psi) = e \leq d$. Then we can write

$$n^d\phi\left(\frac{m}{n}\right) = a_0m^d + a_1m^{d-1}n + \cdots + a_dn^d,$$
$$n^d\psi\left(\frac{m}{n}\right) = b_0m^en^{d-e} + b_1m^{e-1}n^{d-e+1} + \cdots + b_en^d.$$

To ease notation, we let

$$\Phi(m, n) = n^d\phi\left(\frac{m}{n}\right) \quad \text{and} \quad \Psi(m, n) = n^d\psi\left(\frac{m}{n}\right).$$

So we need to find an estimate for $\gcd\big(\Phi(m, n), \Psi(m, n)\big)$ that does not depend on m and n.

Since $\phi(X)$ and $\psi(X)$ have no common roots, they are relatively prime in the Euclidean ring $\mathbb{Q}[X]$. Thus they generate the unit ideal, so we can find polynomials $F(X)$ and $G(X)$ with rational coefficients satisfying

$$F(X)\phi(X) + G(X)\psi(X) = 1. \qquad (**)$$

Let A be a large enough integer so that $AF(X)$ and $AG(X)$ have integer coefficients. Further, let D be the maximum of the degrees of F and G. Note that A and D do not depend on m or n.

Now we evaluate the identity $(**)$ at $X = m/n$ and multiply both sides by An^{D+d}. This gives

$$n^D AF\left(\frac{m}{n}\right) \cdot n^d \phi\left(\frac{m}{n}\right) + n^D AG\left(\frac{m}{n}\right) \cdot n^d \psi\left(\frac{m}{n}\right) = An^{D+d}.$$

Let $\gamma = \gamma(m, n)$ be the greatest common divisor of $\Phi(m, n)$ and $\Psi(m, n)$. We have

$$\left\{n^D AF\left(\frac{m}{n}\right)\right\} n^d \phi\left(\frac{m}{n}\right) + \left\{n^D AG\left(\frac{m}{n}\right)\right\} n^d \psi\left(\frac{m}{n}\right) = An^{D+d}.$$

Since the quantities in braces are integers, we see that γ divides An^{D+d}.

This is not good enough because we need to show that γ divides one fixed number that does not depend on n. We will show that γ actually divides Aa_0^{D+d}, where a_0 is the leading coefficient of $\phi(X)$. To prove this, we observe that since γ divides $\Phi(m, n)$, it certainly divides

$$An^{D+d-1}\Phi(m, n) = Aa_0 m^d n^{D+d-1} + Aa_1 m^{d-1} n^{D+d} + \cdots + Aa_d n^{D+2d-1}.$$

But in the sum, every term after the first one contains An^{D+d} as a factor, and we just proved that γ divides An^{D+d}. It follows that γ also divides the first term $Aa_0 m^d n^{D+d-1}$. Thus

$$\gamma \quad \text{divides} \quad \gcd(An^{D+d}, Aa_0 m^d n^{D+d-1}),$$

and since m and n are relatively prime, we find that γ divides $Aa_0 n^{D+d-1}$. Notice that we have reduced the power of n at the cost of multiplying by a_0.

Now using the fact that γ divides $Aa_0 n^{D+d-2}\Phi(m, n)$ and repeating the above argument shows that γ divides $Aa_0^2 n^{D+d-2}$. The pattern is clear, and eventually we reach the conclusion that γ divides Aa_0^{D+d}, which finishes the proof of (a)

(b) There are two inequalities to be proven. The proof of the upper bound, which is easier than the lower bound, is similar to the proof of Lemma 3.2. We will just prove the lower bound and leave the upper bound for you to do as an exercise.

As usual, it is all right to exclude some finite set of rational numbers when we prove an inequality of this sort. We need merely adjust the constant κ_1 to take care of the finitely many exceptions. So we may assume that the rational number m/n is not a root of ϕ.

If r is any non-zero rational number, it is clear from the definition that $h(r) = h(1/r)$. So reversing the roles of ϕ and ψ if necessary, we may make the same assumption as in (a), namely that ϕ has degree d and ψ has degree $e \leq d$.

Continuing with the notation from (a), the rational number whose height we want to estimate is

$$\xi = \frac{\phi\left(\dfrac{m}{n}\right)}{\psi\left(\dfrac{m}{n}\right)} = \frac{n^d \phi\left(\dfrac{m}{n}\right)}{n^d \psi\left(\dfrac{m}{n}\right)} = \frac{\Phi(m,n)}{\Psi(m,n)}.$$

This gives us an expression for ξ as a quotient of integers, so the height $H(\xi)$ would be the maximum of the integers $|\Phi(m,n)|$ and $|\Psi(m,n)|$ except for the possibility that they may have common factors.

We proved in (a) that there is some integer $R \geq 1$, independent of m and n, so that the greatest common divisor of $\Phi(m,n)$ and $\Psi(m,n)$ divides R. This bounds the possible cancellation, so we find that

$$H(\xi) \geq \frac{1}{R} \max\left\{|\Phi(m,n)|, |\Psi(m,n)|\right\}$$
$$= \frac{1}{R} \max\left\{\left|n^d \phi\left(\frac{m}{n}\right)\right|, \left|n^d \psi\left(\frac{m}{n}\right)\right|\right\}$$
$$\geq \frac{1}{2R}\left(\left|n^d \phi\left(\frac{m}{n}\right)\right| + \left|n^d \psi\left(\frac{m}{n}\right)\right|\right).$$

For the last line, we used the trivial observation that $\max\{a,b\} \geq \frac{1}{2}(a+b)$.

In multiplicative notation, we want to compare $H(\xi)$ to the quantity

$$H\left(\frac{m}{n}\right)^d = \max\{|m|^d, |n|^d\},$$

so we consider the quotient

$$\frac{H(\xi)}{H(m/n)^d} \geq \frac{1}{2R} \cdot \frac{\left|n^d \phi\left(\frac{m}{n}\right)\right| + \left|n^d \psi\left(\frac{m}{n}\right)\right|}{\max\{|m|^d, |n|^d\}}$$

$$= \frac{1}{2R} \cdot \frac{\left|\phi\left(\frac{m}{n}\right)\right| + \left|\psi\left(\frac{m}{n}\right)\right|}{\max\left\{\left|\frac{m}{n}\right|^d, 1\right\}}.$$

This suggests that we look at the function $p(t)$ of the real variable t defined by

$$p(t) = \frac{|\phi(t)| + |\psi(t)|}{\max\{|t|^d, 1\}}.$$

Since ϕ has degree d and ψ has degree at most d, we see that $p(t)$ has a non-zero limit as $|t|$ approaches infinity. This limit is $|a_0|$ if ψ has degree strictly less than d, and it is $|a_0| + |b_0|$ if ψ has degree equal to d. In any case, outside of some closed interval I, the function $p(t)$ is bounded away from 0.

But inside the closed interval I, the function $p(t)$ is continuous, and it never vanishes because by assumption $\phi(X)$ and $\psi(X)$ have no common zeros. And a continuous function on a compact set, such as the closed interval I, actually assumes it maximum and minimum values. In particular, since we know that our function is never equal to zero, its minimum value for $t \in I$ must be positive. This proves that $p(t)$ is bounded away from zero both on I and on the complement of I, and hence there is a constant $C_1 > 0$ so that $p(t) \geq C_1$ for all real numbers t.

We proved earlier that

$$\frac{H(\xi)}{H(m/n)^d} \geq \frac{1}{2R} \cdot p\left(\frac{m}{n}\right),$$

so using the fact that $p(t) \geq C_1$ allows us to conclude that

$$H(\xi) \geq \frac{C_1}{2R} \cdot H\left(\frac{m}{n}\right)^d.$$

The constants C_1 and R depend on ϕ and ψ, but they do not depend on m or n, so taking logarithms gives the desired inequality

$$h(\xi) \geq dh\left(\frac{m}{n}\right) - \kappa_1 \quad \text{with} \quad \kappa_1 = \log(2R/C_1).$$

This concludes the proof of Lemma 3.6. Notice that there are two ideas in the proof. One is to bound the amount of cancellation, and the other is to look at the function

$$\frac{H(\phi(x)/\psi(x))}{H(x)^d}$$

as a function on something compact.   □

And as already noted, this also concludes the proof of Lemma 3.3, which is a special case of Lemma 3.6.   □

3.4   A Useful Homomorphism

To complete the proof of Mordell's theorem, we need to prove Lemma 3.4, which says that the subgroup $2C(\mathbb{Q})$ has finite index inside $C(\mathbb{Q})$. This is the subtlest part of the proof of Mordell's theorem. To ease notation a little bit, we will write Γ for $C(\mathbb{Q})$,

$$\Gamma = C(\mathbb{Q}).$$

Unfortunately, we do not know how to prove Lemma 3.4 for all cubic curves without using some algebraic number theory, and we want to stick to the rational numbers. So we are going to make the additional assumption that the polynomial $f(x)$ has at least one rational root, which amounts to saying that the curve has at least one rational point of order two. The same method of proof works in general if you take a root of the equation $f(x) = 0$ and work in the field generated by that root over the rationals. But ultimately we would need to know some basic facts about the unit group and the ideal class group of this field, topics that we prefer to avoid. So we will prove Lemma 3.4 in the case that $f(x)$ has a rational root x_0. In this section we develop some tools that we need for the proof, and then in the next section we give the proof of Lemma 3.4, thereby completing the proof of Mordell's theorem.

Since $f(x_0) = 0$, and since f is a polynomial with integer coefficients and leading coefficient 1, we know that x_0 is an integer. Making a change of coordinates, we can move the point $(x_0, 0)$ to the origin. This obviously does not affect the group Γ. The new equation again has integer coefficient, and in the new coordinates the curve has the form

$$C : y^2 = f(x) = x^3 + ax^2 + bx,$$

where a and b are integers. Then

$$T = (0,0)$$

is a rational point on C that satisfies $2T = \mathcal{O}$.

The formula for the discriminant of f given in Section 2.3 becomes, in this case,

$$D = b^2(a^2 - 4b).$$

We always assume that our curve is non-singular, which means that $D \neq 0$, or equivalently, neither $a^2 - b$ nor b is zero.

Since we are interested in the index $(\Gamma : 2\Gamma)$, or equivalently in the order of the factor group $\Gamma/2\Gamma$, it is extremely helpful to know that the duplication map $P \mapsto 2P$ can be broken down into two simpler operations. The duplication map is in some sense a map of degree four, since the rational function giving the x-coordinate of $2P$ is of degree four in the x-coordinate of P. We will write the map $P \mapsto 2P$ as a composition of two maps of degree two, each of which will be easier to handle. However, the two maps will not be from C to itself, but rather from C to another curve $\overline{C}$ and then back again to C.

The other curve $\overline{C}$ that we will consider is the curve given by the equation

$$\overline{C} : y^2 = x^3 + \overline{a}x^2 + \overline{b}x,$$

where

$$\overline{a} = -2a \quad \text{and} \quad \overline{b} = a^2 - 4b.$$

For reasons that we will see in a moment, these two curves are intimately related, and it is natural, if you are studying C, to also study $\overline{C}$. One can play C and $\overline{C}$ off against one another, and that is just what we are planning to do.

Suppose that we apply the procedure again and look at

$$\overline{\overline{C}} : y^2 = x^3 + \overline{\overline{a}}x^2 + \overline{\overline{b}}x.$$

Here

$$\overline{\overline{a}} = -2\overline{a} = 4a \quad \text{and} \quad \overline{\overline{b}} = \overline{a}^2 - 4\overline{b} = 4a^2 - 4(a^2 - 4b) = 16b,$$

so the curve $\overline{\overline{C}}$ is the curve

$$\overline{\overline{C}} : y^2 = x^3 + 4ax^2 + 16bx.$$

This is essentially the same as C, we just need to replace x and y with $4x$ and $8y$, respectively, and then divide the equation by 64. Thus the group $\overline{\overline{\Gamma}}$ of rational points on $\overline{\overline{C}}$ is isomorphic to the group Γ of rational points on C.

We are now going to define a map $\phi : C \to \overline{C}$ that will be a group homomorphism and will carry the rational points Γ to the rational points $\overline{\Gamma}$ of $\overline{C}$. And then, by the same procedure, we will define a map $\overline{\phi} : \overline{C} \to \overline{\overline{C}}$. In view of the isomorphism $\overline{\overline{C}} \cong C$, the composition $\overline{\phi} \circ \phi$ gives a homomorphism of C to C that turns out to be the multiplication-by-2 map.

The map $\phi : C \to \overline{C}$ is defined in the following way. If $P = (x, y) \in C$ is a point with $x \neq 0$, then the point $\phi(x, y) = (\overline{x}, \overline{y})$ is given by the formulas

$$\overline{x} = x + a + \frac{b}{x} = \frac{y^2}{x^2} \quad \text{and} \quad \overline{y} = y\left(\frac{x^2 - b}{x^2}\right).$$

To see that ϕ is well-defined, we just have to check that $(\overline{x}, \overline{y})$ satisfies the equation for $\overline{C}$, which is easy:

$$\begin{aligned}
\overline{x}^3 + \overline{a}\,\overline{x}^2 + \overline{b}\,\overline{x} &= \overline{x}\big(\overline{x}^2 - 2a\overline{x} + (a^2 - 4b)\big) \\
&= \frac{y^2}{x^2}\left(\frac{y^4}{x^4} - 2a\frac{y^2}{x^2} + (a^2 - 4b)\right) \\
&= \frac{y^2}{x^2}\left(\frac{(y^2 - ax^2)^2 - 4bx^4}{x^4}\right) \\
&= \frac{y^2}{x^6}\big((x^3 + bx)^2 - 4bx^4\big) \\
&= \left(\frac{y(x^2 - b)}{x^2}\right)^2 \\
&= \overline{y}^2.
\end{aligned}$$

This defines the map ϕ at all points except $T = (0, 0)$ and $\mathcal{O}$. We complete the definition by setting

$$\phi(T) = \overline{\mathcal{O}} \quad \text{and} \quad \phi(\mathcal{O}) = \overline{\mathcal{O}}.$$

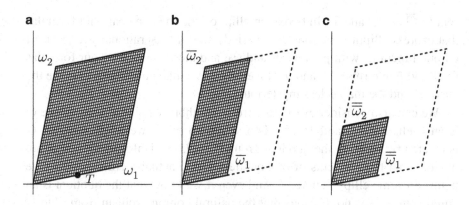

Figure 3.1: The map ϕ described analytically

This ad hoc definition of ϕ looks like magic. We reached into out top hat and out came an amazing map. The reason that we presented ϕ in this way is to emphasize that everything about ϕ follows from a little elementary algebra and arithmetic; there is no need to use any analysis. However, if you are willing to think in terms of complex points and the uniformization of the curve C by the complex variable u, then x and y are elliptic functions of u and you can see ϕ quite clearly. Namely, the complex points on our curve can be represented by the points in the period parallelogram for suitable periods ω_1 and ω_2; see Figure 3.1(a).

If we cut that parallelogram in half by a line parallel to one of the sides, then we get a new parallelogram with sides $\overline{\omega}_1$ and $\overline{\omega}_2$ as in Figure 3.1(b), where $\overline{\omega}_1 = \frac{1}{2}\omega_1$ and $\overline{\omega}_2 = \omega_2$. This parallelogram corresponds to the curve $\overline{C}$. To divide the parallelogram, we had to pick a point of order two on C, which is the point T in the figure. There is a natural map of C onto $\overline{C}$ in which the point

$$u = c_1\omega_1 + c_2\omega_2 \quad \text{is sent to} \quad \overline{u} = c_1\omega_1 + c_2\omega_2 = 2c_1\overline{\omega}_1 + c_2\overline{\omega}_2.$$

Now if we slice the parallelogram the other way, we get $\overline{\overline{C}}$ which has the period parallelogram with sides $\overline{\overline{\omega}}_1$ and $\overline{\overline{\omega}}_2$, where $\overline{\overline{\omega}}_1 = \frac{1}{2}\omega_1$ and $\overline{\overline{\omega}}_2 = \frac{1}{2}\omega_2$; see Figure 3.1(c). Clearly the curve in Figure 3.1(a) is isomorphic to the curve in Figure 3.1(c) via the map $u \mapsto \frac{1}{2}u$, so the elliptic functions with periods $\overline{\overline{\omega}}_1$ and $\overline{\overline{\omega}}_2$ are essentially the same as those with periods ω_1 and ω_2. From an analytic point of view, this is the procedure that we are using.

What is the kernel of ϕ? From the picture it is clear that the kernel of ϕ consists of the two points $\mathcal{O}$ and T, and if you look at the algebraic formula for ϕ that we gave earlier, you will see that the only two points of C that are sent to $\overline{\mathcal{O}}$ are $\mathcal{O}$ and T. In books on elliptic functions one can find formulas that express elliptic functions with periods $\frac{1}{2}\omega_1$ and ω_2 rationally in terms of elliptic functions with periods ω_1 and ω_2, and these are exactly our formulas for $\overline{x}$ and $\overline{y}$ in terms of x and y. Hopefully this explanation helps to make the curve $\overline{C}$ and the map ϕ less mysterious.

We can also consider everything from a highbrow point of view. Since C is an abelian group and $\{\mathcal{O}, T\}$ is a subgroup of C, we might say that $\overline{C}$ is created by forming the quotient group $C/\{\mathcal{O}, T\}$. Unfortunately, it is not obvious that the elements of this quotient group actually correspond to the points on some elliptic curve $\overline{C}$. And even if we know that the quotient is an elliptic curve, it is not obvious that the natural homomorphism from C to $\overline{C}$ is given by rational functions.

However, all of this follows from general theorems on algebraic groups. It is even true that the group of points on an elliptic curve modulo any finite subgroup is again the group of points on an elliptic curve. Granting this, and knowing that any elliptic curve can be written in Weierstrass form, it is not difficult to guess the explicit formulas that we gave earlier.

Both the analytic viewpoint and the "highbrow" approach tell us that the map ϕ is a homomorphism, but we can also prove this directly using explicit formulas. To remind you where we are, and for future reference, we state this as a formal proposition.

Proposition 3.7. *Let C and $\overline{C}$ be elliptic curves given by the equations*

$$C : y^2 = x^3 + ax^2 + bx \quad and \quad \overline{C} : y^2 = x^3 + \overline{a}x^2 + \overline{b}x,$$

where

$$\overline{a} = -2a \quad and \quad \overline{b} = a^2 - 4b.$$

Let $T = (0,0) \in C$.

(a) *There is a homomorphism $\phi : C \to \overline{C}$ defined by*

$$\phi(P) = \begin{cases} \left(\dfrac{y^2}{x^2}, \dfrac{y(x^2 - b)}{x^2} \right), & \text{if } P = (x,y) \neq \mathcal{O}, T, \\ \overline{\mathcal{O}}, & \text{if } P = \mathcal{O} \text{ or } P = T. \end{cases}$$

The kernel of ϕ is $\{\mathcal{O}, T\}$.

(b) *Applying the same process to $\overline{C}$ gives a map $\overline{\phi} : \overline{C} \to \overline{\overline{C}}$. The curve $\overline{\overline{C}}$ is isomorphic to C via the map $(x,y) \mapsto (\frac{1}{4}x, \frac{1}{8}y)$. There is thus a homomorphism $\psi : \overline{C} \to C$ defined by*

$$\psi(\overline{P}) = \begin{cases} \left(\dfrac{\overline{y}^2}{\overline{x}^2}, \dfrac{\overline{y}(\overline{x}^2 - \overline{b})}{\overline{x}^2}\right), & \text{if } \overline{P} = (\overline{x}, \overline{y}) \neq \overline{\mathcal{O}}, \overline{T}, \\ \mathcal{O}, & \text{if } \overline{P} = \overline{\mathcal{O}} \text{ or } \overline{P} = \overline{T}. \end{cases}$$

(c) *The composition $\psi \circ \phi : C \to C$ is the multiplication by two map,*

$$\psi \circ \phi(P) = 2P.$$

Proof. (a) We checked earlier that ϕ maps points of C to points of $\overline{C}$, and once we know that ϕ is a homomorphism, it is obvious that the kernel of ϕ consists of $\mathcal{O}$ and T. So we need to prove that ϕ is a homomorphism. This is somewhat tedious because there are many exceptional cases, so we will do a lot of it and leave a few cases for you.

We have to prove that

$$\phi(P_1 + P_2) = \phi(P_1) + \phi(P_2) \quad \text{for all } P_1, P_2 \in C.$$

Note that the first addition sign is addition on C, whereas the second one is addition on $\overline{C}$.

If P_1 or P_2 is $\mathcal{O}$, there is nothing to prove. If one of P_1 or P_2 is T, say $P_1 = T$, then the formula to be proved is $\phi(T + P) = \phi(P)$. This is not hard to see. Thus using the explicit formula for the addition law, one easily checks that if $P = (x, y)$, then

$$P + T = (x, y) + (0, 0) = \left(\frac{b}{x}, -\frac{by}{x^2}\right).$$

Writing

$$P + T = \big(x(P+T), y(P+T)\big) \quad \text{and} \quad \phi(P+T) = \big(\overline{x}(P+T), \overline{y}(P+T)\big),$$

we find that

$$\overline{x}(P + T) = \left(\frac{y(P+T)}{x(P+T)}\right)^2 = \frac{(-by/x^2)^2}{(b/x)^2} = \frac{y^2}{x^2} = \overline{x}(P).$$

In the same way we compute

$$\overline{y}(P+T) = \frac{y(P+T)(x(P+T)^2 - b)}{x(P+T)^2} = \frac{(-by/x^2)((b/x)^2 - b)}{(b/x)^2} = \overline{y}(P).$$

This shows that $\phi(P + T) = \phi(P)$, except that the argument breaks down if $P = T$. But in that case we obviously have

$$\phi(T + T) = \phi(\mathcal{O}) = \overline{\mathcal{O}} = \overline{\mathcal{O}} + \overline{\mathcal{O}} = \phi(T) + \phi(T).$$

Next we observe that ϕ takes negatives to negatives,

$$\phi(-P) = \phi(x, -y) = \left(\left(\frac{-y}{x} \right)^2, \frac{-y(x^2 - b)}{x^2} \right) = -\phi(x, y) = -\phi(P).$$

So in order to prove that ϕ is a homomorphism, it now suffices to show that if $P_1 + P_2 + P_3 = \mathcal{O}$, then $\phi(P_1) + \phi(P_2) + \phi(P_3) = \overline{\mathcal{O}}$, because once we know this, then

$$\phi(P_1 + P_2) = \phi(-P_3) = -\phi(P_3) = \phi(P_1) + \phi(P_2).$$

Further, from what we have already done, we may assume that none of the points P_1, P_2, or P_3 is equal to $\mathcal{O}$ or T.

From the definition of the group law on a cubic curve, the condition $P_1 + P_2 + P_3 = \mathcal{O}$ is equivalent to the statement that P_1, P_2, and P_3 are colinear, so let $y = \lambda x + \nu$ be the line through them. (If two or three of them coincide, then the line should be appropriately tangent to the curve.) We must show that $\phi(P_1)$, $\phi(P_2)$, and $\phi(P_3)$ are the intersection of some line with $\overline{C}$.

So suppose that P_1, P_2, and P_3 lie on the line $y = \lambda x + \nu$. Note that $\nu \neq 0$, since $\nu = 0$ would mean that the line goes through T, contrary to our assumption that P_1, P_2, P_3 are distinct from T. The line intersecting $\overline{C}$ that we take is

$$y = \overline{\lambda} x + \overline{\nu}, \quad \text{where} \quad \overline{\lambda} = \frac{\nu \lambda - b}{\nu} \quad \text{and} \quad \overline{\nu} = \frac{\nu^2 - a\nu\lambda + b\lambda^2}{\nu}.$$

To check, say, that $\phi(P_1) = (x_1, y_1) = (\overline{x}_1, \overline{y}_1)$ is on the line $y = \overline{\lambda} x + \overline{\nu}$, we just substitute and compute

$$\begin{aligned}
\overline{\lambda}\overline{x}_1 + \overline{\nu} &= \frac{\nu\lambda - b}{\nu} \left(\frac{y_1}{x_1} \right)^2 + \frac{\nu^2 - a\nu\lambda + b\lambda^2}{\nu} \\
&= \frac{(\nu\lambda - b)y_1^2 + (\nu^2 - a\nu\lambda + b\lambda^2)x_1^2}{\nu x_1^2} \\
&= \frac{\nu\lambda(y_1^2 - ax_1^2) - b(y_1 - \lambda x_1)(y_1 + \lambda x_1) + \nu^2 x_1^2}{\nu x_1^2};
\end{aligned}$$

and now using $y_1^2 - ax_2^2 = x_1^3 + bx_1$ and $y_1 - \lambda x_1 = \nu$, we get

$$= \frac{\lambda(x_1^3 + bx_1) - b(y_1 - \lambda x_1) + \nu x_1^2}{x_1^2}$$

$$= \frac{x_1^2(\lambda x_1 + \nu) - by_1}{x_1^2}$$

$$= \frac{(x_1^2 - b)y_1}{x_1^2}$$

$$= \overline{y}_1.$$

The computation for $\phi(P_2)$ and $\phi(P_3)$ is exactly the same.

Notice, however, that strictly speaking it is not enough to show that the three points $\phi(P_1), \phi(P_2), \phi(P_3)$ lie on the line $y = \overline{\lambda}x + \overline{\nu}$. It is enough if $\phi(P_1), \phi(P_2), \phi(P_3)$ are distinct, but in general we really have to show that $\overline{x}(P_1), \overline{x}(P_2), \overline{x}(P_3)$ are the three roots of the cubic $(\overline{\lambda}x + \overline{\nu})^2 = \overline{f}(x)$, whether or not those roots are distinct. We will leave it to you to verify this if there are multiple roots. As an alternative, we might note that ϕ is a continuous map from the complex points of C to the complex points of $\overline{C}$, so once we know that ϕ is a homomorphism for distinct points, we get by continuity that it is a homomorphism in general.

(b) We noted above that the curve $\overline{\overline{C}}$ is given by the equation

$$\overline{\overline{C}} : y^2 = x^3 + 4ax^2 + 16bx,$$

so it is clear that the map $(x, y) \rightarrow (x/4, y/8)$ is an isomorphism from $\overline{\overline{C}}$ to C. From (a) there is a homomorphism $\overline{\phi} : \overline{C} \rightarrow \overline{\overline{C}}$ defined by the same equations that define ϕ, but with $\overline{a}$ and $\overline{b}$ in place of a and b. Since the map $\psi : \overline{C} \rightarrow C$ is the composition of $\overline{\phi} : \overline{C} \rightarrow \overline{\overline{C}}$ with the isomorphism $\overline{\overline{C}} \rightarrow C$, we get immediately that ψ is a well-defined homomorphism from $\overline{C}$ to C.

It remains to verify that $\psi \circ \phi$ is multiplication by two, and that is another tedious computation. A little algebra with the explicit formulas we gave earlier yields

$$2P = 2(x, y) = \left(\frac{(x^2 - b)^2}{4y^2}, \frac{(x^2 - b)(x^4 + 2ax^3 + 6bx^2 + 2abx + b^2)}{8y^3} \right).$$

On the other hand, we have

$$\phi(x, y) = \left(\frac{y^2}{x^2}, \frac{y(x^2 - b)}{x^2} \right), \qquad \psi(\overline{x}, \overline{y}) = \left(\frac{\overline{y}^2}{4\overline{x}^2}, \frac{\overline{y}(\overline{x}^2 - b)}{8\overline{x}^2} \right),$$

so we can compute

$$\psi \circ \phi(x, y) = \psi \left(\frac{y^2}{x^2}, \frac{y(x^2 - b)}{x^2} \right)$$

$$= \left(\frac{\left(\frac{y(x^2 - b)}{x^2} \right)^2}{4 \left(\frac{y^2}{x^2} \right)^2}, \frac{\frac{y(x^2 - b)}{x^2} \left(\left(\frac{y^2}{x^2} \right)^2 - (a^2 - 4b) \right)}{8 \left(\frac{y^2}{x^2} \right)^2} \right)$$

$$= \left(\frac{(x^2 - b)^2}{4y^2}, \frac{(x^2 - b)(y^4 - (a^2 - 4b)x^4)}{8y^3 x^2} \right).$$

Now substituting $y^4 = x^2(x^2 + ax + b)^2$ and doing a little algebra gives the desired result $\psi \circ \phi(x, y) = 2(x, y)$.

A similar computation gives $\phi \circ \psi(\overline{x}, \overline{y}) = 2(\overline{x}, \overline{y})$. Or we can argue as follows. Since ϕ is a homomorphism, we know that

$$\phi(2P) = \phi(P + P) = \phi(P) + \phi(P) = 2\phi(P).$$

We just proved that $2P = \psi \circ \phi(P)$, so we get

$$\phi \circ \psi(\phi(P)) = 2(\phi(P)).$$

Now $\phi : C \to \overline{C}$ is onto as a map of complex points, so for any $\overline{P} \in \overline{C}$ we can find a point $P \in C$ with $\phi(P) = \overline{P}$. Therefore $\phi \circ \psi(\overline{P}) = 2\overline{P}$.

Of course, we have really only proved that $\psi \circ \phi = 2$ for points with $x \neq 0$ and $y \neq 0$ because the formulas that we used are not valid if x or y is zero. So we really should check that $\psi \circ \phi(P) = \mathcal{O}$ in the cases that P is a point of order 2. We will leave that to you to check explicitly, although again we could argue that it must be true by continuity.   □

3.5 Mordell's Theorem

In this section we will complete the proof of Lemma 3.4, and with it the proof of Mordell's theorem. Continuing with the notation from the last section, we recall that we have two curves

$$C : y^2 = x^3 + ax^2 + bx \quad \text{and} \quad \overline{C} : y^2 = x^3 + \overline{a}x^2 + \overline{b}x,$$

where $\bar{a} = -2a$ and $\bar{b} = a^2 - 4b$. Further, we have homomorphisms

$$\phi : C \longrightarrow \overline{C} \quad \text{and} \quad \psi : \overline{C} \longrightarrow C$$

such that the compositions

$$\phi \circ \psi : \overline{C} \longrightarrow \overline{C} \quad \text{and} \quad \psi \circ \phi : C \longrightarrow C$$

are each multiplication by two, and so that the kernel of ϕ consists of the two points $\mathcal{O}$ and $T = (0,0)$ and the kernel of ψ consists of $\overline{\mathcal{O}}$ and $\overline{T} = (0,0)$.

The images of ϕ and ψ are extremely interesting. From the complex point of view, it is obvious that given any point in $\overline{C}$, there is a point in C that maps to it. In other words, on complex points, the map ϕ is onto. But now we examine what happens to the rational points.

It is clear from the formulas that ϕ maps Γ to $\overline{\Gamma}$, but if you are given a rational point in $\overline{\Gamma}$, it is not at all clear if it comes from a rational point in Γ. If we apply the map ϕ to the set of rational points Γ, we get a subgroup of the set of rational points $\overline{\Gamma}$. We denote this group by $\phi(\Gamma)$ and call it the *image of Γ by ϕ*. We make the following three claims which, taken together, provide a good description of the image.

(i) $\overline{\mathcal{O}} \in \phi(\Gamma)$.

(ii) $\overline{T} = (0,0) \in \phi(\Gamma)$ if and only if $\overline{b} = a^2 - 4b$ is a perfect square.

(iii) Let $\overline{P} = (\overline{x}, \overline{y}) \in \overline{\Gamma}$ with $\overline{x} \neq 0$. Then $\overline{P} \in \phi(\Gamma)$ if and only if $\overline{x}$ is the square of a rational number.

Statement (i) is obvious, because $\overline{\mathcal{O}} = \phi(\mathcal{O})$. Let's check statement (ii). From the formula for ϕ we see that $\overline{T} \in \phi(\Gamma)$ if and only if there is a rational point $(x, y) \in \Gamma$ such that $y^2/x^2 = 0$. Note that $x \neq 0$, because $x = 0$ means that $(x, y) = T$ and we know that $\phi(T)$ is $\overline{\mathcal{O}}$, not $\overline{T}$. So $\overline{T} \in \phi(G)$ if and only if there is a rational point $(x, y) \in \Gamma$ with $x \neq 0$ and $y = 0$. Putting $y = 0$ in the equation for Γ gives

$$0 = x^3 + ax^2 + bx = x(x^2 + ax + b).$$

This equation has a non-zero rational root if and only if the quadratic equation $x^2 + ax + b$ has a rational root, which happens if and only if its discriminant $a^2 - 4b$ is a perfect square. This proves statement (ii).

Now we check statement (iii). If $(\overline{x}, \overline{y}) \in \phi(\Gamma)$ is a point with $\overline{x} \neq 0$, then the defining formula for ϕ shows that $\overline{x} = y^2/x^2$ is the square of a rational number. Suppose conversely that $\overline{x} = w^2$ for some rational number w. We want to find a rational point on C that maps to $(\overline{x}, \overline{y})$.

The homomorphism ϕ has two elements in its kernel, $\mathcal{O}$ and T. Thus if $(\overline{x}, \overline{y})$ lies in $\phi(\Gamma)$, there will be two points in Γ that map to it. Let

$$x_1 = \frac{1}{2}\left(w^2 - a + \frac{\overline{y}}{w}\right), \qquad\qquad y_1 = x_1 w,$$

$$x_2 = \frac{1}{2}\left(w^2 - a - \frac{\overline{y}}{w}\right), \qquad\qquad y_2 = -x_2 w.$$

We claim that the points $P_i = (x_i, y_i)$ are on C and that $\phi(P_i) = (\overline{x}, \overline{y})$ for $i = 1, 2$. Since P_1 and P_2 are clearly rational points, this will prove that $(\overline{x}, \overline{y}) = \phi(\Gamma)$.

The most efficient way to check that P_1 and P_2 are on C is to do them together, rather than working with them one at a time. First we compute

$$\begin{aligned}
x_1 x_2 &= \frac{1}{4}\left((w^2 - a)^2 - \frac{\overline{y}^2}{w^2}\right) \\
&= \frac{1}{4}\left((\overline{x} - a)^2 - \frac{\overline{y}^2}{\overline{x}}\right) \\
&= \frac{1}{4}\left(\frac{\overline{x}^3 - 2a\overline{x}^2 + a^2\overline{x} - \overline{y}^2}{\overline{x}}\right) \\
&= b.
\end{aligned}$$

The last line follows because $\overline{y}^2 = \overline{x}^3 - 2a\overline{x}^2 + (a^2 - 4b)\overline{x}$.

To show that $P_i = (x_i, y_i)$ lies on C amounts to showing that

$$\frac{y_i^2}{x_i^2} = x_i + a + \frac{b}{x_i}.$$

Since we just proved that $b = x_1 x_2$, and since from the definition of y_1 and y_2 we have $y_i / x_i = \pm w$, this is the same as showing that

$$w^2 = x_1 + a + x_2.$$

This last equality is obvious from the definition of x_1 and x_2.

It remains to check that $\phi(P_i) = (\overline{x}, \overline{y})$, so we must show that

$$\frac{y_i^2}{x_i^2} = \overline{x} \quad \text{and} \quad \frac{y_i(x_i^2 - b)}{x_i^2} = \overline{y}.$$

The first equality is clear from the definitions $y_i = \pm x_i w$ and $\overline{x} = w^2$. For the second, we use $b = x_1 x_2$ and the definition of y_i to compute

$$\frac{y_1(x_1^2 - b)}{x_1^2} = \frac{x_1 w(x_1^2 - x_1 x_2)}{x_1^2} = w(x_1 - x_2),$$

$$\frac{y_2(x_2^2 - b)}{x_2^2} = \frac{x_2 w(x_2^2 - x_1 x_2)}{x_2^2} = w(x_1 - x_2).$$

So we are left to verify that $w(x_1 - x_2) = \overline{y}$, which is obvious from the definition of x_1 and x_2. This completes the verification of statement (iii).

Recall that our aim is to prove Lemma 3.4, which says that the subgroup 2Γ has finite index inside Γ. As we will see shortly, this will follow if we can prove that both of the indices $(\overline{\Gamma} : \phi(\Gamma))$ and $(\Gamma : \psi(\overline{\Gamma}))$ are finite. In fact, we will now show that

$$(\overline{\Gamma} : \phi(\Gamma)) \leq 2^{s+1} \quad \text{and} \quad (\Gamma : \psi(\overline{\Gamma})) \leq 2^{r+1},$$

where s is the number of distinct prime factors of $\overline{b} = a^2 - 4b$ and r is the number of distinct prime factors of b.

It is clearly enough to prove one of these statements, so we will just prove the second. From statements (i), (ii), and (iii), we know that $\psi(\overline{\Gamma})$ is the set of points $(x, y) \in \Gamma$ such that x is a non-zero rational square, together with $\mathcal{O}$, and also T if b is a perfect square. The idea of the proof is to find a one-to-one homomorphism from the quotient group $\Gamma/\psi(\overline{\Gamma})$ into a finite group.

Let $\mathbb{Q}^*$ be the multiplicative group of non-zero rational numbers, and let $\mathbb{Q}^{*2}$ denote the group of squares of elements of $\mathbb{Q}^*$,

$$\mathbb{Q}^{*2} = \{u^2 : u \in \mathbb{Q}^*\}.$$

We introduce a map α from Γ to $\mathbb{Q}^*/\mathbb{Q}^{*2}$ defined by

$$\alpha(\mathcal{O}) = 1 \quad (\text{mod } \mathbb{Q}^{*2}),$$
$$\alpha(T) = b \quad (\text{mod } \mathbb{Q}^{*2}),$$
$$\alpha(x, y) = x \quad (\text{mod } \mathbb{Q}^{*2}) \quad \text{if } x \neq 0.$$

We claim that α is a homomorphism and that the kernel of α is precisely the image of ψ. Further, we are able to say a lot about the image of α. Because this result is so important, we state it formally and then give the proof. In particular, we want to draw your attention to part (c) of the following proposition. It says that, modulo squares, there are only a finite number of possibilities for the x-coordinate of a point on the curve. This miraculous fact is really the crux of the proof that the index $(\Gamma : 2\Gamma)$ is finite.

Proposition 3.8. (a) *The map $\alpha : \Gamma \to \mathbb{Q}^*/\mathbb{Q}^{*2}$ described above is a homomorphism.*

(b) *The kernel of α is the image $\psi(\overline{\Gamma})$. Hence α induces a one-to-one homomorphism*

$$\Gamma/\psi(\overline{\Gamma}) \longrightarrow \mathbb{Q}^*/\mathbb{Q}^{*2}.$$

(c) *Let $p_1, p_2, \ldots, p_t$ be the distinct primes dividing b. Then the image of α is contained in the subgroup of $\mathbb{Q}^*/\mathbb{Q}^{*2}$ consisting of the elements*

$$\{\pm p_1^{\epsilon_1} p_2^{\epsilon_2} \cdots p_t^{\epsilon_t} : each\ \epsilon_i\ equals\ 0\ or\ 1\}.$$

(d) *The index $(\Gamma : \psi(\overline{\Gamma}))$ is at most 2^{t+1}.*

Proof. (a) First we observe that α sends inverses to inverses, because

$$\alpha(-P) = \alpha(x, -y) = x = \frac{1}{x} \cdot x^2,$$

so

$$\alpha(-P) \equiv \frac{1}{x} = \frac{1}{\alpha(x, y)} = \alpha(P)^{-1} \quad (\mathrm{mod}\ \mathbb{Q}^{*2}).$$

Hence in order to prove that α is a homomorphism, it is enough to show that whenever $P_1 + P_2 + P_3 = \mathcal{O}$, then $\alpha(P_1)\alpha(P_2)\alpha(P_3) \equiv 1 \ (\mathrm{mod}\ \mathbb{Q}^{*2})$.

The triples of points that add to zero consist of the intersections of the curve with a line. If the line is $y = \lambda x + \nu$ and the x-coordinates of the intersections are x_1, x_2, x_3, then we saw in Section 1.4 that x_1, x_2, x_3 are the roots of the equation

$$x^3 + (a - \lambda^2)x^2 + (b - 2\lambda\nu)x + (c - \nu^2) = 0.$$

This is for the cubic $y^2 = x^3 + ax^2 + bx + c$. Thus

$$x_1 + x_2 + x_3 = \lambda^2 - a,$$
$$x_1 x_2 + x_1 x_3 + x_2 x_3 = b - 2\lambda\nu,$$
$$x_1 x_2 x_3 = \nu^2 - c.$$

The last equation is the one that we want. We are looking at a curve with $c = 0$, so we find that

$$x_1 x_2 x_3 = \nu^2 \in \mathbb{Q}^2.$$

Therefore

$$\alpha(P_1)\alpha(P_2)\alpha(P_3) = x_1 x_2 x_3 = \nu^2 \equiv 1 \quad (\mathrm{mod}\ \mathbb{Q}^{*2}).$$

This completes the proof in the case that P_1, P_2, P_3 are distinct from $\mathcal{O}$ and T. We will leave it as an exercise to check the remaining cases. [N.B. Here we cannot argue by "continuity." Even were we to put a topology on $C(\mathbb{Q})$ be using the inclusion of $C(\mathbb{Q})$ into the real points of C, there is no way to put a topology on $\mathbb{Q}^*/\mathbb{Q}^{*2}$ so that the map α is continuous. Up until now, all of the maps that we have looked at have been defined geometrically, but the homomorphism α is completely arithmetic in nature.]

(b) Comparing the definition of α with the description of $\psi(\overline{\Gamma})$ given in statements (i), (ii), and (iii), it is clear that the kernel of α is precisely $\psi(\overline{\Gamma})$.

(c) We want to know what rational numbers x can occur as the x-coordinate of a point in Γ. We know that such points have coordinates of the form $x = m/e^2$ and $y = n/e^3$. Substituting into the equation and clearing denominators gives

$$n^2 = m^3 + am^2e^2 + bme^4 = m(m^2 + ame^2 + be^4).$$

This equation contains the whole secret. It expresses the square n^2 as a product of two integers. If m and $m^2 + ame^2 + be^4$ were relatively prime, then each of them would be plus or minus a square, and so $x = m/e^2$ would be plus or minus the square of a rational number. In the general case, let

$$d = \gcd(m, m^2 + ame^2 + be^4).$$

Then d divides both m and be^4. But m and e are relatively prime, since we assumed that x was written in lowest terms. Therefore d divides b.

Thus the greatest common divisor of m and $m^2 + ame^2 + be^4$ divides b. Since also $n^2 = m(m^2 + ame^2 + be^4)$, we deduce that every prime dividing m appears to an even power except possibly for primes dividing b. Therefore

$$m = \pm(\text{integer})^2 \cdot p_1^{\epsilon_1} \cdot p_2^{\epsilon_2} \cdots p_t^{\epsilon_t},$$

where each ϵ_i is either 0 or 1, and where $p_1, \ldots, p_t$ are the distinct primes dividing b. This proves that

$$\alpha(P) = x = \frac{m}{e^2} \equiv \pm p_1^{\epsilon_1} \cdot p_2^{\epsilon_2} \cdots p_t^{\epsilon_t} \pmod{\mathbb{Q}^{*2}},$$

and thus that the image of α is contained in the indicated set.

If $x = 0$, and hence $m = 0$, then our argument breaks down. But then the definition $\alpha(T) = b \pmod{\mathbb{Q}^{*2}}$ shows that the conclusion is still valid because, up to squares, b can be written in the indicated form.

(d) The subgroup described in (c) has precisely 2^{t+1} elements. On the other hand, (b) says that the quotient group $\Gamma/\psi(\overline{\Gamma})$ maps one-to-one into this subgroup. Hence the index of $\psi(\overline{\Gamma})$ inside Γ is at most 2^{t+1}. $\qquad\square$

It has been a long journey, but we now have all the tools needed to prove Lemma 3.4. Let us remind you what we now know. We have homomorphisms $\phi : \Gamma \to \overline{\Gamma}$ and $\psi : \overline{\Gamma} \to \Gamma$ such that the compositions $\phi \circ \psi$ and $\psi \circ \phi$ are multiplication by two and such that the indices $(\overline{\Gamma} : \phi(\Gamma))$ and $(\Gamma : \psi(\overline{\Gamma}))$ are finite. We want to prove that 2Γ has finite index in Γ. So the following exercise about abelian groups finishes the proof of Lemma 3.4.

Lemma 3.9. *Let A and B be abelian groups, and suppose that $\phi : A \to B$ and $\psi : B \to A$ are homomorphisms satisfying*

$$\psi \circ \phi(a) = 2a \quad \textit{for all } a \in A \quad \textit{and} \quad \phi \circ \psi(b) = 2b \quad \textit{for all } b \in B.$$

Suppose further that $\phi(A)$ has finite index in B and $\psi(B)$ has finite index in A. Then $2A$ has finite index in A. More precisely, the indices satisfy

$$(A : 2A) \leq (A : \psi(B))(B : \phi(A)).$$

Proof. Since $\psi(B)$ has finite index in A, we can find elements $a_1, \ldots, a_n$ representing the finitely many cosets. Similarly, since $\phi(A)$ has finite index in B, we can choose elements $b_1, \ldots, b_m$ representing the finitely many cosets. We claim that the set

$$\{a_i + \psi(b_j) : 1 \leq i \leq n, \ 1 \leq j \leq m\}$$

includes a complete set of representatives for the cosets of $2A$ in A.

To see this, let $a \in A$. We need to show that a can be written as a sum of an element of this set plus an element of $2A$. Since $a_1, \ldots, a_n$ are representatives for the cosets of $\psi(B)$ in A, we can find some a_i so that $a - a_i \in \psi(B)$, say $a - a_i = \psi(b)$. Next, since $b_1, \ldots, b_m$ are representatives for the cosets of $\phi(A)$ inside B, we can find some b_j so that $b - b_j \in \phi(A)$, say $b - b_j = \phi(a')$. Then

$$a = a_i + \psi(b) = a_i + \psi\big(b_j + \phi(a')\big)$$
$$= a_i + \psi(b_j) + \psi(\phi(a'))$$
$$= a_i + \psi(b_j) + 2a',$$

which gives the desired result. $\square$

To celebrate the completion of our proof of Mordell's theorem, we restate the version that we have proven:

Theorem 3.10. Mordell's Theorem (for curves with a rational point of order two) *Let C be a non-singular cubic curve given by an equation*

$$C : y^2 = x^3 + ax^2 + bx,$$

where a and b are integers. Then the group of rational points $C(\mathbb{Q})$ is a finitely generated abelian group.

Proof. We saw in Section 3.1 that Lemmas 3.1, 3.2, 3.3, and 3.4 imply that $C(\mathbb{Q})$ is finitely generated. We proved Lemma 3.1 in Section 3.1, Lemma 3.2 in Section 3.2, Lemma 3.3 in Section 3.3, and Lemma 3.4 (for curves with a rational point of order two) in the current section. □

Mordell's theorem tells us that we can produce all of the rational points on C by starting from some finite set and using geometry, i.e., using the group law. The following question arises: Given a particular cubic curve, how can we find a generating set? Our proof of Mordell's theorem gives us some tools that often allow us to answer this question. We will do a number of examples in the next section. But at present no one knows a procedure that is *guaranteed* to work for all cubic curves!

3.6 Examples and Further Developments

In this section we illustrate Mordell's theorem by working out some numerical examples. First we discuss some consequences of what we have already proven. We have shown that the group Γ of rational points on the curve

$$C : y^2 = x^3 + ax^2 + bx$$

is a finitely generated abelian group. It follows from the fundamental theorem on such groups that Γ is isomorphic, as an abstract group, to a direct sum of infinite cyclic groups and finite cyclic groups of prime power order. We let $\mathbb{Z}$ denote the additive group of integers, and for notational convenience we let $\mathbb{Z}_m$ denote the cyclic group $\mathbb{Z}/m\mathbb{Z}$ of integers modulo m. Then the structure theorem tells us that Γ looks like

$$\Gamma \cong \underbrace{\mathbb{Z} \oplus \mathbb{Z} \oplus \cdots \oplus \mathbb{Z}}_{r \text{ copies}} \oplus \mathbb{Z}_{p_1^{\nu_1}} \oplus \mathbb{Z}_{p_2^{\nu_2}} \oplus \cdots \oplus \mathbb{Z}_{p_s^{\nu_s}}.$$

More naively, this says that there are generators

$$P_1, \ldots, P_r, Q_1, \ldots, Q_s \in \Gamma$$

such that every $P \in \Gamma$ can be written in the form

$$P = n_1 P_1 + \cdots + n_r P_r + m_1 Q_1 + \cdots + m_s Q_s.$$

Here the integers n_i are uniquely determined by P, while the integers m_j are determined modulo $p_j^{\nu_j}$.

The integer r is called the *rank of* Γ. The group Γ is finite if and only if it has rank $r = 0$. The subgroup

$$\mathbb{Z}_{p_1^{\nu_1}} \oplus \mathbb{Z}_{p_2^{\nu_2}} \oplus \cdots \oplus \mathbb{Z}_{p_s^{\nu_s}}$$

corresponds to the elements of finite order in Γ. It has order $p_1^{\nu_1} p_2^{\nu_2} \cdots p_s^{\nu_s}$ and is called the *torsion subgroup of* Γ.

Of course, the points $P_1, \ldots, P_r, Q_1, \ldots, Q_s$ are not unique. There are many possible choices of generators for Γ.

We have already studied how to compute the elements of finite order in Γ in a finite number of steps. It is much harder to get hold of the rank. We want to give some illustrations of how to do this in special cases. First we do a bit more theory, which will help us in doing the computations.

The proof of Mordell's theorem, if we are lucky, allows us to determine the quotient group $\Gamma / 2\Gamma$. From above, the subgroup 2Γ looks like

$$2\Gamma \cong 2\mathbb{Z} \oplus \cdots \oplus 2\mathbb{Z} \oplus 2\mathbb{Z}_{p_1^{\nu_1}} \oplus \cdots \oplus 2\mathbb{Z}_{p_s^{\nu_s}},$$

so the quotient group has the form

$$\Gamma / 2\Gamma \cong \mathbb{Z}/2\mathbb{Z} \oplus \cdots \oplus \mathbb{Z}/2\mathbb{Z} \oplus \mathbb{Z}_{p_1^{\nu_1}}/2\mathbb{Z}_{p_1^{\nu_1}} \oplus \cdots \oplus \mathbb{Z}_{p_s^{\nu_s}}/2\mathbb{Z}_{p_s^{\nu_s}}.$$

Now $\mathbb{Z}/2\mathbb{Z} = \mathbb{Z}_2$ is cyclic of order two, whereas

$$\mathbb{Z}_{p_i^{\nu_i}}/2\mathbb{Z}_{p_i^{\nu_i}} \cong \begin{cases} \mathbb{Z}_2 & \text{if } p_i = 2, \\ 0 & \text{if } p_i \neq 2. \end{cases}$$

Thus

$$(\Gamma : 2\Gamma) = 2^{r + (\text{number of } j \text{ with } p_j = 2)}.$$

On the other hand, let $\Gamma[2]$ denote the subgroup of all $Q \in \Gamma$ such that $2Q = \mathcal{O}$. What does $\Gamma[2]$ look like? We need to know when

$$2(n_1 P_1 + \cdots + n_r P_r + m_1 Q_1 + \cdots + m_s Q_s) = 0.$$

This happens if $n_i = 0$ for every i and $2m_j \equiv 0 \pmod{p_j^{\nu_j}}$ for every j. If p is odd and $2m \equiv 0 \pmod{p^\nu}$, then $m \equiv 0 \pmod{p^\nu}$. However, if $p = 2$ and

$2m \equiv 0 \pmod{p^{\nu}}$, then we only conclude that $m \equiv 0 \pmod{p^{\nu-1}}$. So the order of the subgroup $\Gamma[2]$ is

$$\#\Gamma[2] = 2^{(\text{number of } j \text{ with } p_j = 2)}.$$

Combining these two formulas, we obtain the useful result

$$(\Gamma : 2\Gamma) = 2^r \cdot \#\Gamma[2].$$

This formula holds for any finitely generated abelian group of rank r.

In our case, what are the possibilities for $\#\Gamma[2]$? How many points can we have with $2Q = \mathcal{O}$? Aside from $\mathcal{O}$, these are the points with $y = 0$, so it is clear from the equation for the curve that the answer is

$$\#\Gamma[2] = \begin{cases} 2, & \text{if } a^2 - 4b \text{ is not a square,} \\ 4, & \text{if } a^2 - 4b \text{ is a square.} \end{cases}$$

Now we have only to recall the last step of the proof of Mordell's theorem to get a formula for the rank that makes it computable in some cases if we are lucky. Remember that we have homomorphisms $\phi : \Gamma \to \overline{\Gamma}$ and $\psi : \overline{\Gamma} \to \Gamma$ such that the composition $\psi \circ \phi$ is multiplication by two. Thus

$$(\Gamma : 2\Gamma) = \big(\Gamma : \psi \circ \phi(\Gamma)\big).$$

We have an inclusion of subgroups $\Gamma \supseteq \psi(\overline{\Gamma}) \supseteq 2\Gamma$, and thus

$$(G : 2\Gamma) = \big(\Gamma : \psi(\overline{\Gamma})\big)\big(\psi(\overline{\Gamma}) : \psi \circ \phi(\Gamma)\big).$$

We want to analyze this last index $\big(\psi(\overline{\Gamma}) : \psi \circ \phi(\Gamma)\big)$. We start with an abstract remark. Let A be an abelian group, let B be a subgroup of finite index in A, and let $\psi : A \to A'$ be a homomorphism of A into some group A'. We are interested in the index $\big(\psi(A) : \psi(B)\big)$.

Using the standard isomorphism theorems from elementary group theory, we find that

$$\frac{\psi(A)}{\psi(B)} \cong \frac{A}{B + \ker(\psi)} \cong \frac{A/B}{(B + \ker(\psi))/B} \cong \frac{A/B}{\ker(\psi)/(\ker(\psi) \cap B)}.$$

Hence

$$\big(\psi(A) : \psi(B)\big) = \frac{(A : B)}{\big(\ker(\psi) : \ker(\psi) \cap B\big)}.$$

If you do not like this abstract argument, you can check the equality of indices directly in our case, because $\ker(\psi)$ consists of the two elements $\overline{\mathcal{O}}, \overline{T}$, and thus $\ker(\psi) \cap \phi(\Gamma)$ is either $\overline{\mathcal{O}}$ or $\ker(\psi)$.

We now apply this abstract formula with $A = \overline{\Gamma}$ and $B = \phi(\Gamma)$. This and the formula for $(\Gamma : 2\Gamma)$ that we derived earlier gives

$$(\Gamma : 2\Gamma) = \frac{(\Gamma : \psi(\overline{\Gamma})) \cdot (\overline{\Gamma} : \phi(\Gamma))}{(\ker(\psi) : \ker(\psi) \cap \phi(\Gamma))}.$$

But we have seen that $\overline{T} \in \phi(\Gamma)$ if and only if $\overline{b} = a^2 - 4b$ is a square, so

$$\big(\ker(\psi) : \ker(\psi) \cap \phi(\Gamma)\big) = \begin{cases} 2, & \text{if } \overline{b} \text{ is not a square,} \\ 1, & \text{if } \overline{b} \text{ is a square.} \end{cases}$$

Now everything falls out nicely, and we find that

$$2^r = \frac{(\Gamma : 2\Gamma)}{\#\Gamma[2]} = \frac{(\Gamma : \psi(\overline{\Gamma})) \cdot (\overline{\Gamma} : \phi(\Gamma))}{4}.$$

Of course, each of the indices in the numerator is a power of 2.

How should we compute these indices? Recall the method that we used to prove that they are finite. We found a homomorphism

$$\alpha : \Gamma \longrightarrow \mathbb{Q}^*/\mathbb{Q}^{*2} \qquad \text{defined by} \qquad \begin{cases} \alpha(x, y) = x \pmod{\mathbb{Q}^{*2}}, \\ \alpha(T) = b \pmod{\mathbb{Q}^{*2}}. \end{cases}$$

We showed that the kernel of α equals the image of $\psi(\overline{\Gamma})$, and so the image of α is isomorphic to

$$\alpha(\Gamma) \cong \Gamma/\ker(\alpha) \cong \Gamma/\psi(\overline{\Gamma}).$$

Hence $(\Gamma : \psi(\overline{\Gamma})) = \#\alpha(\Gamma)$.

Similarly, using the analogous homomorphism $\overline{\alpha} : \overline{\Gamma} \to \mathbb{Q}^*/\mathbb{Q}^{*2}$, we find that $(\overline{\Gamma} : \phi(\Gamma)) = \#\overline{\alpha}(\overline{\Gamma})$. This gives the following alternative formula for the rank of Γ:

$$2^r = \frac{\#\alpha(\Gamma) \cdot \#\overline{\alpha}(\overline{\Gamma})}{4}.$$

It is this formula that we use to try to compute the rank.

In order to determine the image of $\alpha(\Gamma)$, we have to find out which rational numbers, modulo squares, can occur as the x-coordinates of points in Γ. The way that we do this is to write

$$x = \frac{m}{e^2} \quad \text{and} \quad y = \frac{n}{e^3}$$

in lowest terms with $e > 0$.

If $m = 0$, then $(x, y) = T$ and $\alpha(T) = b$. Thus $b \pmod{\mathbb{Q}^{*2}}$ is always in $\alpha(\Gamma)$. If $a^2 - 4b$ is a square, say $a^2 - 4b = d^2$, then Γ has two other points of order two, namely

$$\left(\frac{-a+d}{2}, 0\right) \quad \text{and} \quad \left(\frac{-a-d}{2}, 0\right).$$

So if $a^2 - 4b = d^2$, then $\alpha(\Gamma)$ contains $\frac{1}{2}(-a \pm d)$.

Now we look for points with $m, n \neq 0$. These points satisfy

$$n^2 = m^3 + am^2e^2 + bme^4 = m(m^2 + ame^2 + be^4).$$

In Section 3.5 we showed that m and $m^2 + ame^2 + be^4$ are practically relatively prime, so m and $m^2 + ame^2 + be^4$ are both more-or-less squares. Now we do things systematically.

Let $b_1 = \gcd(m, b)$, where we choose the sign so that $mb_1 > 0$. Then we can write

$$m = b_1 m_1 \quad \text{and} \quad b = b_1 b_2 \quad \text{with } \gcd(m_1, b_2) = 1 \text{ and } m_1 > 0.$$

If we substitute into the equation of the curve, we get

$$n^2 = b_1 m_1 (b_1^2 m_1^2 + ab_1 m_1 e^2 + b_1 b_2 e^4) = b_1^2 m_1 (b_1 m_1^2 + am_1 e^2 + b_2 e^4).$$

Thus $b_1^2 \mid n^2$, so $b_1 \mid n$ and we can write $n = b_1 n_1$. Hence

$$n_1^2 = m_1 (b_1 m_1^2 + am_1 e^2 + b_2 e^4).$$

Since $\gcd(b_2, m_1) = 1$ and $\gcd(e, m_1) = 1$, we see that the quantities m_1 and $b_1 m_1^2 + am_1 e^2 + b_2 e^4$ are relatively prime. Their product is a square, and $m_1 > 0$, so we conclude that each of them is a square. Hence we can factor n_1 as $n_1 = MN$ so that

$$M^2 = m_1 \quad \text{and} \quad N^2 = b_1 m_1^2 + am_1 e^2 + b_2 e^4.$$

Eliminating m_1, we obtain

$$N^2 = b_1 M^4 + aM^2 e^2 + b_2 e^4.$$

This tells the whole story. If you have a point $(x, y) \in \Gamma$ with $y \neq 0$, then you can put that point in the form

$$x = \frac{b_1 M^2}{e^2}, \quad y = \frac{b_1 MN}{e^3}. \tag{$*$}$$

Thus modulo squares, the x-coordinate of any point on the curve is one of the values of b_1, and since b_1 is a divisor of the non-zero integer b, there are only a finite number of possibilities for b_1.

It is now very "easy" to find the order of $\alpha(\Gamma)$. We take the integer b and factor it as a product $b = b_1 b_2$ in all possible ways. For each way of factoring, we write down the equation

$$N^2 = b_1 M^4 + aM^2 e^2 + b_2 e^4.$$

Here a, b_1, b_2 are fixed and M, e, N are variables. Then $\alpha(\Gamma)$ consists of $b \pmod{\mathbb{Q}^{*2}}$, together with those $b_1 \pmod{\mathbb{Q}^{*2}}$ such that the equation has a solution with $M \neq 0$.

In addition, the fact that x and y are in lowest terms implies that

$$\gcd(M, e) = \gcd(N, e) = \gcd(b_1, e) = 1,$$

and the assumption that $\gcd(b_2, m_1) = 1$ implies that

$$\gcd(b_2, M) = \gcd(M, N) = 1.$$

All admissible solutions must also satisfy these side conditions. Notice that if we find a solution M, e, N, then we get a point on Γ by the formulas (∗) for x and y.

If you are observant, you will have noticed that we appear to have forgotten two elements of $\alpha(\Gamma)$. We noted above that if $a^2 - 4b$ is a square, say $a^2 - 4b = d^2$, then there are points of order two whose images by α are the values $\frac{1}{2}(-a \pm d) \in \mathbb{Q}^*/\mathbb{Q}^{*2}$. However, notice that there is then a factorization of b given by

$$b = \frac{-a + d}{2} \cdot \frac{-a - d}{2},$$

so in applying the above procedure, we would consider the equation

$$N^2 = \left(\frac{-a \pm d}{2}\right) M^4 + aM^2 e^2 + \left(\frac{-a \mp d}{2}\right) e^4.$$

This equation has the obvious solution $(M, e, N) = (1, 1, 0)$, so our general procedure takes care of these values automatically.

To summarize, in order to determine the order of $\alpha(\Gamma)$, we write down several equations of the form

$$N^2 = b_1 M^4 + aM^2 e^2 + b_2 e^4, \tag{∗∗}$$

one for each factorization $b = b_1 b_2$. We then need to decide whether or not each of these equations has a solution in integers with $M \neq 0$, and each time that we find an equation with a solution (M, e, N), then we get a new point on the curve by the formula

$$x = \frac{b_1 M^2}{e^2}, \quad y = \frac{b_1 M N}{e^3}.$$

The only trouble with all this is that at present, there is no known method for deciding whether an equation of the form (∗∗) has a solution. Except for this "little" difficulty, we now have a method for computing the rank.

We can hope to get some results as follows. For each b_1 and b_2, either exhibit a solution to the equation (∗∗) or show that the equation has no solutions by considering it as a congruence or as an equation in real numbers. We now illustrate this procedure with several examples.

Example 3.11. $\boxed{C : y^2 = x^3 - x, \quad \overline{C} : y^2 = x^3 + 4x}$

We start with a modest example. In this case $a = 0$ and $b = -1$. The first step is to factor b in all possible ways. There are two factorizations:

$$-1 = -1 \times 1 \quad \text{and} \quad -1 = 1 \times -1.$$

Thus b_1 can only be ± 1. Since $\alpha(\mathcal{O}) = 1$ and $\alpha(T) = b = -1$, we see that

$$\alpha(\Gamma) = \left\{ \pm 1 \ (\text{mod } \mathbb{Q}^{*2}) \right\}$$

is a group of two elements.

Next we must compute $\overline{\alpha}(\overline{\Gamma})$, so we need to apply our procedure to the curve $\overline{C} : y^2 = x^3 + 4x$. Now $\overline{b}$ has lots of factorizations, since we can choose

$$b_1 = 1, -1, 2, -2, 4, -4.$$

But $4 \equiv 1 \ (\text{mod } \mathbb{Q}^{*2})$ and $-4 \equiv -1 \ (\text{mod } \mathbb{Q}^{*2})$, so $\overline{\alpha}(\overline{\Gamma})$ consists of at most four elements $\{1, -1, 2, -2\}$. Of course, we always have $\overline{b} \in \overline{\alpha}(\overline{\Gamma})$, but in this case $\overline{b} = 4$ is a square, so that does not help us.

The four equations that we must consider are[2]:

$$\text{(i)} \quad N^2 = M^4 + 4e^4,$$
$$\text{(ii)} \quad N^2 = -M^4 - 4e^4,$$
$$\text{(iii)} \quad N^2 = 2M^4 + 2e^4,$$
$$\text{(iv)} \quad N^2 = -2M^4 - 2e^4.$$

---

[2]There is a subtlety here. The set $\alpha(\Gamma)$ is a subgroup of $\mathbb{Q}^*/\mathbb{Q}^{*2}$, so in principle we need only consider square-free factors b_1 of b, as we have done in this example. However, if we do this, then we may no longer assume that $\gcd(M, N) = 1$ when searching for solutions.

Since $N^2 \geq 0$ and we do not allow solutions with $M = 0$, we see that equations (ii) and (iv) have no solutions in integers. Indeed, they have no solutions in real numbers with $M \neq 0$, since the right-hand side would be strictly negative.

Equation (i) has the obvious solution $(M, e, N) = (1, 0, 1)$, which corresponds to the fact that $1 \in \overline{\alpha}(\overline{\Gamma})$, so that is nothing new. Finally, our theorem tells us that $\#\alpha(\Gamma) \cdot \#\overline{\alpha}(\overline{\Gamma})$ is at least 4, so for this example we know that $\overline{\alpha}(\overline{\Gamma})$ must have order at least two. Thus equation (iii) must have a solution. Of course, we needn't rely on this fancy reasoning, because (iii) has the obvious solution

$$2^2 = 2 \cdot 1^4 + 2 \cdot 1^4.$$

So we conclude that $\overline{\alpha}(\overline{\Gamma})$ has order two. Thus the rank of Γ is zero, and the same for the rank of $\overline{\Gamma}$. This proves that the groups of rational points on C and $\overline{C}$ are both finite, and so all rational points have finite order.

To find the points of finite order, we can use the Nagell–Lutz theorem. Thus if $P = (x, y)$ is a point of finite order in Γ, then either $y = 0$ or y divides $b^2(a^2 - 4b) = 4$. The points with $y = 0$ are $(0, 0)$ and $(\pm 1, 0)$, and it is a simple matter to check that there are no points with $y = \pm 1$, $y = \pm 2$, or $y = \pm 4$. We have thus proven that the group of rational points on the curve $C : y^2 = x^3 - x$ is precisely

$$C(\mathbb{Q}) = \{\mathcal{O}, (0, 0), (1, 0), (-1, 0)\} \cong \mathbb{Z}_2 \oplus \mathbb{Z}_2.$$

So here is the first explicit cubic equation for which we have provably determined all of the rational solutions.

Similarly, the points of finite order in $\overline{\Gamma}$ satisfy either $y = 0$ or y divides $\overline{b}(\overline{a}^2 - 4\overline{b}) = -256$. After some work, one finds four points of finite order,

$$\overline{C}(\mathbb{Q}) = \{\mathcal{O}, (0, 0), (2, 4), (2, -4)\} \cong \mathbb{Z}_4.$$

In this case the group of rational points is a cyclic group of order four, because one easily checks that $(2, 4) + (2, 4) = (0, 0)$.

Example 3.12. $\boxed{C : y^2 = x^3 + x, \quad \overline{C} : y^2 = x^3 - 4x}$
The situation here is a slight variant of the previous example, so we will leave the details to you. Again one finds that the rank is zero. The finite groups of rational points are given by

$$C(\mathbb{Q}) = \{\mathcal{O}, T\} \cong \mathbb{Z}_2,$$
$$\overline{C}(\mathbb{Q}) = \{\mathcal{O}, (0, 0), (2, 0), (-2, 0)\} \cong \mathbb{Z}_2 \oplus \mathbb{Z}_2.$$

As a by-product of the calculation, we get the answer to an interesting question. Any integer solution of the equation $N^2 = M^4 + e^4$ with $e \neq 0$ gives a rational point on the curve C, namely the point $(M^2/e^2, MN/e^3)$. So once we know that Γ has only the two elements $\mathcal{O}$ and $(0,0)$, it follows that the equation $N^2 = M^4 + e^4$ has no solutions in which M, N, e are all non-zero. This means, in particular, that the Fermat equation $Z^4 = X^4 + Y^4$ has no solutions in non-zero integers. Of course, there are more elementary proofs of this fact.

Example 3.13. $\boxed{C : y^2 = x^3 - 5x, \qquad \overline{C} : y^2 = x^3 + 20x}$

For the curve C, we have $a = 0$ and $b = -5$, so the possibilities for b_1 are $1, -1, 5, -5$. The corresponding equations are

$$\begin{aligned}
\text{(i)} \quad & N^2 = M^4 - 5e^4, \\
\text{(ii)} \quad & N^2 = -M^4 + 5e^4, \\
\text{(iii)} \quad & N^2 = 5M^4 - e^4, \\
\text{(iv)} \quad & N^2 = -5M^4 + e^4.
\end{aligned}$$

Note that equations (i) and (ii) are the same as equation (iii) and (iv) with the variables M and e reversed. Since the solutions that we find will satisfy $Me \neq 0$, it is enough to consider the first two equations.

After a little trial-and-error, we find solutions to (i) and (ii):

$$1^2 = 3^4 - 5 \cdot 2^4,$$
$$2^2 = -(1^4) + 5 \cdot 1^4.$$

Hence all b_1's occur, and as a by-product of the method, we can use the formulas

$$x = \frac{b_1 M^2}{e^2}, \quad y = \frac{b_1 M N}{e^3},$$

to get the rational points $(\frac{9}{4}, \frac{3}{8})$ and $(-1, -2)$ on C. This proves that

$$\alpha(\Gamma) = \{\pm 1, \pm 5\} \pmod{\mathbb{Q}^{*2}},$$

which is the Four Group.

What about $\overline{\alpha}(\overline{\Gamma})$? Since $\overline{b} = a^2 - 4b = 20$, the possibilities for $\overline{b}_1$ are

$$\overline{b}_1 = \pm 1, \pm 2, \pm 4, \pm 5, \pm 10, \pm 20.$$

We observe that since $\bar{b}_1 \bar{b}_2 = \bar{b} = 20$, the factors $\bar{b}_1$ and $\bar{b}_2$ have the same sign. If they are negative, then the equation

$$N^2 = \bar{b}_1 M^4 + \bar{b}_2 e^4$$

has no non-zero rational solutions, because it has no non-zero real solutions. So we are down to

$$\bar{\alpha}(\bar{\Gamma}) \subseteq \{1, 2, 4, 5, 10, 20\}.$$

Next we note that

$$\bar{\alpha}(\bar{\mathcal{O}}) = 1 \equiv 4 \pmod{\mathbb{Q}^{*2}},$$
$$\bar{\alpha}(\bar{T}) = \bar{b} = 20 \equiv 5 \pmod{\mathbb{Q}^{*2}},$$

are both in $\bar{\alpha}(\bar{\Gamma})$. How do we eliminate $\bar{b}_1 = 2$ and $\bar{b}_1 = 10$?

We have to decide whether the equation

$$N^2 = 2M^4 + 10e^4$$

has a solution in integers. Looking back at the relative primality conditions satisfied by M, N, e, it is enough to show that there are no solutions with $\gcd(M, 10) = 1$. Suppose that there is such a solution. Since M is relatively prime to 5, we know from Fermat's Little Theorem that $M^4 \equiv 1 \pmod 5$. So reducing the equation modulo 5, we see that N satisfies

$$N^2 \equiv 2 \pmod 5.$$

But this congruence has no solutions, from which we conclude that the equation $N^2 = 2M^4 + 10e^4$ has no solutions in integers with $\gcd(M, 10) = 1$. Therefore $2 \notin \bar{\alpha}(\bar{\Gamma})$.

A similar calculation would show that $10 \notin \bar{\alpha}(\bar{\Gamma})$, but there is an easier way. Since $\bar{\alpha}(\bar{\Gamma})$ is a subgroup of $\mathbb{Q}^*/\mathbb{Q}^{*2}$, and we already know that 5 is in this subgroup and 2 is not, it is immediate that 10 is not. So now we know that

$$\bar{\alpha}(\bar{\Gamma}) = \{1, 5\} \pmod{\mathbb{Q}^{*2}}.$$

Putting all this together, we find that

$$2^r = \frac{\#\alpha(\Gamma) \cdot \#\bar{\alpha}(\bar{\Gamma})}{4} = \frac{4 \cdot 2}{4} = 2,$$

and so the rank of $C(\mathbb{Q})$ is 1.

There is a general principle involved here. In eliminating the equations $N^2 = \bar{b}_1 M^4 + \bar{b}_2 e^4$ with $\bar{b}_1$ and $\bar{b}_2$ negative, we viewed it as an equation in real numbers. This point of view was not helpful in eliminating $\bar{b}_1 = 2$ and $\bar{b}_2 = 10$, but from the point of view of congruences modulo $p = 5$, we saw that there are no solutions to the congruence $N^2 \equiv 2M^4 + 10e^4 \pmod 5$. Thus for the equation $y^2 = x^3 - 5x$, we could settle the whole issue by taking certain equations and looking at them as equations in the real field and as congruences.

Life gets much rougher when we find a curve for which we do our best to eliminate the b_1's by real and congruence considerations, and still there remain some b_1's that we cannot eliminate and for which we cannot find a solution to $N^2 = b_1 M^4 + b_2 e^4$. Such curves do occur in nature, and the problems in such a situation are of a much higher order of difficulty. We exhibit an equation of this sort in the next example, although we will not give a proof.

Example 3.14. $\boxed{C_p : y^2 = x^3 + px}$

It is curious that $y^2 = x^3 + 10x$ has infinitely many rational solutions, whereas $y^2 = x^3 + x$ and $y^2 = x^3 + 4x$ have only a finite number. In general, it is difficult to predict the rank from the equation of the curve. For example, let's look at the curves

$$C_p : y^2 = x^3 + px,$$

where p is a prime. In this case $b = p$ and $\bar{b} = -4p$, and it is not too hard to show that the rank of $C_p(\mathbb{Q})$ is either 0, 1, or 2.

If $p \equiv 7$ or $11 \pmod{16}$, then an argument similar to the ones that we gave earlier can be used to show that C_p has rank 0. Next, if

$$p \equiv 3 \text{ or } 5 \text{ or } 13 \text{ or } 15 \pmod{16},$$

then it is conjectured, but not yet proven, that the rank is always equal to 1. Finally, in the remaining case $p \equiv 1 \pmod 8$, it is believed that the rank is always 0 or 2, never 1. Both of these can occur, since for example the curves C_{73} and C_{89} both have rank 2, whereas the curves C_{17} and C_{41} both have rank 0.

The last two curves give examples of the hard problem mentioned earlier. In trying to compute the rank of C_{17}, for example, one needs to check whether the equation $N^2 = 17M^4 - 4e^4$ has a non-trivial solution in integers. It turns out that there are no such solutions, even though one can check that there are real solutions and also solutions modulo m for every integer m! So the proof that there are no integer solutions is of necessity somewhat indirect.

We cannot resist mentioning one more C_p, studied by Bremner and Cassels [6]. They show that the innocuous looking curve $y^2 = x^3 + 877x$ has rank 1, as it should by the conjecture mentioned earlier. They further show that its group of rational points is generated by the points $T = (0,0)$ and $P = (x_0, y_0)$, where x_0 has the value

$$x_0 = \left(\frac{612776083187947368101}{78841535860683900210} \right)^2.$$

So even cubic curves with comparatively small coefficients may require points of extremely large height to generate the group of rational points.

We have now seen cubic curves whose rational points have rank 0 and 1, and it is not too hard to find examples with rank 2, or 3, or even 4. But it is quite difficult to find curves of very large rank. In fact, it is still an open question as to whether there exist curves of arbitrarily large rank, and even among experts there is no uniform opinion as to whether the answer should be yes or no.

For curves of the form $y^2 = x^3 + bx$, the largest known rank (as of 2015) is the following example of rank 14, constructed by Mark Watkins in 2002:

$$y^2 = x^3 + 4025997743876907010169104272724 83x.$$

Not surprisingly, the value of b has many factors,

$$b = 3^2 \cdot 7 \cdot 11 \cdot 17 \cdot 19 \cdot 23 \cdot 37 \cdot 59 \cdot 71 \cdot 73 \cdot 97 \cdot 127 \cdot 139 \cdot 151 \cdot 263 \cdot 313 \cdot 443 \cdot 733,$$

which leads to many possible factorizations of b and $\overline{b}$.

For elliptic curves that don't necessarily have a rational point of order two, the largest rank (again as of 2015) was constructed by Noam Elkies in 2006. It has rank 28 and is given by the equation

$$y^2 + xy + y = x^3 - x^2 + bx + c$$

with

$$b = -200677624155755265850332082093385427509302303121789565 02,$$
$$c = 34481611795030556467032985690390720374855944359319180361$$
$$2660082962919394487322434 29.$$

3.7 Singular Cubic Curves

As promised earlier, we now briefly look at singular cubic curves. We will show that the rational points on singular cubic curves and on non-singular cubic curves behave completely differently.

Let C be a cubic curve with a singular point $S \in C$. Then any line through S intersects C at S with multiplicity at least two. If there were a second singular point $S' \in C$, then the line connecting S and S' would intersect C at least twice at S and at least twice at S', so L would intersect C at least four times. But a line and a cubic intersect only three times counting multiplicities. Thus a cubic curve can have at most one singular point.

Even if C is singular, we would like to make the points of C into a group, just as we did for non-singular cubics. It turns out that this can be done quite easily provided that we discard the singular point S. So for any cubic curve C we define

$$C_{ns} = \{P \in C : P \text{ is not a singular point}\}.$$

(The subscript stands for "non-singular.") Similarly, we let $C_{ns}(\mathbb{Q})$ denote the subset of C_{ns} consisting of the points with rational coordinates. As usual, we also fix a point $\mathcal{O} \in C_{ns}$ to be the origin. Then to add two points $P, Q \in C_{ns}$, we use the same geometric procedure that worked for non-singular curves. First we draw the line L connecting P and Q and let R be the other intersection point of $L \cap C$. Then we draw the line L' through R and $\mathcal{O}$. The third intersection point of $L' \cap C$ is defined to be the sum $P + Q$. Then one can checks that C_{ns} is an abelian group, and if $\mathcal{O}$ is in $C_{ns}(\mathbb{Q})$, then $C_{ns}(\mathbb{Q})$ is a subgroup of C_{ns}.

This describes the group law geometrically, but we can also give explicit equations. In fact, if we make a change of variables so that the singular cubic curve is given by a Weierstrass equation

$$y^2 = x^3 + ax^2 + bx + c$$

with $\mathcal{O}$ the point at infinity, then all of the formulas for the addition law derived in Section 1.4 are still true. For example, on the singular cubic curve

$$y^2 = x^3$$

with singular point $S = (0,0)$, the addition law becomes

$$(x_1, y_1) + (x_2, y_2) = \left(\frac{\nu^2}{x_1 x_2}, \frac{-\nu^3}{y_1 y_2}\right), \quad \text{where } \nu = \frac{y_1 x_2 - x_1 y_2}{x_2 - x_1}.$$

If C is non-singular, the Mordell–Weil theorem tells us that $C(\mathbb{Q})$ is a finitely generated group. We are now going to describe exactly what the group $C_{ns}(\mathbb{Q})$ looks like in the case that C is singular. The answer and the proof are much easier than the Mordell–Weil theorem. The only slight complication is that there are several different answers, depending on what the singularity looks like.

We observed in Section 1.3 that there are three possible pictures for the singularity S, depending on whether f has a double root or triple root, and if a double root, whether the tangent directions are real or complex. Typical examples with a double root and real tangent directions, respectively complex tangent directions, are the curves

$$C : y^2 = x^3 + x^2 \quad \text{and} \quad C' : y^2 = x^3 - x^2,$$

and a typical example with a cusp is the curve

$$C'' : y^2 = x^3.$$

(See Figures 1.13–1.15.) We saw in Section 1.3 that it is easy to parametrize all of the rational points on C and C''. For the former we used the maps

$$\mathbb{Q} \longrightarrow C(\mathbb{Q}) \qquad\qquad C(\mathbb{Q}) \longrightarrow \mathbb{Q},$$
$$r \longmapsto (r^2 - 1, r^3 - 1) \qquad\qquad (x, y) \longmapsto y/x,$$

which are easily seen to be inverses of one another. Similarly, the map $r \to (r^2, r^3)$ shows that $C''(\mathbb{Q})$ also looks like $\mathbb{Q}$. However, it turns out that if we use slightly different maps, then we actually get group homomorphisms. We describe what happens for C and C'', and we leave C' for you to do in Exercise 3.15.

Theorem 3.15. (a) *Let C be the singular curve $y^2 = x^3 + x^2$. Then the map*

$$\phi : C_{\mathrm{ns}}(\mathbb{Q}) \longrightarrow \mathbb{Q}^*, \qquad \phi(P) = \begin{cases} \dfrac{y - x}{y + x} & \text{if } P = (x, y), \\ 1 & \text{if } P = \mathcal{O}, \end{cases}$$

is a group isomorphism from $C_{\mathrm{ns}}(\mathbb{Q})$ to the multiplicative group of non-zero rational numbers.

(b) *Let C be the singular curve $y^2 = x^3$. Then the map*

$$\phi : C_{\mathrm{ns}}(\mathbb{Q}) \longrightarrow \mathbb{Q}, \qquad \phi(P) = \begin{cases} \dfrac{x}{y} & \text{if } P = (x, y), \\ 0 & \text{if } P = \mathcal{O}, \end{cases}$$

is a group isomorphism from $C_{\mathrm{ns}}(\mathbb{Q})$ to the additive group of all rational numbers.

Proof. (a) First we observe that ϕ is well-defined. The only possible problem would be if we had a point $(x, y) \in C_{\mathrm{ns}}(\mathbb{Q})$ with $y \pm x = 0$. But then the equation of C would imply that

$$x^3 = y^2 - x^2 = (y + x)(y - x) = 0,$$

so $x = 0$, and then also $y = 0$. Since $(0, 0)$ is the singular point on C, we see that $y \pm x \neq 0$ for all points $(x, y) \in C_{ns}$.

Next, if we set

$$t = \frac{y - x}{y + x} \quad \text{and solve for} \quad y = \left(\frac{1 + t}{1 - t}\right) x,$$

then we can substitute into $y^2 = x^3 + x^2$ and solve for x in terms of t,

$$x = \frac{4t}{(1 - t)^2}.$$

This gives a map

$$\psi : \mathbb{Q}^* \longrightarrow C_{ns}(\mathbb{Q}), \qquad \psi(t) = \begin{cases} \left(\dfrac{4t}{(1 - t)^2}, \dfrac{4t(1 + t)}{(1 - t)^3}\right) & \text{if } t \neq 1, \\ \mathcal{O} & \text{if } t = 1. \end{cases}$$

It is easy to check that $\phi(\psi(t)) = t$ and $\psi(\phi(P)) = P$, which proves that ϕ and ψ are inverse maps of sets. It remains to show that they are homomorphisms.

First we check that ψ sends inverses to inverses.

$$\begin{aligned} \psi\left(\frac{1}{t}\right) &= \left(\frac{4t^{-1}}{(1 - t^{-1})^2}, \frac{4t^{-1}(1 + t^{-1})}{(1 - t^{-1})^3}\right) \\ &= \left(\frac{4t}{(1 - t)^2}, -\frac{4t(1 + t)}{(1 - t)^3}\right) \\ &= -\psi(t). \end{aligned}$$

Next let $P_1, P_2, P_3 \in C_{ns}$ be any three points on C_{ns}. We know that their sum is zero if and only if they are colinear. If we use coordinates $P_i = (x_i, y_i)$, then the line through P_1 and P_2 has the equation

$$(x_2 - x_1)(y - y_1) = (y_2 - y_1)(x - x_1).$$

Substituting $(x, y) = (x_3, y_3)$ and multiplying out both sides, we find that the points P_1, P_2, P_3 are colinear if and only if their coordinates satisfy

$$x_1 y_2 - x_2 y_1 + x_2 y_3 - x_3 y_2 + x_3 y_1 - x_1 y_3 = 0. \qquad (*)$$

Now we need to verify that if three elements $t_1, t_2, t_3 \in \mathbb{Q}^*$ satisfy $t_1 t_2 t_3 = 1$, then their images $\psi(t_1), \psi(t_2), \psi(t_3) \in C_{ns}(\mathbb{Q})$ satisfy $\psi(t_1) + \psi(t_2) + \psi(t_3) = \mathcal{O}$. The formula for ψ given above says that

$$\psi(t) = \left(\frac{4t}{(1-t)^2}, \frac{4t(1-t)}{(1-t)^3} \right).$$

Letting $P_1 = \psi(t_1)$, $P_2 = \psi(t_2)$, and $P_3 = \psi(t_3)$, and substituting into the left-hand side of $(*)$, we find after some algebra that

$$x_1 y_2 - x_2 y_1 + x_2 y_3 - x_3 y_2 + x_3 y_1 - x_1 y_3$$
$$= \frac{32(t_1 - t_2)(t_1 - t_3)(t_2 - t_3)(t_1 t_2 t_3 - 1)}{(1 - t_1)^3 (1 - t_2)^3 (1 - t_3)^3}.$$

This proves that

$$t_1 t_2 t_3 = 1 \implies \psi(t_1), \psi(t_2), \text{ and } \psi(t_3) \text{ are colinear}$$
$$\implies \psi(t_1) + \psi(t_2) + \psi(t_3) = \mathcal{O},$$

at least provided that t_1, t_2, and t_3 are distinct and not equal to 1. The remaining cases can be dealt with similarly, or we could define the group law on all of the real points in C_{ns} and argue that because $\psi : \mathbb{R}^* \to C_{ns}(\mathbb{R})$ is a homomorphism for distinct points, it is a homomorphism for all points by continuity.

(b) The proof for this curve is similar to the proof for (a), but easier, so we leave it for you as an exercise.   □

The Mordell–Weil theorem tells us that if C is a non-singular cubic curve, then the group $C(\mathbb{Q})$ is finitely generated. On the other hand, it is easy to see that the groups $(\mathbb{Q}^*, *)$ and $(\mathbb{Q}, +)$ are not finitely generated. So Theorem 3.15 implies that the group of rational points $C_{ns}(\mathbb{Q})$ on a singular cubic curve is not finitely generated, at least for the two curves covered in the theorem. In the exercises we explain how to show that $C_{ns}(\mathbb{Q})$ is not finitely generated for all singular cubic curves. So the rational points on singular and non-singular cubic curves behave quite differently, and further, the rational points on the singular curves form groups such as $\mathbb{Q}^*$ and $\mathbb{Q}$ with which we are very familiar. We hope that this explains why we have devoted most of our attention to studying rational point on the more interesting and mysterious non-singular cubic curves.

Exercises

3.1. (a) Prove that the set of rational numbers x with height $H(x)$ less than κ contains at most $2\kappa^2 + \kappa$ elements.

(b) * Let $R(\kappa)$ be the set of rational numbers x with height $H(x)$ less than κ. Prove that

$$\lim_{\kappa \to \infty} \frac{\#R(\kappa)}{\kappa^2} = \frac{12}{\pi^2}.$$

3.2. Let $P_1 = (x_1, y_1)$ and $P_2 = (x_2, y_2)$ be points on the non-singular cubic curve

$$y^2 = x^3 + ax^2 + bx + c,$$

where a, b, and c are integers. Let

$$P_3 = (x_3, y_3) = P_1 + P_2 \quad \text{and} \quad P_4 = (x_4, y_4) = P_1 - P_2.$$

(a) Derive formulas for the quantities $x_3 + x_4$ and $x_3 x_4$ in terms of x_1 and x_2. (Note that you should be able to eliminate y_1 and y_2 from these formulas.)

(b) Prove that there is a constant κ, which depends only on a, b, c, so that for all rational points P_1 and P_2,

$$h(P_1 + P_2) + h(P_1 - P_2) \le 2h(P_1) + 2h(P_2) + \kappa.$$

Notice that this greatly strengthens the inequality given in Lemma 3.2.

(c) Prove that if κ is replaced by a suitably large negative number, then the opposite inequality in (b) is true. In other words, prove that there is a constant κ, depending only a, b, c, so that for all rational points P_1 and P_2,

$$-\kappa \le h(P_1 + P_2) + h(P_1 - P_2) - 2h(P_1) - 2h(P_2) \le \kappa.$$

(*Hint.* In (b), replace P_1 and P_2 by $P_1 + P_2$ and $P_1 - P_2$ and use the lower bound $h(2P) \ge 4h(P) - \kappa_0$ provided by Lemma 3.3.)

(d) Prove that for any integer m there is a constant κ_m, depending on a, b, c, m, so that for all rational points P,

$$-\kappa_m \le h(mP) - m^2 h(P) \le \kappa_m.$$

3.3. * Let C be a rational cubic curve given by the usual Weierstrass equation.

(a) Prove that for any rational point $P \in C(\mathbb{Q})$, the limit

$$\hat{h}(P) = \lim_{n \to \infty} \frac{1}{4^n} h(2^n P)$$

exists. The quantity $\hat{h}(P)$ is called the *canonical height* of P. (*Hint.* Use exercise 3.2 to prove that the sequence $4^{-n} h(2^n P)$ is Cauchy.)

(b) Prove that there is a constant κ, depending only on a, b, c, so that for all rational points P we have

$$-\kappa \le \hat{h}(P) - h(P) \le \kappa.$$

(c) Prove that for every integer m and every rational point P,

$$\hat{h}(mP) = m^2 \hat{h}(P).$$

(d) Prove that $\hat{h}(P) = 0$ if and only if P is a point of finite order.

3.4. Prove the upper bound in Lemma 3.6 in Section 3.3 whose proof was omitted in the text.

3.5. Let $\alpha : \Gamma \to \mathbb{Q}^*/\mathbb{Q}^{*2}$ be the map defined in Section 3.5 by the rule

$$\alpha(\mathcal{O}) = 1 \quad (\mathrm{mod} \ \mathbb{Q}^{*2}),$$
$$\alpha(T) = b \quad (\mathrm{mod} \ \mathbb{Q}^{*2}),$$
$$\alpha(x, y) = x \quad (\mathrm{mod} \ \mathbb{Q}^{*2}) \quad \text{if } x \neq 0.$$

Prove that if $P_1 + P_2 + P_3 = \mathcal{O}$, then $\alpha(P_1)\alpha(P_2)\alpha(P_3) \equiv 1 \ (\mathrm{mod} \ \mathbb{Q}^{*2})$. (Except for a few trivial cases, this completes the proof that α is a homomorphism.)

3.6. Let A and B be abelian groups, and let $\phi : A \to B$ and $\psi : B \to A$ be homomorphisms. Suppose that there is an integer $m \geq 2$ so that

$$\psi \circ \phi(a) = ma \quad \text{for all } a \in A,$$
$$\phi \circ \psi(b) = mb \quad \text{for all } b \in B.$$

Suppose further that $\phi(A)$ has finite index in B and $\psi(B)$ has finite index in A.
(a) Prove that mA has finite index in A and that the index satisfies the inequality

$$(A : mA) \leq (A : \psi(B))(B : \phi(A)).$$

(b) Give an example to show that it is possible for the inequality in (a) to be a strict inequality. More generally, show that the ratio

$$\frac{(A : \psi(B))(B : \phi(A))}{(A : mA)}$$

is an integer and give a good description of what this ratio represents.

3.7. This exercise describes a variant of the Nagell–Lutz theorem that often simplifies calculations on curves with a rational point of order two.
(a) Let C be a non-singular cubic curve given in Weierstrass form by an equation

$$y^2 = x^3 + ax^2 + bx,$$

where a and b are integers. Let $P = (x, y) \in C(\mathbb{Q})$ be a point of finite order with $y \neq 0$. Prove that x divides b and that the quantity

$$x + a + \frac{b}{x}$$

is a perfect square. (Note that if this quantity is a square, say equal to N^2, then (x, xN) is a rational point on C, but that such such a point need not have finite order. So this exercise gives a necessary condition for P to have finite order, but not a sufficient condition.)

(b) Let p be a prime. Prove that the only points of finite order on the curve $C : y^2 = x^3 + px$ are $\mathcal{O}$ and $T = (0,0)$.

(c) ** Let $D \neq 0$ be an integer. Prove that the points of finite order on the curve $y^2 = x^3 + Dx$ are as described in the following table:

$$\{P \in C(\mathbb{Q}) : P \text{ has finite order}\} \cong \begin{cases} \mathbb{Z}/4\mathbb{Z} & \text{if } D = 4d^4 \text{ for some } d, \\ \mathbb{Z}/2\mathbb{Z} \oplus \mathbb{Z}/2\mathbb{Z} & \text{if } D = -d^4 \text{ for some } d, \\ \mathbb{Z}/2\mathbb{Z} & \text{otherwise.} \end{cases}$$

3.8. For prime p, let C_p be the cubic curve $y^2 = x^3 + px$ discussed in Section 3.6.
(a) Prove that the rank of C_p is either 0, 1, or 2.
(b) If $p \equiv 7 \pmod{16}$, prove that C_p has rank 0.
(c) If $p \equiv 3 \pmod{16}$, prove that C_p has rank either 0 or 1.

3.9. Using the method developed in Section 3.6, find the rank of each of the following curves.
(a) $y^2 = x^3 + 3x$
(b) $y^2 = x^3 + 5x$
(c) $y^2 = x^3 + 7x$
(d) ** $y^2 = x^3 + 17x$
(e) $y^2 = x^3 + 73x$
(f) * $y^2 = x^3 - 82x$
In each case, if the rank is positive, find points in $C(\mathbb{Q})$ that generate $C(\mathbb{Q})/2C(\mathbb{Q})$.

3.10. (a) Let C be the singular cubic curve $y^2 = x^3$. Prove that the group law on C_{ns} is given by the formula

$$(x_1, y_1) + (x_2, y_2) = \left(\frac{\nu^2}{x_1 x_2}, \frac{-\nu^3}{y_1 y_2} \right), \quad \text{where } \nu = \frac{y_1 x_2 - x_1 y_2}{x_2 - x_1}.$$

(b) Let C be the singular cubic curve $y^2 = x^3 + x^2$. Find a formula for the group law on C_{ns} similar to the formula in (a).

3.11. Let C be the singular cubic curve $y^2 = x^3$. Prove that the map

$$\phi : C_{\text{ns}}(\mathbb{Q}) \longrightarrow \mathbb{Q}, \quad \phi(P) = \begin{cases} \dfrac{x}{y} & \text{if } P = (x, y), \\ 0 & \text{if } P = \mathcal{O}, \end{cases}$$

is a group isomorphism from $C_{\text{ns}}(\mathbb{Q})$ to the additive group of all rational numbers.

3.12. Let $P_1 = (x_1, y_1)$, $P_2 = (x_2, y_2)$, and $P_3 = (x_3, y_3)$ be three points in the plane. Prove that P_1, P_2, and P_3 are colinear if and only if

$$\det \begin{pmatrix} x_1 & y_1 & 1 \\ x_2 & y_2 & 1 \\ x_3 & y_3 & 1 \end{pmatrix} = 0.$$

3.13. (a) Prove that additive group of rational numbers $(\mathbb{Q}, +)$ is not a finitely generated group.

(b) Prove that the multiplicative group of non-zero rational numbers $(\mathbb{Q}^*, *)$ is not a finitely generated group.

3.14. Let C be the cubic curve given by an equation

$$y^2 = x^3 + ax^2 + bx + c$$

with $a, b, c \in \mathbb{Q}$. Suppose that C is singular, and let $S = (x_0, y_0)$ be the singular point.

(a) Prove that x_0 and y_0 are in $\mathbb{Q}$.

(b) Prove that the change of coordinates $x = X + x_0$ and $y = Y$ gives a new equation for C of the form

$$Y^2 = X^3 + AX^2 \quad \text{for some } A \in \mathbb{Q}.$$

(c) Suppose that $A = B^2$ for some non-zero $B \in \mathbb{Q}$. Prove that $C_{\text{ns}}(\mathbb{Q})$ is isomorphic, as a group, to the multiplicative group $\mathbb{Q}^*$ of non-zero rational numbers.

3.15. This is a continuation of the previous exercise. Let $A \in \mathbb{Q}$ be a non-zero rational number that is not a perfect square, i.e., $\sqrt{A} \notin \mathbb{Q}$.

(a) Let H be the conic $u^2 - Av^2 = 1$. If (u_1, v_1) and (u_2, v_2) are two points in $H(\mathbb{Q})$, we define their product by the formula

$$(u_1, v_1) * (u_2, v_2) = (u_1 u_2 + A v_1 v_2, u_1 v_2 + u_2 v_1).$$

Prove that with this operation, $H(\mathbb{Q})$ is an abelian group.

(b) Prove that $H(\mathbb{Q})$ is not a finitely generated group.

(c) Let C be the singular cubic curve $y^2 = x^3 + Ax^2$. Prove that the map

$$\phi : C_{\text{ns}}(\mathbb{Q}) \longrightarrow H(\mathbb{Q}), \qquad \phi(P) = \begin{cases} \left(\dfrac{y^2 + Ax^2}{x^3}, \dfrac{-2y}{x^2} \right) & \text{if } P = (x, y), \\ (1, 0) & \text{if } P = \mathcal{O}, \end{cases}$$

is an isomorphism of groups. Deduce that $C_{\text{ns}}(\mathbb{Q})$ is not a finitely generated group.

(*Hint.* If you have studied field theory, it might help to reformulate this problem in terms of the field $K = \mathbb{Q}(\sqrt{A})$. Show that the product formula in (a) comes from identifying points (u, v) on H with numbers $u + v\sqrt{A}$, and use this to prove that $H(\mathbb{Q})$ is isomorphic to a certain subgroup of K^*. Then check that the map in (c) becomes $(x, y) \mapsto (y - x\sqrt{A})/(y + x\sqrt{A})$.)

3.16. Let

$$\phi(X) = a_0 X^d + a_1 X^{d-1} \cdots + a_d \quad \text{and} \quad \psi(X) = b_0 X^d + b_1 X^{d-1} \cdots + b_d$$

be polynomials of degree $d \geq 2$ with integer coefficients and no common complex roots. We use ϕ and ψ to define a rational function

$$F(X) = \frac{\phi(X)}{\psi(X)} : \mathbb{Q} \cup \{\infty\} \longrightarrow \mathbb{Q} \cup \{\infty\}$$

by setting

$$F(\alpha) = \begin{cases} \phi(\alpha)/\psi(\alpha) & \text{if } \alpha \neq \infty \text{ and } \psi(\alpha) \neq 0, \\ \infty & \text{if } \alpha \neq \infty \text{ and } \psi(\alpha) = 0, \\ a_0/b_0 & \text{if } \alpha = \infty. \end{cases}$$

For $n \geq 1$, we write $F^n = F \circ F \circ \cdots \circ F$ for the n'th iterate of F, and we say that a point $\alpha \in \mathbb{Q}$ is *preperiodic for F* if there are integers $n > m \geq 1$ such that

$$F^n(\alpha) = F^m(\alpha).$$

In other words, α is preperiodic if applying F repeatedly to α eventually comes back to some point that we've already seen. Prove that

$$\{\alpha \in \mathbb{Q} : \alpha \text{ is preperiodic for } F\}$$

is a finite set. This special case of a theorem of Northcott is a basic result in the field of arithmetic dynamics. (*Hint.* Use Lemma 3.6. It may be easier to first prove that there are only finitely many points satisfying $F^n(\alpha) = \alpha$ for some $n \geq 1$. These are called *periodic points*.)

Chapter 4

Cubic Curves over Finite Fields

4.1 Rational Points over Finite Fields

In this chapter we look at cubic equations over a finite field, the field of
integers modulo p. We denote this field by $\mathbb{F}_p$. Of course, now we cannot
visualize things, but we can look at polynomial equations

$$C : F(x, y) = 0$$

with coefficients in $\mathbb{F}_p$ and ask for solutions (x, y) with $x, y \in \mathbb{F}_p$. More
generally, we can look for solutions $x, y \in \mathbb{F}_q$, where $\mathbb{F}_q$ is an extension field
of $\mathbb{F}_p$ containing $q = p^e$ elements. We call such a solution a point on the
curve C. If the coordinates x and y of a solution lie in $\mathbb{F}_p$, we call it a *rational
point*.

If we have a cubic curve that is non-singular, then we can define an
addition law on it, and the points form an abelian group. There is no need
to use any pictures, since the procedures and formulas that we described in
Chapter 1 make perfect sense for any field.

For example, consider the curve

$$y^2 = x^3 + ax^2 + bx + c$$

for some $a, b, c \in \mathbb{F}_p$. This curve is non-singular if and only if $p \neq 2$ and the
discriminant

$$D = -4a^3c + a^2b^2 + 18abc - 4b^3 - 27c^2$$

© Springer International Publishing Switzerland 2015   117
J.H. Silverman, J.T. Tate, *Rational Points on Elliptic Curves*,
Undergraduate Texts in Mathematics, DOI 10.1007/978-3-319-18588-0_4

of the cubic is not zero as an element of $\mathbb{F}_p$. Given points $P_1 = (x_1, y_1)$ and $P_2 = (x_2, y_2)$, we define the sum $P_1 + P_2$ by the usual rules. Ignoring a few exceptional cases (namely $P_1 = \mathcal{O}$, $P_2 = \mathcal{O}$, and $P_1 + P_2 = \mathcal{O}$), we take $y = \lambda x + \nu$ to be the line through P_1 and P_2, so

$$\lambda = \begin{cases} \dfrac{y_2 - y_1}{x_2 - x_1} & \text{if } x_1 \neq x_2, \\ \dfrac{3x_1^2 + 2ax_1 + b}{2y_1} & \text{if } P_1 = P_2, \end{cases}$$

and we let $\nu = y_1 - \lambda x_1 = y_2 - \lambda x_2$. Then $P_3 = (x_3, y_3) = P_1 + P_2$ is given by the formulas

$$x_3 = \lambda^2 - a - x_1 - x_2 \quad \text{and} \quad y_3 = -\lambda x_3 - \nu.$$

All of this makes perfect sense if $a, b, c, x_1, y_1, x_2, y_2$ are in the finite field $\mathbb{F}_p$. Of course, it would be a lot of work to verify that this addition law defines a group, since there are a lot of special cases to check. In particular, the associative law would require lengthy calculations. But we have given you explicit formulas with which to work, so if you have any doubts, feel free to do the necessary checking.

If C is a curve given by an equation of the form

$$C : F(x, y) = 0,$$

we denote the set of rational points by

$$C(\mathbb{F}_p) = \{(x, y) : x, y \in \mathbb{F}_p \text{ and } F(x, y) = 0\}.$$

Actually, just as with our cubic curves, we may also include one or more points "at infinity." These extra points come from making F into a homogeneous polynomial of three variables. We will see an example in the next section.

Before doing more general theory, let's look at an example. Consider the curve

$$y^2 = x^3 + x + 1$$

over the field $\mathbb{F}_5$. How can we find the rational points? Since x and y are supposed to be in $\mathbb{F}_5$, we can just take each of the five possibilities for x, put them into the polynomial $x^2 + x + 1$, and check if the result is a square in $\mathbb{F}_5$. Doing this, we find nine points, including the point $\mathcal{O}$ at infinity:

$$C(\mathbb{F}_5) = \{\mathcal{O}, (0, \pm 1), (2, \pm 1), (3, \pm 1), (4, \pm 2)\}.$$

Thus $C(\mathbb{F}_5)$ is an abelian group of order nine, so it is either a cyclic group of order nine or a product of two cyclic groups of order three. We can determine which one by starting to make a group table. Let $P = (0, 1) \in C(\mathbb{F}_5)$. Then using the formulas given earlier, we compute

$$2P = (4, 2), \quad 3P = (2, 1), \quad 4P = (3, -1), \ldots .$$

Hence $C(\mathbb{F}_5)$ is a cyclic group of order nine. The two points of order three in $C(\mathbb{F}_5)$ are $(2, \pm 1)$, and all of the other non-zero points have order nine.

As this example makes clear, there is never a problem about the group $C(\mathbb{F}_p)$ being finitely generated. Since there are only a finite number of possibilities for x and y, the group $C(\mathbb{F}_p)$ is a finite group. A natural question is to ask for its size. Or if not an exact formula, can we at least give an estimate for the number of points in $C(\mathbb{F}_p)$?

To get an idea of what might be true, let's consider some simpler cases. First, how many points are there on a straight line? If the line is $y = ax + b$, we can take any value for x and then the value for y is determined. So that gives p points. But we really want to count projective points, and a line always has one additional point "at infinity." (In homogeneous coordinates, the line has the equation $Y = aX + bZ$, so it contains the extra point $[1, a, 0]$. See Appendix A, Sections A.1 and A.2.) Thus a line has $p + 1$ points.

Next we might look at a conic C, which is the set of solution $x, y \in \mathbb{F}_p$ to a quadratic equation

$$ax^2 + bxy + cy^2 + dx + ey + f = 0.$$

In Section 1.1 we discussed the solutions to such equations with x and y in the field of rational numbers $\mathbb{Q}$, and everything that we said there works equally well if we replace $\mathbb{Q}$ by a finite field $\mathbb{F}_p$. Further, it turns out that if C is non-singular, then $C(\mathbb{F}_p)$ is never empty, so for a non-singular C, there are always exactly $p + 1$ points in $C(\mathbb{F}_p)$.

We now turn our attention to the curve C given by the equation

$$C : y^2 = f(x),$$

where $f(x)$ is a polynomial with coefficients in $\mathbb{F}_p$. How many points would we expect C to have? We suppose that $p \neq 2$. As we observed earlier, among the non-zero elements $1, 2, \ldots, p - 1$ of the field $\mathbb{F}_p$, half of them are squares (the quadratic residues) and half of them are non-squares (the quadratic non-residues).

Now think of substituting the different values $x = 0, 1, \ldots, p - 1$ into the equation $y^2 = f(x)$. If $f(x) = 0$, there is only the one solution $y = 0$.

If $f(x) \neq 0$, then for half the possible non-zero values of $f(x)$, there are two solutions for y, and for the other possible values of $f(x)$, there are no solutions y. So if the $f(x)$'s were randomly distributed among the squares and the non-squares, we would expect again to get approximately $p+1$ roots. Of course, this does not constitute a proof. But intuitively, each value for x yields either one solution (if $f(x) = 0$), or else it has a 50 % chance of producing two solutions and a 50 % chance of producing no solutions. So the p possible values for x should give approximately p solutions, and then including the point $\mathcal{O}$ at infinity gives $p+1$ points. Thus the number of solutions should look like

$$\#C(\mathbb{F}_p) = p + 1 + (\text{error term}),$$

where we expect the "error term" to be fairly small compared to p.

It turns out that this is true. As long as the polynomial $f(x)$ has distinct roots, there is no tendency for the values of $f(x)$ to be squares or non-squares. So it is true that the number of points on a curve does not differ too much from the number of points on a line. These rough remarks are made precise by the following theorem.

Theorem 4.1 (Hasse–Weil Theorem). *If C is a non-singular irreducible curve of genus g defined over a finite field $\mathbb{F}_p$, then the number of points on C with coordinates in $\mathbb{F}_p$ is equal to $p + 1 - \epsilon$, where the "error term" ϵ satisfies $|\epsilon| \leq 2g\sqrt{p}$.*

It would take us too far afield to actually define the genus, but that will not matter. Let us just say that whenever you have a curve $F(x, y) = 0$, there is a non-negative integer g associated to it called its genus, and as long as the curve is not too singular, the genus increases as the degree of F increases. For example, if p does not divide n, then the Fermat curve $x^n + y^n = 1$ has genus equal to $\frac{1}{2}(n - 1)(n - 2)$. In particular, the cubic curve $x^3 + y^3 = 1$ that we will study in Section 4.2 is a curve of genus 1. More generally, any non-singular curve given by a cubic equation is a curve of genus 1, so an alternative title for this book would have been "Rational Points on Curves of Genus 1"! (But that might have sounded too forbidding to the uninitiated.)

For an elliptic curve C over a finite field $\mathbb{F}_p$, the Hasse–Weil theorem gives the estimate

$$-2\sqrt{p} \leq \#C(\mathbb{F}_p) - p - 1 \leq 2\sqrt{p}.$$

The Hasse–Weil theorem is also called the Riemann hypothesis for curves over finite fields, because there is an alternative way to state it that is analogous to the famous, and as yet unsolved, Riemann hypothesis. The theorem was conjectured by Emil Artin in his thesis and was proven by Hasse [21] in the case $g = 1$, i.e., for elliptic curves. Weil [58] subsequently proved it for curves of arbitrary genus g, and an amazingly deep generalization in higher dimensions was proposed by Weil [59] and proven by Deligne [13].

For some special cubic curves, the result is due to Gauss. In the next section we give Gauss' proof of one of these special cases.

4.2 A Theorem of Gauss

In the last section we stated, without proof, an estimate for the number of solutions to a cubic equation over a finite field. Certain special cases of that theorem were proved by Gauss. In this section we discuss one of those cases, the cubic Fermat curve

$$x^3 + y^3 = 1.$$

This comes from Gauss' *Disquistiones Arithmeticae*, Article 358. It is the first non-trivial case of the theorem ever treated. If you want, you can read about it in Latin in the *Disquistiones*. (It's easy Latin. Or you can read it in the language of your choice – there are several translations available.)

We take the curve in homogeneous form

$$x^3 + y^3 + z^3 = 0$$

and consider solutions in the projective sense. That is, we do not count the trivial solution $(0,0,0)$, and we identify a solution (x, y, z) with all of its non-zero multiples (ax, ay, az). With these conventions, we can now state the theorem of Gauss.

Theorem 4.2 (Gauss). *Let M_p be the number of projective solutions to the equation*

$$x^3 + y^3 + z^3 = 0$$

with x, y, z in the finite field $\mathbb{F}_p$.
(a) If $p \not\equiv 1 \pmod 3$, then $M_p = p + 1$.
(b) If $p \equiv 1 \pmod 3$, then there are integers A and B such that

$$4p = A^2 + 27B^2.$$

The numbers A and B are unique up to changing their signs, and if we fix the sign of A so that $A \equiv 1 \pmod 3$, then

$$M_p = p + 1 + A.$$

Note that if $p \equiv 1 \pmod 3$, then the equation $4p = A^2 + 27B^2$ implies that $A^2 \equiv 1 \pmod 3$. So $A \equiv \pm 1 \pmod 3$, and replacing A by $-A$ if necessary, we can always make $A \equiv 1 \pmod 3$.

Since $B^2 > 0$, it follows that $A^2 = 4p - 27B^2 < 4p$, and thus $|A| < 2\sqrt{p}$. Since the genus in this case is $g = 1$, the Hasse–Weil theorem says that we should have $|M_p - p - 1| \leq 2\sqrt{p}$. But $M_p - p - 1 = A$, so Gauss' theorem is indeed a special case of the Hasse–Weil theorem.

Before beginning the proof of Gauss' theorem, we make a few remarks about the field $\mathbb{F}_p$. This field consists of p elements, $0, 1, \ldots, p - 1$. The multiplicative group $\mathbb{F}_p^*$ of $\mathbb{F}_p$ consists of the non-zero elements $1, 2, \ldots, p - 1$, with the group operation being multiplication.

The multiplicative group $\mathbb{F}_p^*$ is a *cyclic* group of order $p - 1$. Why is it cyclic? Well, if G is a non-cyclic finite abelian group, and if ℓ is the least common multiple of the orders of its elements, then we have a strict inequality $\ell < \#G$, and every element of G satisfies the equation $x^\ell = 1$. Taking $G = \mathbb{F}_p^*$, this would mean that the polynomial $x^\ell - 1$ has more than ℓ solutions in $\mathbb{F}_p^*$. But over a field, a polynomial never has more roots than its degree. Hence the multiplicative group of a finite field is cyclic. More generally, if K is any field and if $G \subset K^*$ is a finite subgroup of the multiplicative group of K, then G is cyclic. You may have run across this fact when K is the field of complex numbers and G is a finite group of roots of unity.

Using this elementary fact about $\mathbb{F}_p^*$, the first part of Gauss' theorem is easy.

Proof of Gauss' theorem. (a) For this part, we assume that

$$p \not\equiv 1 \pmod 3.$$

Then 3 does not divide the order $p - 1$ of the cyclic group $\mathbb{F}_p^*$. It follows that the map $x \to x^3$ is an isomorphism from $\mathbb{F}_p^*$ to itself.

For example, if $p = 5$, then in $\mathbb{F}_5$ we have

$$1^3 = 1, \quad 2^3 = 3, \quad 3^3 = 2, \quad 4^3 = 4.$$

And of course, $0^3 = 0$. So in the case that $p \not\equiv 1 \pmod 3$, every element of $\mathbb{F}_p$ has a unique cube root. Thus the number of solutions of $x^3 + y^3 + z^3 = 0$ is equal to the number of solutions of the linear equation $x + y + z = 0$. This is the equation of a line in the projective plane, so it has exactly $p + 1$ points rational over $\mathbb{F}_p$. Therefore $M_p = p + 1$. So the case that $p \not\equiv 1 \pmod 3$ is extremely easy.

(b) Now we consider the case that $p \equiv 1 \pmod 3$. Let us write

$$p = 3m + 1.$$

Since 3 divides the order of the group $\mathbb{F}_p^*$, the map $x \to x^3$ is a homomorphism of $\mathbb{F}_p^*$ to itself that is neither one-to-one nor onto. For example, if $p = 13$, then the cubes in $\mathbb{F}_{13}^*$ are

$$1^3 = 1, \quad 2^3 = 8, \quad 3^3 = 1, \quad 4^3 = 12, \quad 5^3 = 8, \quad 6^3 = 8,$$
$$7^3 = 5, \quad 8^3 = 5, \quad 9^3 = 1, \quad 10^3 = 12, \quad 11^3 = 5, \quad 12^3 = 12.$$

The image of the homomorphism $x \to x^3$ is a subgroup of $\mathbb{F}_p^*$ which we denote by R, so

$$R = \{x^3 : x \in \mathbb{F}_p^*\}.$$

The subgroup R has index 3 inside $\mathbb{F}_p^*$. The kernel of the map $x \to x^3$ consists of three elements $1, u, u^2$ satisfying $u^3 = 1$. Thus for $p = 13$, we have $R = \{\pm 1 \pm 5\}$, and the kernel of the cubing map consists of the numbers $1, 3, 9 \in \mathbb{F}_p^*$.

The elements of R are called *cubic residues*. We will let S and T denote the other two cosets of R in $\mathbb{F}_p^*$. For example, if we take any $s \in \mathbb{F}_p^*$ that is not in R, then we could take

$$S = sR = \{sr : r \in R\} \quad \text{and} \quad T = s^2R = \{s^2r : r \in R\}.$$

Continuing with our example of $p = 13$, we can choose $s = 2$, and then $S = 2R = \{\pm 2, \pm 10\}$ and $T = 4R = \{\pm 4, \pm 7\}$.

In general the field $\mathbb{F}_p$ is a disjoint union

$$\mathbb{F}_p = \{0\} \cup R \cup S \cup T.$$

The number of elements in each of the sets R, S, and T is $m = \frac{p-1}{3}$. Notice also that $(-1) = (-1)^3$ is a cube, so $R = -R$, $S = -S$, and $T = -T$. In other words, if $r \in R$, then $-r \in R$, and similarly for S and T. Thinking in terms of R, S and T is the key to finding the number of solutions of $x^3 + y^3 + z^3 = 0$.

We want to express the number of solutions M_p in terms of R, S, and T. It's a question of counting. We need to introduce a symbol. Suppose that X, Y, Z are subsets of the field $\mathbb{F}_p$. We let $[XYZ]$ denote the number of triples (x, y, z) such that

$$x \in X, \quad y \in Y, \quad z \in Z, \quad \text{and} \quad x + y + z = 0.$$

What is the number of solutions M_p in terms of this symbol? We first consider the solutions of $x^3 + y^3 + z^3 = 0$ in which x, y, and z are all non-zero. The number of ways of writing zero as a sum of three non-zero cubes is obviously $[RRR]$. But for each non-zero cube, there are three possible field elements which give that cube. Thus there are $27[RRR]$ solutions (x, y, z) of $x^3 + y^3 + z^3 = 0$ with $xyz \neq 0$. But we have agreed not to distinguish proportional solutions (x, y, z) and (ax, ay, az). There are $p - 1$ choices for the multiplier a. Thus there are

$$\frac{27[RRR]}{p - 1} = \frac{9[RRR]}{m}$$

projective solutions of $x^3 + y^3 + z^3 = 0$ in which none of x, y, z is zero.

How many solutions are there if one of the coordinates is zero, say $z = 0$. Then neither x nor y can be zero, because we do not allow $(0, 0, 0)$. So we can pick any non-zero value for x, and once we do that, then there are three possible values for y, namely the solutions of $y^3 = -x^3$. This has three solutions because, as we noted earlier, the group $\mathbb{F}_p^*$ has an element u of order 3. So for a given x, the equation $y^3 = -x^3$ has the three solutions $y = -x$, $y = -ux$, and $y = -u^2x$. Thus there are $3(p-1)$ triples $(x, y, 0)$ such that $x^3 + y^3 = 0$. Similarly for $y = 0$ and $z = 0$, so there are $9(p - 1)$ triples (x, y, z) such that $x^3 + y^3 + z^3 = 0$ and one of x, y, z is zero. Since we do not distinguish proportional triples, we must divide by the $p - 1$ possible multipliers, and so we conclude that there are $\frac{9(p-1)}{p-1} = 9$ projective solutions with one coordinate zero.

Combining these two calculations, we have shown that

$$M_p = \frac{9[RRR]}{m} + 9 = 9 \left(\frac{[RRR]}{m} + 1 \right).$$

The symbol $[XYZ]$ has many marvelous properties that are easy to verify, such as the following, where for any a, we write $aX = \{ax : x \in X\}$.

$$[XY(Z \cup W)] = [XYZ] + [XYW] \quad \text{if } Z \cap W = \emptyset.$$
$$[XYZ] = [aX, aY, aZ] \quad \text{for any } a \neq 0.$$
$$[XYZ] = [XZY] = [YXZ] = [YZX] = [ZXY] = [ZYX].$$

Thus, since $\mathbb{F}_p = \{0\} \cup R \cup S \cup T$ is a disjoint union and $[RR\mathbb{F}_p] = m^2$, we have

$$[RR\{0\}] + [RRR] + [RRS] + [RRT] = m^2.$$

Now fix elements $s \in S$ and $t \in T$. Since

$$[RRS] = [sR, sR, sS] = [SST] \quad \text{and} \quad [RRT] = [tR, tR, tT] = [TTS],$$

we obtain

$$[RR\{0\}] + [RRR] + [SST] + [TTS] = m^2. \qquad (*)$$

Again using $\mathbb{F}_p = \{0\} \cup R \cup S \cup T$ and the obvious fact that $[\mathbb{F}_p TS] = m^2$, we similarly get

$$[\{0\}TS] + [RTS] + [STS] + [TTS] = m^2. \qquad (**)$$

Now $[\{0\}TS] = 0$ because $-S = S$ and $S \cap T = \emptyset$. Also $[RR\{0\}] = m$, because $-R = R$. So if we subtract $(**)$ from $(*)$, we get

$$m + [RRR] = [RTS],$$

and so we have the beautiful formula

$$M_p = 9\frac{[RTS]}{m}.$$

Now we just have to find a clever method of getting $[RST]$. What we are going to do is look at some complex numbers called *cubic Gauss sums*. These complex numbers that we use in the proof are gadgets for keeping track of information about the sets R, S, and T, and in particular they will allow us to relate sums of elements of R, S, and T to products of the associated Gauss sums.

We recall a little bit about the p'th roots of unity. (See Figure 4.1.) Let

$$\zeta = e^{2\pi i/p}.$$

The complex p'th roots of unity are then $1 = \zeta^0, \zeta, \zeta^2, \ldots, \zeta^{p-1}$. Further, we know that $\zeta^a = \zeta^b$ if and only if $a \equiv b \pmod{p}$, which tells us that ζ^a makes sense if a is an element of our finite field $\mathbb{F}_p$. Further, if $a, b \in \mathbb{F}_p$, then $\zeta^{a+b} = \zeta^a \zeta^b$.

We define three complex numbers $\alpha_1, \alpha_2, \alpha_3$ as certain sums of powers of ζ,

$$\alpha_1 = \sum_{r \in R} \zeta^r, \qquad \alpha_2 = \sum_{s \in S} \zeta^s, \qquad \alpha_3 = \sum_{t \in T} \zeta^t.$$

The complex numbers $\alpha_1, \alpha_2, \alpha_3$ are thus each a sum of m different p'th roots of unity. They are called *cubic Gauss sums*. It turns out that they are the three roots of a polynomial equation having integer coefficients. Out next task is to find the equation of that polynomial.

To do this, we multiply together two of the α_i's, say $\alpha_2\alpha_3$. Thus

$$\alpha_2\alpha_3 = \sum_{s \in S} \zeta^s \cdot \sum_{t \in T} \zeta^t = \sum_{s \in S,\, t \in T} \zeta^{s+t} = \sum_{x \in \mathbb{F}_p} N_x \zeta^x,$$

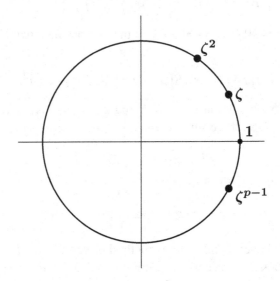

Figure 4.1: The p'th roots of unity

where N_x is the number of pairs (s, t) with $s \in S$ and $t \in T$ satisfying $s + t = x$. We observe that for $r \in R$, we have

$$N_x = [ST\{-x\}] = [rS, rT, \{-rx\}] = [S, T, \{-rx\}] = N_{rx},$$

which shows that N_x depends only on the coset R, S, or T in which x lies. Thus

$$mN_x = [S, T, Rx] = \begin{cases} [STR] & \text{if } x \in R, \\ [STS] & \text{if } x \in S, \\ [STT] & \text{if } x \in T. \end{cases}$$

Define integers a, b, c by

$$[STR] = ma, \quad [STS] = mb, \quad [STT] = mc.$$

Then

$$M_p = 9a$$

and

$$\alpha_2\alpha_3 = a\alpha_1 + b\alpha_2 + c\alpha_3.$$

A similar calculation gives

$$\alpha_3\alpha_1 = a\alpha_2 + b\alpha_3 + c\alpha_1,$$
$$\alpha_1\alpha_2 = a\alpha_3 + b\alpha_1 + c\alpha_2.$$

From now on you can relax because everything else is merely substituting one formula into another until we find an expression for the integer a. Since

$$0 = \zeta^p - 1 = (\zeta - 1)(\zeta^{p-1} + \zeta^{p-2} + \cdots + \zeta + 1)$$

and $\zeta \neq 1$, we have $\zeta^{p-1} + \zeta^{p-2} + \cdots + \zeta + 1 = 0$. Hence

$$\alpha_1 + \alpha_2 + \alpha_3 = \sum_{x \in R \cup S \cup T} \zeta^x = \sum_{x=1}^{p-1} \zeta^x = -1,$$

since the three α_i's include all powers of ζ except ζ^0. Now summing the three formulas for the $\alpha_i\alpha_j$'s, we find that

$$\alpha_1\alpha_2 + \alpha_1\alpha_3 + \alpha_2\alpha_3 = (a + b + c)(\alpha_1 + \alpha_2 + \alpha_3) = -(a + b + c).$$

But

$$\begin{aligned}
m(a + b + c) &= [STR] + [STS] + [STT] \\
&= [ST(R \cup S \cup T)] \\
&= [ST\mathbb{F}_p] - [ST\{0\}] \\
&= m^2,
\end{aligned}$$

so we find that

$$\alpha_1\alpha_2 + \alpha_1\alpha_3 + \alpha_2\alpha_3 = -m.$$

This also allows us to compute the sum of the squares of the α_i's as

$$\alpha_1^2 + \alpha_2^2 + \alpha_3^2 = (\alpha_1 + \alpha_2 + \alpha_3)^2 - 2(\alpha_1\alpha_2 + \alpha_1\alpha_3 + \alpha_2\alpha_3) = 1 + 2m.$$

Our next task is to find $\alpha_1\alpha_2\alpha_3$. To get this quantity, we write

$$\alpha_1(\alpha_2\alpha_3) = \alpha_1(a\alpha_1 + b\alpha_2 + c\alpha_3),$$
$$\alpha_2(\alpha_3\alpha_1) = \alpha_2(a\alpha_2 + b\alpha_3 + c\alpha_1),$$
$$\alpha_3(\alpha_1\alpha_2) = \alpha_3(a\alpha_3 + b\alpha_1 + c\alpha_2).$$

Summing these and using the known facts

$$\alpha_1^2 + \alpha_2^2 + \alpha_3^2 = 1 + 2m \quad \text{and} \quad \alpha_1\alpha_2 + \alpha_1\alpha_3 + \alpha_2\alpha_3 = -m,$$

we get

$$3\alpha_1\alpha_2\alpha_3 = a(1 + 2m) + (b + c)(-m) = a + km,$$

where we have introduced a new letter

$$k = 2a - b - c = 3a - m.$$

So if we can find a value for k, then we will also have computed

$$M_p = 9a = 3(k + m) = 3k + p - 1.$$

Let's stop for a moment and review what we are doing. The sets R, S, and T are defined multiplicatively in terms of cubing, whereas the symbol $[RTS]$ tells us how many times the sum of three things is zero. We are mixing up multiplication and addition and counting, and out of that mixture we have concocted three complex numbers $\alpha_1, \alpha_2, \alpha_3$ and three integers a, b, c and various algebraic relations among them. Now all that we are doing is manipulating those relations until we get what we want because we know that $9a$ is our answer for the number of points on the curve.

Using the values of $\alpha_1 + \alpha_2 + \alpha_3$, $\alpha_1\alpha_2 + \alpha_1\alpha_3 + \alpha_2\alpha_3$, and $\alpha_1\alpha_2\alpha_3$ that we have computed, we see that the complex numbers $\alpha_1, \alpha_2, \alpha_3$ are the roots of the polynomial

$$F(t) = (t - \alpha_1)(t - \alpha_2)(t - \alpha_3) = t^3 + t^2 - mt - \frac{a + km}{3}.$$

Let D_F be the discriminant of F. Using our formula for the $\alpha_i\alpha_j$'s, we can calculate a square root of D_F as

$$\begin{aligned}
\sqrt{D_F} &= (\alpha_1 - \alpha_2)(\alpha_1 - \alpha_3)(\alpha_2 - \alpha_3) \\
&= \alpha_2\alpha_3(\alpha_2 - \alpha_3) + \alpha_3\alpha_1(\alpha_3 - \alpha_1) + \alpha_1\alpha_2(\alpha_1 - \alpha_2) \\
&= (a\alpha_1 + b\alpha_2 + c\alpha_3)(\alpha_2 - \alpha_3) + (a\alpha_2 + b\alpha_e + c\alpha_1)(\alpha_3 - \alpha_1) \\
&\qquad\qquad\qquad\qquad\qquad + (a\alpha_3 + b\alpha_1 + c\alpha_2)(\alpha_1 - \alpha_2) \\
&= (b - c)(\alpha_1^2 + \alpha_2^2 + \alpha_3^2 - \alpha_1\alpha_2 - \alpha_1\alpha_3 - \alpha_2\alpha_3) \\
&= (b - c)(1 + 3m) \\
&= (b - c)p.
\end{aligned}$$

Put

$$\beta_i = 1 + 3\alpha_i \quad \text{for } i = 1, 2, 3.$$

Then we find that

$$\beta_1 + \beta_2 + \beta_3 = 0,$$
$$\beta_1\beta_2 + \beta_1\beta_3 + \beta_2\beta_3 = -3p,$$
$$\beta_1\beta_2\beta_3 = (3k - 2)p.$$

The polynomial whose roots are $\beta_1, \beta_2, \beta_3$ is

$$G(t) = (t - \beta_1)(t - \beta_2)(t - \beta_3) = t^3 - 3pt - (3k - 2)p.$$

Let $A = 2k - 2$. Then as noted earlier, the number of solutions M_p is given by the formula

$$M_p = 3k + p - 1 = p + 1 + A.$$

This is the A referred to in the statement of Gauss' theorem. We just need to show that it has all of the necessary properties.

Let D_G be the discriminant of the polynomial $G(t)$. From the formula for the discriminant of a cubic, we have

$$D_G = -4(-3p)^3 - 27(Ap)^2 = 4 \cdot 27p^3 - 27A^2p^2.$$

On the other hand, since $\beta_i - \beta_j = 3(\alpha_i - \alpha_j)$, we have

$$D_G = 27^2 D_F.$$

Thus

$$4 \cdot 27p^3 - 27A^2p^2 = D_G = 27^2 D_F = 27^2(b - c)^2 p^2.$$

Canceling $27p^2$, we find that

$$4p = A^2 + 27B^2$$

with

$$B = b - c \quad \text{and} \quad A = 3k - 2 \equiv 1 \pmod 3.$$

So magically we obtain the result that $4p$ can be written in the from $4p = A^2 + 27B^2$ with $A \equiv 1 \pmod 3$ and $M_p = p + 1 + A$.

It remains to show that A is uniquely determined by the two conditions $4p = A^2 + 27B^2$ and $A \equiv 1 \pmod 3$. One can argue conceptually or do it with formulas. In keeping with the first part of the proof, we will do it with formulas. So suppose that we have another representation $4p = A_1^2 + 27B_1^2$. Then

$$4p(B_1^2 - B^2) = (A^2 + 27B^2)B_1^2 - (A_1^2 + 27B_1^2)B^2$$
$$= (AB_1 + A_1B)(AB_1 - A_1B).$$

Since p divides the product on the left-hand side, it divides one of the factors on the right, say $p \mid (AB_1 - A_1B)$.

Now we multiply the two formulas for $4p$ to get

$$16p^2 = A^2A_1^2 + 27B^2A_1^2 + 27B_1^2A^2 + 27^2B^2B_1^2,$$

so that

$$16p^2 - (AA_1 + 27BB_1)^2 = 27(AB_1 - A_1B)^2.$$

Since p divides $AB_1 - A_1B$, we see that

$$16 - \left(\frac{AA_1 + 27BB_1}{p}\right)^2 = 27\left(\frac{AB_1 - A_1B}{p}\right)^2.$$

Well now, something is fishy, because the left-hand side is at most 16, whereas the right-hand side is 27 times the square of an integer. So both sides must be zero. In particular, $AB_1 - A_1B = 0$, so if we let

$$\lambda = \frac{A_1}{A} = \frac{B_1}{B}, \quad \text{then} \quad A_1 = \lambda A \quad \text{and} \quad B_1 = \lambda B.$$

Substituting into $A^2 + 27B^2 = 4p = A_1^2 + 27B_1^2$ gives $\lambda^2 = 1$, so $\lambda = \pm 1$. Finally, the assumption that $A \equiv A_1 \equiv 1 \pmod 3$ forces $\lambda = 1$, which proves uniqueness and completes the proof of Gauss' theorem. $\qquad\square$

We illustrate the theorem with some examples. To find the number of points M_p, we just have to solve the equation $4p = A^2 + 27B^2$. For small-ish p, this is not too hard. Here is a short table with a few values.

p	A	B	$M_p = A + p + 1$
7	1	1	9
13	-5	1	9
19	7	1	27
31	4	2	36
4027	-104	14	3924

M_p is always divisible by 9. This is because the group of points on the curve $x^3 + y^3 + z^2 = 0$ has nine points of order three, corresponding to the solutions where one of x, y or z is zero and the other two are cube roots of 1 and -1. Note that in the field $\mathbb{F}_p$, there are three distinct cube roots of 1, so we get nine distinct projective points on the curve. We will leave it to you to check that these nine points form a subgroup that is isomorphic to $\mathbb{Z}/3\mathbb{Z} \oplus \mathbb{Z}/3\mathbb{Z}$, which implies that M_p is divisible by 9. Of course, all of this is only for the case that $p \equiv 1 \pmod 3$.

So now we have this crazy method for computing the number of points on the curve. Take $4p$ and write it as $A^2 + 27B^2$. We know that we can do it. If we actually want to compute A and B, it helps to note that M_p is divisible by 9, so $A \equiv -p - 1 \pmod 9$. And in looking for B, we can think of the formula $4p = A^2 + 27B^2$ as a congruence modulo some small primes. This gives us some information, a kind of sieve, with congruences that B must satisfy. This somewhat eases the quest for A and B.

There is a famous problem concerning the roots $\alpha_1, \alpha_2, \alpha_3$. Letting $\zeta = e^{2\pi i/p}$ be the usual p'th root of unity, we have the well-defined complex number

$$\alpha_1 = \sum_{r \in R} \zeta^r = \frac{1}{3} \sum_{x \in \mathbb{F}_p^*} \zeta^{x^3}.$$

In fact, since ζ^{-r} is the complex conjugate of ζ^r and $-R = R$, we see that α_1 is actually a real number

$$\alpha_1 = \frac{1}{3} \sum_{n=1}^{(p-1)/2} \left(\zeta^{n^3} + \zeta^{-n^3} \right) = \frac{2}{3} \sum_{n=1}^{(p-1)/2} \cos \left(\frac{2\pi n^3}{p} \right).$$

Similarly, both α_2 and α_3 are real. For a given prime p, we can compute the α_i's easily by writing $4p = A^2 + 27B^2$ and using the fact that the α_i's are the roots of the polynomial

$$F(t) = t^3 + t^2 - \frac{p-1}{3}t - \frac{p(A+3) - 1}{27}.$$

Since $D_F \neq 0$, the α_i's are distinct.

> **Question:** For which primes p is α_1 the smallest of the three roots?

The primes $p \equiv 1 \pmod 3$ are mysteriously divided into three types, those types for which α_1 is the smallest root, the middle root, and the largest root of the equation $F(t) = 0$. Let's call these Class 1, Class 2, and Class 3. Kummer [28] made a table for all primes less than 500. He found that there are 7 primes of Class 1, 14 primes of Class 2, and 24 primes of Class 3. Based on this evidence, he suggested that maybe primes fall into the three classes in the ratio 1-to-2-to-3. When early computers became available in the 1950s, Emil Artin suggested this problem to von Neumann and Goldstine to try out on the MANIAC computer. This is a good problem to test on a machine because there is a built-in check. On the one hand, it can compute α_1 directly as a sum of cosines, while on the other hand, it can search for A and B and use them to get the polynomial $F(t)$. Then it can substitute α_1 into $F(t)$ and see if it gets (approximately) zero. They computed for all primes less than 2000 and found that Kummer's table is correct – he had not made any mistakes. This is quite a feat, since when p is around 500, you have to add up 133 cosines to get α_1.

However, tables of this sort for small primes can be quite misleading, and Kummer's guess turned out to be wrong. What is true is that the primes $p \equiv 1 \pmod{3}$ are equally distributed among the three types. This beautiful result was proven by Heath-Brown and Patterson [22]. The proof, which is extremely difficult, uses tools from number theory, geometry, and analysis.

Suppose that we take a non-singular cubic curve with integer coefficients, say

$$ax^3 + bx^2y + cxy^2 + dy^3 + ex^2 + fxy + gy^2 + hx + iy + j = 0,$$

and suppose that we read it as a congruence modulo p for various primes p. If we ask for a formula for the number of solutions M_p, then it is only for some very special cubics that we get an answer like the one that we obtained for $x^3 + y^3 = 1$. In general, the behavior of M_p as a function of p is quite complicated, but a beautiful conjecture of Shimura and Taniyama, which was further refined by Weil, says that the collection of M_p's can be used to form a certain kind of holomorphic function called a *modular form* that has wonderful transformation properties. The semi-stable case of this Modularity Conjecture, which was the case required to prove Fermat's Last Theorem, was proven by Andrew Wiles [60] (with some assistance from Richard Taylor [53]) in 1995, and after several further years of intense work, a proof of the full conjecture was completed by Breuil, Conrad, Diamond, and Taylor [7] in 2001. We briefly discuss the modularity conjecture, and its relation to Fermat's last theorem, in Section 6.6.

We conclude this section by describing another unexpected pattern in the distribution of the M_p's. Since $|M_p - p - 1| \leq 2\sqrt{p}$, we can define an angle θ_p between 0 and π by the condition

$$\cos \theta_p = \frac{M_p - p - 1}{2\sqrt{p}}.$$

We also recall the standard notation $\pi(X)$ for the number of primes less than or equal to X. The prime number theorem says that $\pi(X)$ is asymptotic to $X/\log X$, which means that

$$\lim_{X\to\infty} \frac{\pi(X)}{X/\log X} = 1.$$

A conjecture of Sato and Tate which has recently been proven describes how the angles θ_p are distributed.

Theorem 4.3 (Conjectured by Sato and Tate). *Assume that the cubic curve does not have complex multiplication.*[1] *Then for any fixed angles $0 \le \alpha \le \beta \le \pi$, we have*

$$\lim_{X\to\infty} \frac{\#\{p \le X : \alpha \le \theta_p \le \beta\}}{\pi(X)} = \frac{2}{\pi} \int_{\alpha}^{\beta} \sin^2 t\, dt.$$

Thus the angles θ_p, which determine the number of solutions M_p by the formula

$$M_p = p + 1 + 2\sqrt{p}\cos\theta_p,$$

are distributed in the interval $[0, \pi]$ according to a $\sin^2$ distribution. The Sato–Tate conjecture was proven for an important class of cubic curves by Clozel, Harris, Shepherd-Barron, and Taylor [12, 19, 52], and building on their work, the tools to establish the full conjecture for all cubic curves were developed by a number of mathematicians and appeared in the papers [3, 10, 11, 44]

4.3 Points of Finite Order Revisited

Let C be a cubic curve, given as usual by a Weierstrass equation

$$C : y^2 = x^3 + ax^2 + bx + c$$

with integer coefficients a, b, c. In Chapters 2 and 3 we studied the group of rational points $C(\mathbb{Q})$ on this curve, and in particular we showed that this group is finitely generate (Mordell's theorem) and that the points of finite order have integer coordinates (Nagell–Lutz theorem).

In the present chapter we have been looking at curves with coefficients in a finite field $\mathbb{F}_p$. Suppose that we write $z \to \tilde{z}$ for the map "reduction modulo p,"

---

[1]We define and study cubic curves that do have complex multiplication in Chapter 6.

$$\mathbb{Z} \longrightarrow \mathbb{Z}/p\mathbb{Z} = \mathbb{F}_p, \qquad z \longmapsto \tilde{z}.$$

Then we can take the equation for C, which has integer coefficients, and we can reduce those coefficients modulo p to get a new curve with coefficients in $\mathbb{F}_p$,

$$\tilde{C} : y^2 = x^3 + \tilde{a}x^2 + \tilde{b}x + \tilde{c}.$$

When will the curve $\tilde{C}$ be non-singular? It will be non-singular if $p \geq 3$ and the discriminant

$$\tilde{D} = -4\tilde{a}^3\tilde{c} + \tilde{a}^2\tilde{b}^2 + 18\tilde{a}\tilde{b}\tilde{c} - 4\tilde{b}^3 - 27\tilde{c}^2$$

is non-zero. But reduction modulo p from $\mathbb{Z}$ to $\mathbb{F}_p$ is a homomorphism, so $\tilde{D}$ is just the reduction modulo p of the discriminant D of the cubic $x^3 + ax^2 + bx + c$. In other words, the reduced curve $\tilde{C}$ (mod p) is non-singular provided that $p \geq 3$ and p does not divide the discriminant D.

Having reduced the curve C, it is natural to try taking points in $C(\mathbb{Q})$ and reducing them modulo p to get points on $\tilde{C}$. We can do this provided that the coordinates of the point have no p in their denominator. In particular, if a point has integer coordinates, then we can reduce that point modulo p for any prime p. That is, if $P = (x, y) \in C(\mathbb{Q})$ is a point that happens to have integer coordinates, then x and y satisfy the relation

$$y^2 = x^3 + ax^2 + bx + c,$$

among integers, so we can reduce this relation modulo p to get the equation

$$\tilde{y}^2 = \tilde{x}^3 + \tilde{a}\tilde{x}^2 + \tilde{b}\tilde{x} + \tilde{c}.$$

This last equation says that $\tilde{P} = (\tilde{x}, \tilde{y})$ is a point in $\tilde{C}(\mathbb{F}_p)$. So we get a map from the points in $C(\mathbb{Q})$ with integer coordinates to $\tilde{C}(\mathbb{F}_p)$.

We proved in Section 2.4 that aside from $\mathcal{O}$, all points of finite order in $C(\mathbb{Q})$ have integer coordinates. This was the hard part of the Nagell–Lutz theorem. We are going to study the collection of points of finite order. This is called the *torsion subgroup of* $C(\mathbb{Q})$, and we will denote it by

$$\Phi = \{P = (x, y) \in C(\mathbb{Q}) : P \text{ has finite order}\} \cup \{\mathcal{O}\}.$$

The set Φ is a subgroup of $C(\mathbb{Q})$, since if P_1 and P_2 are points of finite order, then so are $P_1 + P_2$ and $P_1 - P_2$. To see this, we may suppose that $m_1 P_1 = \mathcal{O}$ and $m_2 P_2 = \mathcal{O}$ for some positive integers m_1 and m_2, and then we clearly have $m_1 m_2 (P_1 \pm P_2) = \mathcal{O}$.

Since Φ consists of points with integer coordinates, together with $\mathcal{O}$, we can define a *reduction modulo p map*

$$\Phi \longrightarrow \tilde{C}(\mathbb{F}_p), \qquad P \longmapsto \tilde{P} = \begin{cases} (\tilde{x}, \tilde{y}) & \text{if } P = (x, y), \\ \tilde{\mathcal{O}} & \text{if } P = \mathcal{O}. \end{cases}$$

Now Φ is a subgroup of $C(\mathbb{Q})$, so it is a group, and provided that p does not divide $2D$, we know that $\tilde{C}(\mathbb{F}_p)$ is a group. So we have a map from the group $C(\mathbb{Q})$ to the group $\tilde{C}(\mathbb{F}_p)$, and we next want to check that this map is a homomorphism. (For a more general description of the reduction modulo p map $C(\mathbb{Q}) \to \tilde{C}(\mathbb{F}_p)$ and a proof that it is a homomorphism, see Exercise 4.12 and Appendix A.5.)

First we show that negatives go to negatives. Thus

$$\widetilde{-P} = \widetilde{(x, -y)} = (\tilde{x}, -\tilde{y}) = -\tilde{P}.$$

So it suffices to show that if $P_1 + P_2 + P_3 = \mathcal{O}$, then $\tilde{P}_1 + \tilde{P}_2 + \tilde{P}_3 = \tilde{\mathcal{O}}$. As usual, there are some special cases to check.

If any of P_1, P_2, or P_3 equals $\mathcal{O}$, then the result that we want follows from the fact that negatives go to negatives. So we may assume that P_1, P_2, and P_3 are not equal to $\mathcal{O}$. We write their coordinates as

$$P_1 = (x_1, y_1), \quad P_2 = (x_2, y_2), \quad P_3 = (x_3, y_3).$$

From the definition of the group law on C, the condition $P_1 + P_2 + P_3 = \mathcal{O}$ is equivalent to saying that P_1, P_2, and P_3 lie on a line. Let

$$y = \lambda x + \nu$$

be the line through P_1, P_2, P_3. (If two or three of the points coincide, then the line has to satisfy certain tangency conditions.)

Our explicit formula for adding points says that

$$x_3 = \lambda^2 - a - x_1 - x_2 \quad \text{and} \quad y_3 = \lambda x_3 + \nu.$$

Since x_1, x_2, x_3, y_3 and a are all integers, we see that λ and ν are also integers. This fact is what we need because now we can reduce λ and ν modulo p.

Substituting the equation of the line into the equation of the cubic, we know that the equation

$$x^3 + ax^2 + bx + c - (\lambda x + \nu)^2 = 0$$

has x_1, x_2, x_3 as its three roots. In other words, we have the factorization

$$x^3 + ax^2 + bx + c - (\lambda x + \nu)^2 = (x - x_1)(x - x_2)(x - x_3).$$

This is the relation that ensures that $P_1 + P_2 + P_3 = \mathcal{O}$, regardless of whether the points are distinct.

Reducing this last equation modulo p, we obtain

$$x^3 + \tilde{a}x^2 + \tilde{b}x + \tilde{c} - (\tilde{\lambda}x + \tilde{\nu})^2 = (x - \tilde{x}_1)(x - \tilde{x}_2)(x - \tilde{x}_3).$$

Of course, we can also reduce the equations $y_i = \lambda x_i + \nu$ to get

$$\tilde{y}_i = \tilde{\lambda}\tilde{x}_i + \tilde{\nu} \quad \text{for } i = 1, 2, 3.$$

This means that the line $y = \tilde{\lambda}x + \tilde{\nu}$ intersects the curve $\tilde{C}$ at the three points $\tilde{P}_1$, $\tilde{P}_2$, and $\tilde{P}_3$. Further, if two of the points $\tilde{P}_1, \tilde{P}_2, \tilde{P}_3$ are the same, say $\tilde{P}_1 = \tilde{P}_2$, then the line is tangent to $\tilde{C}$ at $\tilde{P}_1$, and similarly, if all three coincide, then the line has a triple order contact with $\tilde{C}$. Therefore

$$\tilde{P}_1 + \tilde{P}_2 + \tilde{P}_3 = \tilde{\mathcal{O}},$$

which completes the proof that the reduction modulo p map is a homomorphism from Φ to $\tilde{C}(\mathbb{F}_p)$.

Now, lo and behold, we observe that this homomorphism is one-to-one. Why is this true? Because a non-zero point $(x, y) \in \Phi$ is sent to the reduced point $(\tilde{x}, \tilde{y}) \in \tilde{C}(\mathbb{F}_p)$, and that reduced point is clearly not $\tilde{\mathcal{O}}$. So the kernel of the reduction map consists only of $\mathcal{O}$, and hence the map is one-to-one. This means that Φ looks like a subgroup of $\tilde{C}(\mathbb{F}_p)$ for every prime p such that p is relatively prime to $2D$. As we will see, this often allows us to determine Φ with very little work. But before giving some examples, we restate formally the theorem that we have just finished proving.

Theorem 4.4 (Reduction Modulo p Theorem). *Let C be a non-singular cubic curve*

$$y^2 = x^3 + ax^2 + bx + c$$

with integer coefficients a, b, c, and let D be the discriminant

$$D = -4a^3c + a^2b^2 + 18abc - 4b^3 - 27c^2.$$

Let $\Phi \subseteq C(\mathbb{Q})$ be the subgroup consisting of all points of finite order. For any prime p, let $P \to \tilde{P}$ be the reduction modulo p map

$$\Phi \longrightarrow \tilde{C}(\mathbb{F}_p), \qquad P \longmapsto \tilde{P} = \begin{cases} (\tilde{x}, \tilde{y}) & \text{if } P = (x, y), \\ \tilde{\mathcal{O}} & \text{if } P = \mathcal{O}. \end{cases}$$

If p does not divide $2D$, then the reduction modulo p map is an isomorphism of Φ onto a subgroup of $\tilde{C}(\mathbb{F}_p)$.

How can we use this theorem to determine the points of finite order? We give three examples to illustrate how it is used.

Example 4.5. $\boxed{C : y^2 = x^3 + 3}$

The discriminant for this curve is $D = -243 = -3^5$, so there is a one-to-one homomorphism $\Phi \to \tilde{C}(\mathbb{F}_p)$ for all primes $p \geq 5$. But it is easy to check that

$$\#\tilde{C}(\mathbb{F}_5) = 6 \quad \text{and} \quad \#\tilde{C}(\mathbb{F}_7) = 13.$$

Thus $\#\Phi$ divides both 6 and 13, so $\#\Phi = 1$. In other words, the curve C has no rational points of finite order other than $\mathcal{O}$. In particular, this means that the point $(1, 2) \in C(\mathbb{Q})$ has infinite order, so C has infinitely many rational points. (We mention that an alternative way to see that the point $P = (1, 2)$ is not a torsion point is to compute $2P = \left(-\frac{23}{16}, -\frac{11}{64}\right)$. The coordinates of $2P$ are not integers, so Nagell–Lutz tells us that $2P$, and hence also P, are not torsion points.)

It is worth comparing this method for determining Φ with the procedure given by the Nagell–Lutz theorem. Using Nagell–Lutz, we would need to check that there are no rational points on C with y-coordinate in the set

$$\{\pm 1, \pm 3, \pm 9, \pm 27, \pm 81, \pm 243\}.$$

(Using the stronger form of Nagell–Lutz would reduce our task to checking $y \in \{\pm 1, \pm 3, \pm 9\}$.) Clearly $y = \pm 1$ gives no rational points. But if y is divisible by 3, then the equation $y^2 = x^3 + 3$ shows that x must also be divisible by 3. Then $3 = y^2 - x^3$ means that 3 would be divisible by 9, which is absurd. So using the Nagell–Lutz theorem, we have again proven that $\#\Phi = 1$. We will let you decide which method you think was more efficient for computing Φ for this curve.

Example 4.6. $\boxed{C : y^2 = x^3 + x}$

Here the discriminant $D = -4$ is quite small, so it might be easiest to use the Nagell–Lutz theorem, but we will use the reduction theorem to illustrate how it works. We have a one-to-one map $\Phi \to \tilde{C}(\mathbb{F}_p)$ for all primes $p \geq 3$. A little computation gives the values

$$\#\tilde{C}(\mathbb{F}_3) = 4, \quad \#\tilde{C}(\mathbb{F}_5) = 4, \quad \#\tilde{C}(\mathbb{F}_7) = 8.$$

In fact, it is not hard to check that $\#\tilde{C}(\mathbb{F}_p)$ is divisible by 4 for every prime $p \geq 3$.

But suppose that we look at the actual groups.

$$\tilde{C}(\mathbb{F}_3) = \big\{\tilde{O}, (0,0), (2,1), (2,2)\big\},$$
$$\tilde{C}(\mathbb{F}_5) = \big\{\tilde{O}, (0,0), (2,0), (3,0)\big\}.$$

We know that a point in $\tilde{C}$ has order two if and only if its y-coordinate is zero. So

$$\tilde{C}(\mathbb{F}_3) \cong \mathbb{Z}/4\mathbb{Z} \quad \text{and} \quad \tilde{C}(\mathbb{F}_5) \cong \mathbb{Z}/2\mathbb{Z} \oplus \mathbb{Z}/2\mathbb{Z}.$$

The reduction theorem says that Φ looks like a subgroup of both of these groups, so the only possibilities are that Φ is trivial or cyclic of order two. Since $(0,0) \in C(\mathbb{Q})$ is a point of order two, we conclude that $\Phi = \{O, (0,0)\}$.

Example 4.7. $\boxed{C : y^2 = x^3 - 43x + 166}$

The discriminant is $D = -425984 = -2^{15} \cdot 13$. Starting to apply the Nagell–Lutz theorem, we soon find the point $P = (3, 8)$, which might be a point of finite order. Using the doubling formula, we can easily compute the x-coordinates of $2P$, $4P$, and $8P$, which turn out to be

$$x(P) = 3, \quad x(2P) = -5, \quad x(4P) = 11, \quad x(8P) = 3.$$

Thus $x(8P) = x(P)$, so $8P = \pm P$, which shows that P is a point of finite order.

Next we use the reduction theorem. Since $2D$ is relatively prime to 3, we know that Φ is a subgroup of $\tilde{C}(\mathbb{F}_3)$. It is easy to check that $\#\tilde{C}(\mathbb{F}_3) = 7$, so Φ must have order 1 or 7. Since Φ contains the point $P = (3, 8)$, we conclude that Φ has order 7. Therefore the points of finite order in $C(\mathbb{Q})$ form a cyclic group of order 7, and $(3, 8)$ generates this subgroup. Computing the multiples of $(3, 8)$, we find that the group of points of finite order is

$$\Phi = \big\{O, (3, \pm 8), (-5, \pm 16), (11, \pm 32)\big\}.$$

4.4   A Factorization Algorithm Using Elliptic Curves

In the section we are going to discuss the classical problem of factoring integers. The fundamental theorem of arithmetic says that every integer can be written as a product of primes in an essentially unique way. So suppose that we are given a large positive integer n and asked to factor it into primes. First, n itself might be prime, in which case we're done. How can we check? We will see below that it is not difficult to compute $2^k \pmod{n}$, even if n and k are very large. If n is prime, then Fermat's little theorem says that $2^{n-1} \equiv 1 \pmod{n}$. So if we compute $2^{n-1} \pmod{n}$ and find that it is not equal to 1, then we know that n is composite. Suppose that this happens. Then we have conclusively proven that n is composite without having any idea what the factors are!

Warning. The converse to Fermat's little theorem is not true. In fact, there are composite numbers n such that

$$a^{n-1} \equiv 1 \pmod{n}$$

for all a that are relatively prime to n, the smallest such number being 561. Numbers with this property are called *Carmichael numbers*. So we cannot use Fermat's theorem to prove that a number is prime, but only (frequently) to prove that a number is not prime.[2]

Suppose that we are given a number n which we know is composite. If n factors as $n = n_1 n_2$, then the smaller factor is at most $\sqrt{n}$. So this gives a method that is guaranteed to factor n. First we check if $2 \mid n$. If it does, we have found a factor. If not, then we check if $3 \mid n$, then if $4 \mid n$, then if $5 \mid n$, etc. And by the time we get up to $\sqrt{n}$, we are guaranteed to find a factor. Of course, this procedure is wildly inefficient. For example, suppose that n has around 100 digits, and suppose that every second we can check one million possible divisors. Then we will certainly find a factor of n in no more that 3.2×10^{37} years. And even if we make our calculation a million times faster, it could still take us around 3.2×10^{31} years. So we clearly need to find a better procedure.

Why do we want to be able to factor large numbers? From a purely mathematical point of view, the fundamental theorem of arithmetic is a beautiful

---

[2]In theory and in practice, there are many methods that are used to check if a number is prime or composite. If you are interested in learning more about this topic, look up "primality testing," or more specifically the "Miller–Rabin test" and the "Agrawal–Kayal–Saxena (AKS) test."

theorem, so it is natural to want to be able to compute the factorizations that it describes. But there is also a practical reason to factor large numbers. In the 1970s mathematicians devised new sorts of codes based on so-called trapdoor functions built around the problem of factoring large integers. The most famous of these, the RSA cryptosystem, is briefly described in Exercise 4.25. Its security relies on the fact that it is generally easy to check if a number is composite, even though it may be quite hard to actually find a factor. So if you encrypt a message using a composite integer n, then an adversary will be able to read your message if she can factor n. Thus the question of how easy it is to factor large integers is of great interest to governments and to businesses if they want to be sure that their communications remain private.

Before we discuss the problem of factorization, we consider two other computational number theory problems for which there are very efficient algorithms.

Raising to Powers Modulo n

Suppose that we are given three positive integers a, k, and n, and that we want to compute

$$a^k \pmod{n}.$$

This means that we want to find an integer b satisfying

$$b \equiv a^k \pmod{n} \quad \text{and} \quad 0 \le b < n.$$

How long will it take us to compute b? The obvious method is to compute $a_2 = a \cdot a$, reduce a_2 modulo n, compute $a_3 = a_2 \cdot a$, reduce a_3 modulo n, and so on. When we get to a_k we will have our answer, at the cost of k operations, where each operation consists of one multiplication and one reduction modulo n. Is there a better way?

The answer is that there is a much better way, which we illustrate for the exponent $k = 1000$. The first step is to write k as a sum of powers of 2, that is, write k to the base 2. Thus

$$1000 = 2^3 + 2^5 + 2^6 + 2^7 + 2^8 + 2^9.$$

Then we observe that a^{1000} may be written as

$$a^{1000} = a^{2^3} \cdot a^{2^5} \cdot a^{2^6} \cdot a^{2^7} \cdot a^{2^8} \cdot a^{2^9}.$$

For any exponent i, we can use successive squaring to compute a^{2^i} in only i multiplications. Thus we let $A_0 = a$ and calculate

$$A_1 \equiv A_0 \cdot A_0 \equiv a^2 \pmod{n}$$

$$A_2 \equiv A_1 \cdot A_1 \equiv a^4 \pmod{n}$$
$$A_3 \equiv A_2 \cdot A_2 \equiv a^8 \pmod{n}$$
$$\vdots \qquad \vdots$$
$$A_9 \equiv A_8 \cdot A_8 \equiv a^{2^9} \pmod{n}.$$

Then
$$a^{1000} = A_3 \cdot A_5 \cdot A_6 \cdot A_7 \cdot A_8 \cdot A_9 \pmod{n}.$$

So it takes nine operations to get the A_i's, and then six more operations to get a^{1000}. This is much better than the 1000 operations required by our original method. And if k is much larger, say $k \approx 10^{100}$, the savings are enormous.

In general, to compute $a^k \pmod{n}$, we write

$$k = k_0 + k_1 \cdot 2 + k_2 \cdot 2^2 + k_3 \cdot 2^3 + \cdots + k_r \cdot 2^r$$

with each k_i equal to 0 or 1. Next we make a table of values[3]

$$A_0 \equiv a, \quad A_1 \equiv A_0^2, \quad A_2 \equiv A_1^2, \quad \ldots, \quad A_r \equiv A_{r-1}^2,$$

all calculations being done modulo n. Finally we get a^k as

$$a^k \equiv (\text{product of the } A_i\text{'s for which } k_i = 1) \pmod{n}.$$

It takes r operations to compute the A_i's, and then at most r operations to get a^k. So the speed of the algorithm depends on the size of r. We may assume that $k_r = 1$, since otherwise k has a shorter binary expansion. Then

$$k = k_0 + k_1 \cdot 2 + k_2 \cdot 2^2 + k_3 \cdot 2^3 + \cdots + k_r \cdot 2^r \geq 2^r,$$

so
$$r \leq \log_2 k.$$

We have proved the following result.

Proposition 4.8. *It is possible to compute $a^k \pmod{n}$ in at most $2 \log_2 k$ operations, where each operation consists of one multiplication and one reduction modulo n.*

---

[3]In practice, one can use more efficient bookkeeping to avoid storing the whole table of values; see Exercise 4.23.

The logarithm function grows very slowly, so this provides a practical method for computing $a^k \pmod{n}$ even for very large k. For example, if $k = 10^{100}$, then the computation takes fewer than 700 steps.

Computing Greatest Common Divisors

Let a and b be positive integers. How can we compute the greatest common divisor of a and b, that is, the largest integer that divides both a and b? If we can factor a and b into primes, then it is easy, but if a and b are large, this may not be feasible.

An efficient way to compute $\gcd(a, b)$ is the *Euclidean algorithm*, which many of you have probably already seen. The idea is to use division with remainder. Thus first we divide a by b to get a quotient q and a remainder r. In other words,

$$a = bq + r \quad \text{with } 0 \le r < b.$$

Next we divide b by r, and so on. This leads to a sequence of equations

$$
\begin{aligned}
a &= bq_1 + r_2 & &\text{with } 0 \le r_2 < b, \\
b &= r_2 q_2 + r_3 & &\text{with } 0 \le r_3 < r_2, \\
r_2 &= r_3 q_3 + r_4 & &\text{with } 0 \le r_4 < r_3, \\
&\ \ \vdots & &\ \ \vdots \\
r_{n-1} &= r_n q_n + r_{n+1} & &\text{with } 0 \le r_{n+1} < r_n, \\
r_n &= r_{n+1} q_{n+1}.
\end{aligned}
$$

(If you let $r_0 = a$ and $r_1 = b$, the numbering system of the r_i's and q_i's will make more sense.) Since

$$b = r_1 > r_2 > r_3 > \cdots$$

and the r_i's are non-negative integers, we eventually get to zero, say $r_{n+2} = 0$. Then it is not hard to check that

$$\gcd(a, b) = r_{n+1}.$$

How many steps does the Euclidean algorithm take in order to compute $\gcd(a, b)$? We claim that the successive remainders satisfy the estimate

$$r_{i+1} < \frac{1}{2} r_{i-1}.$$

So every two steps cuts the remainder at least in half, and the algorithm terminates when we reach a remainder of zero. Switching a and b if necessary, we may assume that $a \ge b$, and at the first step we have $r_2 < b$. Hence

$$r_4 < \frac{1}{2}b, \qquad r_6 < \frac{1}{2}r_4 < \frac{1}{4}b, \qquad r_8 < \frac{1}{2}r_6 < \frac{1}{8}b, \qquad \dots \qquad r_{2i} < \frac{1}{2^{i-1}}b.$$

But r_{2i} is a non-negative integer, so as soon as $2^{i-1} \geq b$, we get $r_{2i} < 1$, which means that $r_{2i} = 0$. In other words,

$$i \geq 1 + \log_2 b = \log_2(2b) \quad \text{implies that} \quad r_{2i} = 0.$$

So the Euclidean algorithm takes at most $2\log_2(2b)$ steps to compute the greatest common divisor of a and b. And again, since the logarithm function grows so slowly, the Euclidean algorithm is practical even for very large values of a and b.

Now we verify the claim that

$$r_{i+1} < \frac{1}{2}r_{i-1}.$$

If $r_i \leq \frac{1}{2}r_{i-1}$, then we are done, since we know that $r_{i+1} < r_i$. On the other hand, suppose that $r_i > \frac{1}{2}r_{i-1}$. We know that

$$r_{i-1} = r_i q_i + r_{i+1} \quad \text{with } 0 \leq r_{i+1} < r_i,$$

so using our assumption that $r_i > \frac{1}{2}r_{i-1}$, we find that

$$r_{i+1} = r_{i-1} - r_i q_i < r_{i-1} - \frac{1}{2}r_{i-1}q_i = r_{i-1}\left(1 - \frac{1}{2}q_i\right).$$

Since the r_i's are strictly decreasing, we must have $q_i \geq 1$, and since the r_i's are non-negative, we must have $q_i \leq 1$, so $q_i = 1$. This gives the desired inequality $r_{i+1} < \frac{1}{2}r_{i-1}$. We have proven the following result.

Proposition 4.9. *Let a and b be positive integers. The Euclidean algorithm computes the greatest common divisor of a and b in at most*

$$2\log_2 \max\{2a, 2b\} \text{ operations,}$$

where each operation is one division with remainder.

Now we turn to the difficult problem of factoring integers. We saw earlier that if n is composite, then it is always possible to factor n in no more than $\sqrt{n}$ steps, but that usually takes far too long. We start by describing a factorization algorithm due to Pollard [36]. Pollard's method does not work for all n's, but when it does work, it is fairly efficient. And more importantly, it is the prototype for the elliptic curve factorization algorithm that we discuss later in this section.

The idea underlying Pollard's algorithm is not difficult. Suppose that n happens to have a prime factor p such that $p - 1$ is a product of small primes. Then

$$a^{p-1} \equiv 1 \pmod{p},$$

so p divides $\gcd(a^{p-1} - 1, n)$.

Of course, initially we do not know the value of p, so we cannot compute $a^{p-1} - 1$. Instead we choose an integer

$$k = 2^{e_2} \cdot 3^{e_3} \cdot 5^{e_5} \cdots r^{e_r},$$

where $2, 3, 5, \ldots, r$ are the first few primes and $e_1, e_2, \ldots, e_r$ are small positive integers. Then we compute

$$\gcd(a^k - 1, n).$$

In doing this computation, we only need the value of $a^k - 1$ modulo n, so using Propositions 4.8 and 4.9, we can compute $\gcd(a^k - 1, n)$ in no more than about $2 \log_2(2kn)$ operations. This can be done quite easily even if k and n are as large as 10^{1000}.

Now suppose that we are lucky and n has a prime factor p satisfying $p - 1 \mid k$. Then p will divide $a^k - 1$,[4] so

$$\gcd(a^k - 1, n) \geq p > 1.$$

If $\gcd(a^k - 1, n) \neq n$, then this gcd value is a non-trivial factor of n, so we can factor n into two pieces and repeat the procedure on each piece. On the other hand, if the gcd equals n, then we can choose a new a and try again. So the idea is to compute $\gcd(a^k - 1, n)$. If it is strictly between 1 and n, then we have factored n, if it equals n, then we choose a new a, and if it equals 1, then we choose a larger k.

We illustrate with an example. Let

$$n = 246082373.$$

The first thing to do is to check that n is not itself prime. This follows from the computation $2^{n-1} \equiv 180137693 \pmod{246082373}$. So now we know that n is composite and we want to find a factor.

---

[4]If a and n are not relatively prime, then Fermat's little theorem cannot be used. But in the unlikely event that $\gcd(a, n) > 1$, the gcd is already a non-trivial factor of n.

We take

$$a = 2 \quad \text{and} \quad k = 5! = 120 = 2^3 \cdot 3 \cdot 5.$$

Writing 120 in binary as $120 = 2^3 + 2^4 + 2^5 + 2^6$, the fast powering algorithm allows us to rapidly compute

$$2^{120} = 2^{2^3} \cdot 2^{2^4} \cdot 2^{2^5} \cdot 2^{2^6} \equiv 153677509 \pmod{246082373}.$$

Then the equally fast Euclidean algorithm gives

$$\gcd(2^{120} - 1, n) = \gcd(153677508, 246082373) = 1.$$

So the algorithm fails, and n has no prime factors p such that $p-1$ divides 120.

But all is not lost, we can just go back and choose a larger k. For our new k we take

$$k = 7! = 5040 = 2^4 \cdot 3^2 \cdot 5 \cdot 7.$$

Then

$$2^{5040} = 2^{2^4} \cdot 2^{2^5} \cdot 2^{2^7} \cdot 2^{2^8} \cdot 2^{2^9} \cdot 2^{2^{12}} \equiv 101220672 \pmod{246082373},$$

and the Euclidean algorithm yields

$$\gcd(2^{5040} - 1, n) = \gcd(101220671, 246082373) = 2521,$$

so we have found a non-trivial factor of n.

More precisely, we have factored n as

$$n = 246082373 = 2521 \cdot 97613.$$

It is easy to check that each of the factors is prime, so this gives the complete factorization of n. Of course, we do not mean to suggest that Pollard's algorithm is needed to factor a small number such as $n = 246082373$. But this example illustrates the salient features of the algorithm.

There is one more issue. How should we choose the exponent k? We want k to be divisible by a lot of small primes to small powers, and we want to be able to compute $a^k \pmod{n}$ efficiently. Taking k to be successive factorials works well. Thus we take $k = 1!, 2!, 3!, \ldots$. The reason that this is especially convenient is because, having computed the value of $a^{d!} \pmod{n}$, we can compute the next value as

$$a^{(d+1)!} \equiv (a^{d!})^{d+1} \pmod{n}.$$

A formula due to Sterling says that $d!$ is roughly equal to $(d/e)^d$, so comput-ing $a^{d!}$ with fast powering takes (roughly) at most $2d\log_2(d)$ steps, which is quite reasonable if d is not too large. Table 4.1 summarizes Pollard's $p-1$ algorithm. (We mention that for added efficiency, it's probably best to only evaluate the gcd in Step 4 every m'th time through the loop for some appro-priately chosen value of m.)

Let $n \geq 2$ be a composite integer to be factored.

Step 1: Set $a = 2$ (or any other convenient value).

Step 2: Loop $d = 2, 3, 4, \ldots$ up to a specified bound.

    Step 3: Replace a with $a^d \pmod{n}$.

    Step 4: Compute $g = \gcd(a - 1, n)$.

    Step 5: If $1 < g < n$, then **success**, return the value of g.

    Step 6: If $g = n$, go to **Step 1** and choose a new a.

Step 7: Increment d and loop again at **Step 2**.

Table 4.1: Pollard's $p - 1$ factorization algorithm

Notice that Pollard's algorithm should eventually stop, since eventually $d!$ is divisible by $p - 1$ for some prime $p \mid n$, and for that d, we have $a^{d!} \equiv 1 \pmod{p}$. So for that d the gcd in Step 4 is greater than 1, and the algorithm terminates unless we are very unlucky and the gcd turns out to be n. However, if $p - 1$ is not a product of small primes for some prime divisor of n, then the algorithm is not practical for large values of n. The algorithm only works in a "reasonable" amount of time if it happens that n has a prime divisor p satisfying

$$p - 1 = \text{product of small primes to small powers.}$$

Now we are ready to describe Lenstra's idea [30] for using elliptic curves to create an algorithm that (conjecturally) does not have this defect. Pollard's algorithm is based on the fact that the non-zero elements in $\mathbb{Z}/p\mathbb{Z}$ form a group $(\mathbb{Z}/p\mathbb{Z})^*$ of order $p - 1$, so if $p - 1 \mid k$, then $a^k = 1$ in the group. Lenstra's idea is to replace the group $(\mathbb{Z}/p\mathbb{Z})^*$ by the group of points on an elliptic curve $C(\mathbb{F}_p)$, and to replace the integer a by a point $P \in C(\mathbb{F}_p)$. As in Pollard's algorithm, we choose an integer k composed of a product of small primes, say $k = d!$. Then, if it happens that the number of elements in $C(\mathbb{F}_p)$ divides k, then we will have $kP = \mathcal{O}$ in $C(\mathbb{F}_p)$. And just as before, the fact that $kP = \mathcal{O}$ generally allows us to find p, which is a non-trivial factor of n.

What is the advantage of Lenstra's algorithm? If we use only one curve C with integer coefficients and consider its reductions modulo various primes, then there is no advantage. For a single curve C, we win if there is some prime p dividing n such that $\#\tilde{C}(\mathbb{F}_p)$ is a product of small primes. Similarly, we win using Pollard's algorithm if there is a prime p dividing n such that $p - 1$ is a product of small primes. But suppose now that we do not win. Using Pollard's algorithm, not winning means losing and going home, the game is over. But with Lenstra's algorithm, there is a new flexibility that allows us to continue playing. Namely, we are free to choose a new elliptic curve and start over again. Since $\#\tilde{C}(\mathbb{F}_p)$ varies considerably for a fixed prime p and varying curve C, our odds of eventually winning are fairly good.

Now we take these vague comments and turn them into an explicit algorithm. We noted in Section 4.1 that if C is a non-singular cubic curve with coefficients in $\mathbb{F}_p$, then

$$\#C(\mathbb{F}_p) = p + 1 - \epsilon_p \quad \text{with } |\epsilon_p| \le 2\sqrt{p}.$$

Further, Birch [5] has shown that as C varies over all cubic curves modulo p, the numbers ϵ_p are quite well spread out over the interval from $-2\sqrt{p}$ to $2\sqrt{p}$. So it is quite likely (but not yet rigorously proven) that we will fairly rapidly run across a curve C for which $\#C(\mathbb{F}_p)$ is a equal to a product of small primes.

So we choose an elliptic curve E with mod n coefficients and a point $P \in E$ with mod n coordinates and compute kP with $k = 1!, 2!, 3!, \ldots$. This raises several issues.

First, for a given b and c modulo n, how do we find even one solution (x_1, y_1) to the congruence

$$y^2 \equiv x^3 + bx + c \pmod{n}?$$

This appears to be difficult if we don't know how to factor n. However, we're content to use a random elliptic curve, so rather than fixing b and c, we simply take random values for b, x_1, and y_1, and set

$$c \equiv y_1^2 - x_1^3 - bx_1 \pmod{n}.$$

Second, how can we efficiently compute kP if k is large. Clearly not as a k-fold sum $P + P + \cdots + P$. Instead we use the same binary expansion trick that we used to compute a^k. First write

$$k = k_0 + k_1 \cdot 2 + k_2 \cdot 2^2 + k_3 \cdot 2^3 + \cdots + k_r \cdot 2^r,$$

with each k_i either 0 or 1. As before, we can do this with $r \leq \log_2 k$. Next we compute

$$P_0 = P$$
$$P_1 = 2P_0 = 2P$$
$$P_2 = 2P_1 = 2^2 P$$
$$P_3 = 2P_2 = 2^3 P$$
$$\vdots \qquad \vdots$$
$$P_r = 2P_{r-1} = 2^r P.$$

Finally we calculate

$$kP = (\text{sum of } P_i\text{'s for which } k_i = 1).$$

This allows us to compute kP in fewer than $2\log_2 k$ steps of doubling and adding points.

Note, however, that we do not want to compute the coordinates of kP as rational numbers, because the numerators and denominators would have approximately k^2 digits. Even for relatively small values of k, such as $k = 41!$, this leads to numbers with more digits than there are elementary particles in the known universe. So it is much better to perform all computations modulo n.

But n is not prime, so how can we use the formulas for addition and doubling? Let's consider the problem of adding two points $Q_1 = (x_1, y_1)$ and $Q_2 = (x_2, y_2)$, where x_1, y_1, x_2, y_2 are integers modulo n and we want to perform all computations modulo n. Our formula for $Q_3 = Q_1 + Q_2$ says that

$$x_3 = \lambda^2 - x_1 - x_2 \quad \text{and} \quad y_3 = -\lambda x_3 - (y_1 - \lambda x_1), \quad \text{where} \quad \lambda = \frac{y_2 - y_1}{x_2 - x_1}.$$

The difficulty lies in computing λ, because the ring $\mathbb{Z}/n\mathbb{Z}$ is not a field, so $x_2 - x_1$ might not have an inverse. When we try to compute the inverse of $x_2 - x_1$ modulo n, there are three possible outcomes:

(1) $\boxed{\gcd(x_2 - x_1, n) = 1}$ In this case $x_2 - x_1$ has an inverse in $\mathbb{Z}/n\mathbb{Z}$, so we can calculate Q_3 modulo n. (Note that if $\gcd(a, n) = 1$, then an adaptation of the Euclidean algorithm gives a solution to the equation $ax \equiv 1 \pmod{n}$. So if the inverse exists, then there is a fast way to find it; see Exercises 4.18 and 4.24.)

(2) $\boxed{1 < \gcd(x_2 - x_1, n) < n}$ In this case we cannot find Q_3, but we don't care because the integer $\gcd(x_2 - x_1, n)$ is a non-trivial factor of n. So the algorithm can be terminated here.

(3) $\boxed{\gcd(x_2 - x_1, n) = n}$ If this case occurs, then we have been very unlucky. We could try a smaller value of k, or just go back to the beginning and choose a new curve.

Similarly, to double a point $Q = (x, y)$ modulo n, we need to compute the ratio

$$\lambda = \frac{f'(x)}{2y} = \frac{2x^2 + 2ax + b}{2y} \pmod{n}.$$

So we get the same three alternatives: either we can compute $2Q$ modulo n, or we get a non-trivial factor of n, or $\gcd(y, n) = n$ and we have to start with a new curve.

Lenstra's elliptic curve factorization is summarized in Table 4.2. The description includes all of the essential underlying features of the algorithm, although in practice there are many ways to make it more efficient.

We illustrate Lenstra's algorithm by factoring

$$n = 1715761513.$$

Let $n \geq 2$ be a composite integer to be factored.
Step 1: Check that $\gcd(n, 6) = 1$ and that n is not a perfect power.
Step 2: Choose random integers b, x_1, and y_1 modulo n.
Step 3: Set $P = (x_1, y_1)$ and $c \equiv y_1^2 - x_1^3 - bx_1 \pmod{n}$.
Step 4: Let E be the elliptic curve $E : y^2 = x^3 + bx + c$.
Step 5: Loop $d = 2, 3, 4, \ldots$ up to a specified bound $d_{\max}$.
Step 6: Compute $Q = dP \pmod{n}$ and set $P = Q$.
Step 7: If the computation in **Step 6** fails,   then we have found a divisor $g > 1$ of n.
Step 8: If $g < n$, then **success**, return the value of g
Step 9: If $g = n$, go to **Step 2** to pick a new curve and point.
Step 10: Increment d and, if $d \leq d_{\max}$, loop again at **Step 5**.
Step 11: Go to **Step 2** to pick a new curve and point.

Table 4.2: Lenstra's elliptic curve factorization algorithm

The first thing to check is that n is not prime. Using the square-and-multiply scheme described earlier, we easily calculate that

$$2^{1715761512} \equiv 114094409 \pmod{1715761513}.$$

Applying Fermat's little theorem, this proves that n is not prime, so now we search for a factor.

The first step of Lenstra's algorithm says to check that n is not a perfect power. Using a calculator, we compute each of

$$\sqrt{n}, \quad \sqrt[3]{n}, \quad \sqrt[4]{n}, \quad \sqrt[5]{n}, \dots, \sqrt[31]{n} \approx 1.9855.$$

None of them are integers, so n is not a perfect power.

Step 2 says to choose random integers b, x_1, and y_1 modulo n. We will take

$$x_1 = 2 \quad \text{and} \quad y_1 = 3, \quad \text{so} \quad P = (2, 3).$$

For b we will use various values until we find one that works, and for a given b, we take $c = 1 - 2b$. Also, to make it clearer what's happening as the algorithm progresses, we write P_d for the point computed in Step 6 during the d'th loop. Thus

$$P_d = dP_{d-1} = \dots = d!P_1.$$

Let's start with $b = 1$ and $c = -1$, so we are looking at the curve and initial point

$$C : y^2 = x^3 + x - 1 \quad \text{and} \quad P = P_1 = (2, 3) \in C.$$

We'll take $d_{\max} = 20$, so we iterate the d-loop from Step 5 to Step 10, with $d = 2, 3, \dots, 20$. The points that we find are listed in Table 4.3.

So now we know that on the curve $y^2 = x^3 + x - 1$ considered modulo $n = 1715761513$, the point $P = (2, 3)$ satisfies

$$20!P = 2432902008176640000(2, 3) = (693588502, 858100579).$$

What does this tell us about the factors of n? Nothing! The whole point of Lenstra's algorithm is that it gives us a factor of n precisely when the addition law breaks down. So if we are actually able to compute $d!P \pmod{n}$, then we have to continue with either a larger multiplier d or a new point P and curve C.

So suppose we stick with this point and curve and take d up to 50? Then we find that

d	P_d
1	$(2, 3)$
2	$(524260463, 1437744601)$
3	$(1580374945, 1281688384)$
4	$(102166583, 726409659)$
5	$(1230754737, 656248933)$
6	$(1439423743, 261453828)$
7	$(649350388, 251146533)$
8	$(850659306, 148388675)$
9	$(859697522, 1168641628)$
10	$(1393637669, 651726681)$

d	P_d
11	$(535090466, 120781551)$
12	$(621168269, 1626584297)$
13	$(1562301880, 1546127470)$
14	$(1506757996, 1569723892)$
15	$(1234029292, 1672539306)$
16	$(1276800395, 664055804)$
17	$(1547160202, 159566783)$
18	$(495807207, 511034411)$
19	$(1226239889, 1164547094)$
20	$(693588502, 858100579)$

Table 4.3: An example of factoring using elliptic curves

$$P_{50} = 50!P = (321131143, 586731948) \pmod{1715761513}.$$

This doesn't help, so maybe it's time to try a new curve. We stick with the point $P = (2, 3)$, but now we take $b = 2$ and $c = -3$, and we compute up to $d_{\max} = 20$. Again we hit no obstacles to computing $20!P$ modulo n, nor is there a problem with $b = 3$ or $b = 4$. But when we try $b = 5$, we hit the jackpot. Everything goes smoothly as we compute up to

$$16!P = (962228801, 946564039) \pmod{1715761513}$$
$$\text{on the curve} \quad y^2 = x^3 + 5x - 9.$$

But see what happens when we try to compute $17!P$. Letting

$$Q = 16!P = (962228801, 946564039),$$

we have to compute $17Q$, which we do via the double-and-add formula

$$17!P = 17Q = 2 \cdot 2 \cdot 2 \cdot 2 \cdot Q + Q.$$

First we compute $2^i Q$ modulo n for $i = 0, 1, 2, 3, 4$,

$$Q = (962228801, 946564039)$$
$$2Q = (731126553, 1349251536)$$
$$4Q = (731200636, 806528011)$$
$$8Q = (108793287, 1488256803)$$
$$16Q = (505708443, 718251590).$$

Then to compute $17Q$, we need to add Q to $16Q$. This involves finding the inverse, modulo n, of the difference of the x-coordinates of Q and $16Q$, so we need to invert

$$x(16Q) - x(Q) = 505708443 - 962228801 = -456520358 \quad \text{modulo } n.$$

But when we use the Euclidean algorithm to compute the gcd of this quantity and n, we find that

$$\gcd\big(x(16Q) - x(Q), n\big) = \gcd(-456520358, 1715761513) = 26927.$$

This gives a non-trivial factor of n, and indeed in this case it gives the complete prime factorization of n,

$$n = 1715761513 = 26927 \cdot 63719.$$

With hindsight, we can see why this choice of elliptic curve managed to factor n. The curve $C : y^2 = x^3 + 5x - 9$ has the property that

$$\#C(\mathbb{F}_{26927}) = 2^4 \cdot 3^2 \cdot 11 \cdot 17 \quad \text{and} \quad \#C(\mathbb{F}_{63719}) = 2^2 \cdot 3 \cdot 5303.$$

So $17!\tilde{P} = \tilde{O}$ in $C(\mathbb{F}_{26927})$, since the order of the group $C(\mathbb{F}_{26927})$ divides $17!$, but not surprisingly, we have $17!\tilde{P} \neq \tilde{O}$ in $C(\mathbb{F}_{63719})$, since the orders of most points in $C(\mathbb{F}_{63719})$ are multiples of 5303.

Of course, as with the example that we did for Pollard's $p - 1$ algorithm, there is no need to use elliptic curves to factor the comparatively small number $n = 1715761513$. Our aim is simply to illustrate the basic operation of Lenstra's algorithm.

4.5   Elliptic Curve Cryptography

The 1970s saw a revolutionary advance in the field of cryptography with the introduction of public key cryptosystems by Diffie, Hellman, Merkle, Rivest, Shamir, Adelman, and others. A cryptosystem allows two parties, typically called "Bob" and "Alice", to exchange information over an insecure communication channel in such a way that their adversary, "Eve", is unable to determine the information. Mathematically, one may view a basic cryptosystem as an injective function

$$f : \{\text{messages}\} \longrightarrow \{\text{encrypted messages}\}.$$

Bob encrypts his message m by computing $c = f(m)$ and sending the value of c to Alice, who decrypts the message by computing

$$f^{-1}(c) = f^{-1}\big(f(m)\big) = m.$$

In classical *private key cryptosystems*, anyone who knows how to compute the function f is also easily able to compute f^{-1}. So it is essential that the private key, which is the function f, be a closely guarded secret known only to Bob and Alice. In particular, before they can communicate securely, Bob and Alice need to agree on a secret key f that is unknown to Eve.

But suppose that Bob and Alice have never met, and that their only means of communication is via email or text messaging. Even if their personal adversary Eve doesn't have the resources to monitor their communications, there are likely to be other agencies that do. Public key cryptography solves this problem. In a *public key cryptosystem*, Alice can publish her encryption key f, and despite the fact that Eve and a host of criminal enterprises and initialed agencies know Alice's public encryption key f, they are unable to compute the inverse function f^{-1} required to decrypt messages.

It was a brilliant idea to conceive that public key cryptography might be possible, as Diffie and Hellman did in 1976, but even knowing the concept, it's far from clear how one might actually construct a public key cryptosystem. Indeed, the Diffie–Hellman paper did not give an example. Various public key cryptosystems have been proposed, and many broken, over the subsequent decades. The best known, which you've probably seen, is the RSA system. In this system, Alice's public key is a large number N that is a product of two large primes, $N = pq$, and it is believed that in order to decrypt messages, Eve needs to find p and q. (For a brief reminder of how RSA works, see Exercise 4.25.) So one says that the security of RSA relies on the difficulty of factoring large numbers. The most powerful factorization method currently known is called the number field sieve. The time that it takes to factor N is (more-or-less) proportional to $e^{c\sqrt[3]{\log N}}$ for a small constant c.[5] At present, it is considered infeasible to factor numbers N that satisfy $N \geq 2^{2048} \approx 10^{617}$.

Other public key cryptosystems rely on the difficulty of the so-called *discrete logarithm problem* (DLP), which asks the following: Let p be a prime, and let a and b be non-zero numbers modulo p.

DLP: Find an integer m that solves the congruence $a^m \equiv b \pmod{p}$.

It is clear why this is called a logarithm problem, since if we didn't work modulo p, then m would simply be the logarithm of b to the base a. If p is

---

[5] If the two prime factors of $N = pq$ are of approximately the same size, then the number field sieve is faster than the elliptic curve factorization method described in Section 4.4. But if p is significantly smaller than q, then the elliptic curve method may be faster, since it takes roughly $e^{c\sqrt{\log p}}$ steps to factor N.

large, this is a hard problem, with the best solution method (called the *index calculus* for reasons that we do not discuss) taking time roughly proportional to $e^{c\sqrt[3]{\log p}}$. So as with RSA, DLP-based public key cryptosystems generally use numbers satisfying $p \geq 2^{2048}$. See Exercise 4.27 for a brief description of a DLP-based system called Elgamal.

Our modest goal in this section is to discuss how public key cryptosystems can be created using a hard problem on elliptic curves, and to explain why these elliptic curve systems appear to have practical advantages over RSA and DLP cryptosystems. This not being a text on cryptography, we do not want to enter too deeply into the details, and we acknowledge that we will be sweeping under the rug a great number of important issues that affect the security of such systems. There are many texts, such as [24] and [27], where the interested reader can learn more about public key cryptography in general, and elliptic curve cryptography in particular.

The first step is to note that there is a version of the (discrete) logarithm problem in any group. Thus if G is a given group and $a, b \in G$ are elements of G, we may ask for an exponent m solving the formula $a^m = b$ in the group G. Taking $G = \mathbb{F}_p^*$, the multiplicative group of the field $\mathbb{F}_p$, gives the DLP described earlier. Taking $G = \mathbb{R}^*$ (and allowing $m \in \mathbb{R}$) gives classical logarithms that have been studied since the seventeenth century. And as you have undoubtedly guessed, if we take $G = C(\mathbb{F}_p)$ to be the group of mod p points on an elliptic curve, then we have the *elliptic curve discrete logarithm problem*, which is abbreviated as the ECDLP. Since the group law on an elliptic curve is written additively, the ECDLP in $C(\mathbb{F}_p)$ is the following:[6]

ECDLP: Given $P, Q \in C(\mathbb{F}_p)$, find an integer m so that $mP = Q$.

Example 4.10. Consider the elliptic curve

$$C : y^2 = x^3 + x^2 + x + 1 \quad \text{over the field} \quad \mathbb{F}_{97}.$$

The points $P = (7, 20)$ and $Q = (17, 46)$ are in $C(\mathbb{F}_{97})$. The ECDLP asks for an integer m such that $mP = Q$. One way to solve this problem is to compute $2P, 3P, 4P, \ldots$ until eventually finding that $47P = Q$. A faster method is to use what is called a "collision algorithm." Here one makes two lists, say $P, 2P, 3P, \ldots$ and $Q - 10P, Q - 20P, Q - 30P, \ldots$, until finding a point that appears on both lists, say $aP = Q - 10bP$. Then $(a + 10b)P = Q$, so we can take $m = a + 10b$. (Here we choose 10 because 10 is close to $\sqrt{97}$.) Thus

---

[6]We will always assume that P and Q are chosen so that there is such an m.

$$P = (7, 20), \ 2P = (71, 70), \ 3P = (17, 51), \ 4P = (69, 40),$$
$$5P = (52, 75), \ 6P = (84, 26), \ 7P = (8, 87), \dots ,$$
$$Q - 10P = (1, 2), \ Q - 20P = (61, 96), \ Q - 30P = (80, 93),$$
$$Q - 40P = (8, 87), \ Q - 50P = (17, 46), \dots .$$

Looking at the lists, we see the collision $7P = (8, 87) = Q - 40P$, so $47P = Q$.

In general working over $\mathbb{F}_p$, one takes n to be approximately $\sqrt{p}$ and makes lists kP and $Q - nkP$ for $k = 1, 2, 3 \dots$ Under suitable hypotheses, one can show that a collision will occur for some $k < n$; see Exercise 4.28. So using the collision method only requires about $2\sqrt{p}$ additions on the curve, as opposed to the naive method of computing $P, 2P, 3P, \dots$ until finding Q, which on average takes $\frac{1}{2}p$ additions. For more about collision algorithms, including a brilliant idea due to Pollard that achieves the same result without having to store long lists of data, see [24, §§2.7, 5.4, 5.5].

The Elgamal cryptosystem, and another important cryptographic construction called Diffie–Hellman key exchange (Exercise 4.26), can be formulated using the discrete logarithm problem in almost any group. So why should we use an elliptic curve group $C(\mathbb{F}_p)$, where the group law is so complicated, rather than the multiplicative group $\mathbb{F}_p^*$, where the group law is simply multiplication modulo p? The answer lies in the differing degrees of difficulty of the discrete logarithm problem in different groups. As an extreme example, consider the discrete logarithm problem in the cyclic group $\mathbb{Z}_n$. Solving $ma \equiv b \pmod{n}$ for m using the Euclidean algorithm takes at most $2 \log n$ steps, so it is very easy to find m even if n is enormous. On the other hand, as we noted earlier, solving the DLP in $\mathbb{F}_p^*$ currently takes around $e^{c\sqrt[3]{\log p}}$ steps, so is infeasible if $p > 2^{2048}$.

In the mid-1980s, Neal Koblitz and Victor Miller (independently) suggested that the discrete logarithm problem on elliptic curves might be much more difficult than on $\mathbb{F}_p^*$. Using a collision algorithm as described in Example 4.10, one can solve the discrete logarithm problem in any group G in roughly $2\sqrt{o(G)}$ steps; cf. Exercise 4.28. And despite decades of study, no one has found a better algorithm to solve the ECDLP on general elliptic curves, although faster methods are known in certain special cases. So as of 2015, the best known algorithms take a small multiple of $\sqrt{p}$ to solve the ECDLP in $C(\mathbb{F}_p)$. And if p is large, then $\sqrt{p}$ steps take much longer than the $e^{c\sqrt[3]{p}}$ steps required to solve the DLP in $\mathbb{F}_p^*$ using the index calculus. The upshot is that instead of using a prime $p > 2^{2048}$, it suffices for ECDLP-based cryptosystems to take roughly $p > 2^{200}$.

Why does this matter? Suppose that Alice wants to put her public key on her credit card, or that an airline wants to use a public key in the bar code on your printed airline ticket, or that a manufacturer wants to put a public key on a computer chip in your car and your refrigerator and your microwave.[7] On such constrained devices, every bit stored and every bit transmitted is expensive. An RSA key, or an Elgamal key using $\mathbb{F}_p^*$, requires around 2000 bits, while an Elgamal key using an elliptic curve $C(\mathbb{F}_p)$ requires only around 200 bits. That's a huge savings, and explains why elliptic curve cryptography is used in many real-world situations.

What might make us believe that the ECDLP in $C(\mathbb{F}_p)$ is harder than the DLP in $\mathbb{F}_p^*$? One explanation comes from comparing the natural homomorphisms

$$R_p^* \longrightarrow \mathbb{F}_p^* \quad \text{and} \quad C(\mathbb{Q}) \longrightarrow C(\mathbb{F}_p),$$

where $R_p = \{a/b \in \mathbb{Q} : p \nmid b\}$ is the local ring that we used in Section 2.4. The index calculus, which is the strongest method known for solving the DLP, uses this homomorphism and the fact that R_p^* is infinitely generated with many "small" generators. By way of contrast, Mordell's theorem tells us that the group $C(\mathbb{Q})$ is finitely generated, so it appears that an elliptic curve index calculus cannot even get started.

Unfortunately, the preceding paragraph must be viewed as a mix of philosophy and marketing! It's a disconcerting fact that we currently don't know, in the sense of having proofs, that integer factorization or the DLP or the ECDLP is hard in an appropriately rigorous sense. For all that anyone knows, it may be possible to factor N in time proportional to a small power of $\log N$, or to solve the ECDLP in $C(\mathbb{F}_p)$ in time proportional to a small power of $\log p$. The question of rigorously classifying which mathematical problems can be solved in polynomial time, which problems require exponential time, and which problems lie between, is a fundamental research topic in computer science and complexity theory.

Finally, we would be remiss without a quick mention of quantum computers, amazing devices that are under development, but which no one knows when, or even if, will ever be built. What is known is that a working quantum computer with enough quantum bits will be able to factor N and to solve the DLP and ECDLP in polynomial time. So quantum computers, if they're ever constructed, are likely to sound the death knell on the use of elliptic curves in cryptography. But it's unlikely that they, or any other discovery or invention

---

[7]These are all actual real-world applications, although some use something called a digital signature, rather than a public key cryptosystem.

or device, will ever dissuade people from studying the beautiful mathematical theory of elliptic curves.

Exercises

4.1. Let $p \neq 2$ be a prime, let $a, b, c, d \in \mathbb{F}_p$ satisfy $acd \neq 0$, and let C be the conic given by the homogeneous equation

$$C : ax^2 + bxy + cy^2 = dz^2.$$

(a) If $b^2 \neq 4ac$, prove that $\#C(\mathbb{F}_p) = p + 1$.
(b) If $b^2 = 4ac$, prove that either

$$\#C(\mathbb{F}_p) = 1 \quad \text{or} \quad \#C(\mathbb{F}_p) = 2p + 1.$$

Give examples for $p = 3$ to show that both possibilities can occur. More generally, show that both possibilities occur for all odd primes.

4.2. Compute the group $C(\mathbb{F}_p)$ for the curve

$$C : y^2 = x^3 + x + 1$$

and the primes $p = 3, 7, 11$, and 13.

4.3. Let $p \geq 3$ be a prime, and let $m \geq 1$ be an integer that is relatively prime to $p - 1$.
(a) Prove that the map $x \mapsto x^m$ is an isomorphism of $\mathbb{F}_p^*$ to itself.
(b) Prove that the equation

$$x^m + y^m + z^m = 0$$

has exactly $p + 1$ projective solutions with $x, y, z \in \mathbb{F}_p$.
(c) ** Suppose instead that m divides $p - 1$. Let M_p be the number of projective solutions to the equation given in (b). Prove that M_p satisfies the inequality

$$|M_p - p - 1| \leq (m - 1)(m - 2)\sqrt{p}.$$

This problem is a little easier if you take m to be a prime, so you might want to try that case first. We mention that the Fermat curve $x^m + y^m + z^m = 0$ has genus $\frac{1}{2}(m - 1)(m - 2)$, so (c) is a special case of the Hasse–Weil theorem.

4.4. Let p be an odd prime and let $\zeta \in \mathbb{C}$ be a root of the equation

$$x^{p-1} + x^{p-2} + \cdots + x + 1 = 0.$$

Thus ζ is a primitive p'th root of unity, i.e., it satisfies $\zeta \neq 1$ and $\zeta^p = 1$. We define the set of *quadratic residues* R in $\mathbb{F}_p^*$ by

$$R = \{x^2 : x \in \mathbb{F}_p^*\}.$$

(a) Prove that R is a subgroup of $\mathbb{F}_p^*$ of index 2. We denote the other coset of R in $\mathbb{F}_p^*$ by N and call it the set of *quadratic non-residues*.

(b) Prove that $-1 \in R$ if and only if $p \equiv 1 \pmod 4$.

(c) Define quadratic Gauss sums by the formulas

$$\alpha = \sum_{r \in R} \zeta^r \quad \text{and} \quad \beta = \sum_{n \in N} \zeta^n.$$

Prove that $\alpha + \beta = -1$.

(d) * Prove that

$$\alpha\beta = \begin{cases} -\frac{p-1}{4} & \text{if } p \equiv 1 \pmod 4, \\ \frac{p+1}{4} & \text{if } p \equiv 3 \pmod 4. \end{cases}$$

Deduce that

$$2\alpha + 1 = \begin{cases} \pm\sqrt{p} & \text{if } p \equiv 1 \pmod 4, \\ \pm\sqrt{-p} & \text{if } p \equiv 3 \pmod 4. \end{cases}$$

(e) Fix $\zeta = e^{2\pi i/p}$ and compute the value of α for some small values of p. Use your computation to make a conjecture about the correct sign for $2\alpha + 1$.

(f) ** Prove that your conjecture in (e) is correct.

4.5. Let C be the cubic curve given by the equation

$$C : y^2 = x^3 + x + 1.$$

(a) For each prime $p < 1000$, compute the number of points

$$M_p = \#C(\mathbb{F}_p)$$

on C over the field $\mathbb{F}_p$. Don't forget to include the point $\mathcal{O}$. Also compute the angles θ_p determined by the conditions

$$\cos\theta_p = \frac{M_p - p - 1}{2\sqrt{p}} \quad \text{and} \quad 0 \le \theta_p \le \pi.$$

(b) Compare the quantities

$$\frac{\#\{p \le 1000 : \alpha \le \theta_p \le \beta\}}{\pi(1000)} \quad \text{and} \quad \frac{2}{\pi} \int_\alpha^\beta \sin^2(t)\, dt$$

for various values of α and β. (The number of primes less than 1000 is $168 = \pi(1000)$.) How well do your computations support the conclusion of Theorem 4.3?

4.6. This exercise describes a special case of a theorem that was originally proven by Eichler and Shimura. The modularity theorem of Wiles et al. says that a similar statement is true for every elliptic curve given by an equation with rational coefficients. See Section 6.6 for further material on the modularity theorem.

(a) Let C be the cubic curve given by the equation

$$C : y^2 = x^3 - 4x^2 + 16.$$

As usual, let $M_p = \#C(\mathbb{F}_p)$ be the number of points on C over the field $\mathbb{F}_p$. Calculate M_p by hand for all primes $3 \leq p \leq 13$, or use a computer and calculate M_p for all primes $p < 100$ (or even $p < 1000$).

(b) Let $F(q)$ be the formal power series given by the infinite product

$$F(q) = q \prod_{n=1}^{\infty} (1 - q^n)^2 (1 - q^{11n})^2$$
$$= q - 2q^2 - q^3 + 2q^4 + q^5 + 2q^6 - 2q^7 + \cdots .$$

Let N_n be the coefficient of q^n in $F(q)$,

$$F(q) = \sum_{n=1}^{\infty} N_n q^n.$$

Calculate N_n by hand for $n \leq 13$, or use a computer and calculate N_n for all $n < 100$ (or even $n < 1000$).

(c) For each prime p, compute the sum $M_p + N_p$ of the quantities that you calculated in (a) and (b). Formulate a conjecture as to what this value should be in general.

(d) ** Prove that your conjecture in (c) is correct.

(e) If we replace the indeterminate q by the quantity $e^{2\pi i z}$, we obtain a function

$$\Phi(z) = F(e^{2\pi i z}) = e^{2\pi i z} \prod_{n=1}^{\infty} (1 - e^{2\pi i n z})^2 (1 - e^{2\pi i 11 n z})^2.$$

Prove that $\Phi(z)$ is holomorphic in the upper half plane

$$\mathfrak{H} = \{z = x + iy \in \mathbb{C} : y > 0\},$$

and that

$$\lim_{y \to \infty} \Phi(x + iy) = 0.$$

(f) ** Prove that for every prime p except $p = 11$, the function $\Phi(z)$ satisfies the relation

$$N_p \Phi(z) = \Phi(pz) + \sum_{j=0}^{p} \Phi\left(\frac{z+j}{p}\right) \quad \text{for all } z \in \mathfrak{H}.$$

(g) ** Prove that if a, b, c, d are integers satisfying

$$ad - bc = 1 \quad \text{and} \quad c \equiv 0 \pmod{11},$$

then

$$\Phi\left(\frac{az + b}{cz = d}\right) = (cz + d)^2 \Phi(z) \quad \text{for all } z \in \mathfrak{H}.$$

The identities in (f) and (g) are two of the amazing properties enjoyed by the function $\Phi(z)$. It is called a modular form of weight 2 for the congruence subgroup $\Gamma_0(11)$. And the formula that you found in (c) and (d) implies that the coefficients of the modular form $\Phi(z)$ completely determine the number of points in $C(\mathbb{F}_p)$ for all primes p.

4.7. Let b and c be integers satisfying

$$b \equiv 11 \pmod{15} \quad \text{and} \quad c \equiv 4 \pmod{15}.$$

Assume further that $4b^3 + 27c^2 \neq 0$, and let C be the elliptic curve

$$C : y^2 = x^3 + bx + c.$$

Find all points of finite order in $C(\mathbb{Q})$.

4.8. Let $p \equiv 3 \pmod 4$ be a prime, and let $b \in \mathbb{F}_p^*$.
(a) Show that the equation

$$v^2 = u^4 - 4b$$

has $p - 1$ solutions (u, v) with $u, v \in \mathbb{F}_p$.
(b) Show that if (u, v) is a solution of the equation in (a), then

$$\phi(u, v) = \left(\frac{u^2 + v}{2}, \frac{u(u^2 + v)}{2} \right)$$

is a point on the elliptic curve

$$C : y^2 = x^3 + bx.$$

(c) Prove that the curve C defined in (b) satisfies $\#C(\mathbb{F}_p) = p + 1$.
(d) ** What does $\#C(\mathbb{F}_p)$ look like if $p \equiv 1 \pmod 4$?

4.9. Let b be a non-zero integer that is *fourth power free*. (This means that $p^4 \nmid b$ for all primes p.) Let C be the elliptic curve

$$C : y^2 = x^3 + bx,$$

and let $\Phi \subseteq C(\mathbb{Q})$ be the subgroup consisting of all points of finite order.

(a) Prove that $\#\Phi$ divides 4.

(b) More precisely, show that Φ is given by the following table:

$$\Phi \cong \begin{cases} \mathbb{Z}/4\mathbb{Z} & \text{if } b = 4, \\ \mathbb{Z}/2\mathbb{Z} \oplus \mathbb{Z}/2\mathbb{Z} & \text{if } -b \text{ is a square}, \\ \mathbb{Z}/2\mathbb{Z} & \text{otherwise.} \end{cases}$$

4.10. Let $p \equiv 2 \pmod 3$ be a prime, and let $c \in \mathbb{F}_p^*$. Prove that the curve

$$C : y^2 = x^3 + c$$

satisfies $\#C(\mathbb{F}_p) = p + 1$.

4.11. Let c be a non-zero integer that is *sixth power free*. (This means that $p^6 \nmid b$ for all primes p.) Let C be the elliptic curve

$$C : y^2 = x^3 + c,$$

and let $\Phi \subseteq C(\mathbb{Q})$ be the subgroup consisting of all points of finite order.

(a) Prove that $\#\Phi$ divides 6.

(b) More precisely, show that Φ is given by the following table:

$$\Phi \cong \begin{cases} \mathbb{Z}/6\mathbb{Z} & \text{if } c = 1, \\ \mathbb{Z}/3\mathbb{Z} & \text{if } c \neq 1 \text{ is a square, or if } c = -432, \\ \mathbb{Z}/2\mathbb{Z} & \text{if } c \neq 1 \text{ is a cube}, \\ \{\mathcal{O}\} & \text{otherwise.} \end{cases}$$

4.12. Let C be a cubic curve given by a Weierstrass equation

$$y^2 = x^3 + ax^2 + bx + c$$

with integer coefficients. Let $p \geq 3$ be a prime that does not divide the discriminant, so when we reduce C modulo p we get a non-singular cubic curve $\tilde{C}$ with coefficients in $\mathbb{F}_p$. We define a general *reduction modulo p map* from $C(\mathbb{Q})$ to $\tilde{C}(\mathbb{F}_p)$ as follows. Let

$$P = (x, y) = \left(\frac{u}{d^2}, \frac{v}{d^3} \right) \in C(\mathbb{Q})$$

with $\gcd(u, d) = \gcd(v, d) = 1$. If p does not divide d, then we choose an integer e satisfying $de \equiv 1 \pmod p$ and set

$$\tilde{P} = (\tilde{u}\tilde{e}^2, \tilde{v}\tilde{e}^3) \in \tilde{C}(\mathbb{F}_p).$$

And if p does divide d, then we set $\tilde{P} = \tilde{\mathcal{O}}$. Prove that this map is a group homomorphism from $C(\mathbb{Q})$ to $\tilde{C}(\mathbb{F}_p)$ and that its kernel is the subgroup $C(p)$ that we discussed in Sections 2.4 and 2.5. Conclude that there is a one-to-one homomorphism

$$C(\mathbb{Q})/C(p) \longrightarrow \tilde{C}(\mathbb{F}_p).$$

Since we proved in Chapter 2 that $C(p) \cap \Phi = \emptyset$, this immediately implies the reduction theorem (Theorem 4.4) and provides a useful generalization.

4.13. (a) Prove that $561 = 3 \cdot 11 \cdot 17$ is a Carmichael number, that is, prove that if a is any integer that is relatively prime to 561, then

$$a^{560} \equiv 1 \pmod{561}.$$

(This can, of course, be checked on a computer by trying every a value. But with a little thought, you should be able to verify it by hand in just a few lines.)

(b) Fix an integer $a \geq 2$. Prove that there are infinitely many composite numbers m such that $a^{m-1} \equiv 1 \pmod{m}$. One says that m is a *pseudo-prime to the base a*.

4.14. Use the square-and-multiply method described in Section 4.4 to compute the following powers.
(a) $17^{5386} \pmod{26}$.
(b) $2^{35687} \pmod{38521}$.

4.15. Prove that the Euclidean algorithm described in Section 4.4 correctly computes the greatest common divisor of a and b.

4.16. Use the Euclidean algorithm to compute $\gcd(a, b)$ for the following pairs of integers. Write out each of the intermediate equations and compare the number of steps required to the upper bound $2 \log_2(2b)$.
(a) $a = 1187319, \quad b = 438987$.
(b) $a = 4152983, \quad b = 298936$.

4.17. If $a > b > 0$, we proved that the Euclidean algorithm computes $\gcd(a, b)$ in no more than $2 \log_2(2b)$ steps.
(a) Suppose that we revise the Euclidean algorithm as follows. Each time that we do a division with remainder $r_{i-1} = r_i q_i + r_{i+1}$, we choose the remainder to satisfy $-\frac{1}{2}|r_i| < r_{i+1} \leq \frac{1}{2}|r_i|$. Prove that the algorithm still computes $\gcd(a, b)$, but now in no more than $\log_2(2b)$ steps.
(b) Using the revised version of the Euclidean algorithm described in (a), prove that the r_i's satisfy

$$|r_{i+2}| \leq \frac{1}{5}|r_i|.$$

Deduce that the revised algorithm computes $\gcd(a, b)$ in no more than $2 \log_5(5b)$ steps. How large does b have to be before this bound is better than the bound in (a)?
(c) Compute $\gcd(4152983, 298936)$ using the revised algorithm in (a). Compare the actual number of steps with the upper bound $2 \log_5(5b)$ from (b).

4.18. If $\gcd(a, b) = 1$, then we know that there exist integers a' and b' satisfying

$$aa' + bb' = 1.$$

The Euclidean algorithm described in Section 4.4 provides a sequence of quotients $q_1, \ldots, q_{n+1}$ and remainders $r_0, \ldots, r_{n+1}$ that arise when computing $\gcd(a, b)$. Explain how to use the q_i's and r_i's to find a' and b'. Note that this gives a (moderately) efficient way to find the inverse of a modulo b, which is needed in the implementation of Lenstra's algorithm. For a more efficient method that's well-suited for computers, see Exercise 4.24.

4.19. Let $n = 246082373$
(a) Write $n - 1$ in the form

$$n - 1 = k_0 + k_1 \cdot 2 + k_2 \cdot 2^2 + \cdots + k_r \cdot 2^r$$

with each k_i either 0 or 1 and with $k_r = 1$.
(b) Use successive squaring to make a table of values $2^{2^i} \pmod{n}$ for $0 \le i \le r$.
(c) Use the binary expansion in (a) and the table in (b) to compute $2^{n-1} \pmod{n}$. Use your answer to deduce that n is not prime.

4.20. Let $n = 7591548931$.
(a) Calculate $2^{n-1} \pmod{n}$ and deduce that n is not prime.
(b) Use Pollard's $p - 1$ factorization algorithm (Figure 4.1) to factor n. What is the smallest value of d such that $\gcd(2^{d!} - 1, n)$ returns a non-trivial factor p? What is the prime factorization of $p - 1$?

4.21. Let $n = 199843247$. Using the point $P = (1, 1)$ and the elliptic curve

$$C : y^2 = x^3 + bx - b,$$

for each $b = 1, 2, \ldots$, compute $20!P \pmod{n}$ until some computation does not work, and use that failure to factor n.

4.22. Let C be the curve $x^3 + y^3 + z^3 = 0$ that we studied in Section 4.2. We stated there that if $p \equiv 1 \pmod 3$, then M_p is divisible by 9, and we justified this by indicating that the group of points $C(\mathbb{F}_p)$ contains a subgroup of order 9. This exercise sketches an alternative proof that does not use the group law.
   Prove that there is an element $u \in \mathbb{F}_p^*$ satisfying $u \neq 1$ and $u^3 = 1$. Then observe that each solution (x, y, z) with $xyz \neq 0$ leads to 27 points by taking $(u^i x, u^j y, u^k z)$ with $i, j, k \in \{0, 1, 2\}$. Prove that if we only want to count projective points, then we need to divide by 3. Finally, prove that there are exactly 9 (projective) solutions satisfying $xyz = 0$. Conclude that M_p is divisible by 9.

4.23. In Section 4.4 we described a square-and-multiply algorithm for computing large powers a^k of a number and an analogous double-and-add algorithm for computing large multiples kP of a point on an elliptic curve. These algorithms, as we presented them, require a fair amount of storage. Prove that the algorithm described in Table 4.4 computes kP while using very little storage.

4.24. Show that the algorithm described in Table 4.5, which is quite efficient and easily implemented on a computer, computes $g = \gcd(a, b)$ and a pair of integers u and v satisfying $au + bv = g$.

Input a point $P \in C(\mathbb{F}_p)$ and an integer $k \geq 1$.
Step 1: Set $Q = P$ and $R = \mathcal{O}$.
Step 2: Loop while $k > 0$.
Step 3: If $k \equiv 1 \pmod 2$, set $R = R + Q$.
Step 4: Set $Q = 2Q$ and $k = \lfloor k/2 \rfloor$.
Step 5: If $k > 0$, go to **Step 2**.
Step 6: Return the point R, which equals kP.

Table 4.4: An efficient double-and-add algorithm

Input integers $a > 0$ and $b > 0$.
Step 1: Set $u = 1$ and $g = a$ and $x = 0$ and $y = b$
Step 2: If $y = 0$, then set $v = (g - au)/b$ and return (g, u, v).
Step 3: Divide g by y with remainder, so $g = qy + t$ with $0 \leq t < y$.
Step 4: Set $s = u - qx$.
Step 5: Set $u = x$ and $g = y$.
Step 6: Set $x = s$ and $y = t$.
Step 7: Go to **Step 2**.

Table 4.5: An efficient extended gcd algorithm

4.25 (RSA Cryptosystem). Let p and q be distinct odd primes, let $N = pq$, and let e be an integer that is relatively prime to $(p-1)(q-1)$. Bob encrypts a message $m \in \mathbb{Z}/N\mathbb{Z}$ by computing

$$c \equiv m^e \pmod N$$

and sending c to Alice.

(a) Assuming that Alice knows p and q, show how she can use these values to efficiently find an integer f satisfying

$$a^{ef} \equiv a \pmod N$$

for all $a \in \mathbb{Z}/N\mathbb{Z}$. Hence Alice can decrypt Bob's message by computing $c^f \bmod N$.

(b) Prove that if Eve knows N and $(p-1)(q-1)$, then she, too, can find a value of f as in (a). Prove further that Eve can use the values of N and $(p-1)(q-1)$ to easily compute p and q. So knowing how to factor N is equivalent to knowing the values of N and $(p-1)(q-1)$.

4.26 (Diffie–Hellman Key Exchange). Suppose that Bob and Alice are content to exchange some random information that neither knows in advance, as long as they can keep their information secret from their adversary Eve. This might be useful, for example, if they then use the exchanged information as the secret key for a private key cryptosystem. We describe a method to perform such a *key exchange*.

  (i) Bob and Alice agree on a (large) finite group G, for example G might be $\mathbb{F}_p^*$ or $C(\mathbb{F}_p)$. They also pick an element $g \in G$. It is assumed that Eve knows G and g.

 (ii) Alice picks a secret number a and Bob picks a secret number b.

(iii) Alice computes $A = g^a$ and sends it to Bob, while Bob computes $B = g^b$ and sends it to Alice. It is assumed that Eve reads their communication, so she knows the values of A and B.

(iv) Alice computes B^a and Bob computes A^b. (Note that Alice knows a and Bob knows b, but Eve knows neither a nor b.)

Prove the following statements:

 (a) The quantities that Alice and Bob compute in Step (iv) are the same, so they have indeed exchanged a piece of information.

 (b) If Eve can solve the DLP in G, then she can find a and b, and hence can compute Alice and Bob's shared information.

 (c) ** Is there an efficient way for Eve to compute the shared information that doesn't require knowing a and b? (This is currently an open problem.)

Explain why it might be advantageous to use an elliptic curve group $C(\mathbb{F}_p)$, instead of $\mathbb{F}_p^*$, for Diffie–Hellman key exchange.

4.27 (Elgamal Cryptosystem). This exercise describes a public key cryptosystem based on discrete logarithms.

  (i) Bob and Alice agree on a (large) finite group G, for example G might be $\mathbb{F}_p^*$ or $C(\mathbb{F}_p)$. They also pick an element $g \in G$. It is assumed that Eve knows G and g.

 (ii) Alice picks a secret number a and computes $A = g^a$. Her private key is the number a and her public key is the group element A.

(iii) Bob picks a message $m \in G$ to send to Alice. He also chooses a random integer k. He computes the two group elements $c_1 = g^k$ and $c_2 = mA^k$ and sends them to Alice. It is assumed that Eve reads the communication, so she knows the values of c_1 and c_2.

(iv) Alice computes $c_2 c_1^{-a}$ in the group G.

Prove the following statements:

 (a) The quantity that Alice computes in Step (iv) is indeed Bob's message m.

 (b) If Eve can solve the DLP in G, then she can find a, and hence can compute Bob's message.

 (c) If Eve can figure out the value of Bob's random number k, then she can easily compute his message.

 (d) ** Is there an efficient way for Eve to compute Bob's message that doesn't require knowing the value of a and/or k? (This is currently an open problem.)

4.28 (Shank's Babystep-Giantstep Algorithm). Let G be a finite group of order N, and let $a, b \in G$ be elements for which we want to solve the DLP, i.e., we want to find an m such that $a^m = b$. (We always assume that such an m exists.) Prove that the algorithm described in Table 4.6 has the following properties:
  (a) There is always at least one element that appears in both List_1 and List_2, i.e., there is always a collision.
  (b) The number m computed in Step (5) is a solution to the DLP for a and b, i.e., it satisfies $a^m = b$.

Input elements a and b of a group G of order N.

Step 1: Let $n = \lceil \sqrt{N} \rceil$ be the smallest integer that is greater than $\sqrt{N}$.

Step 2: Compute a list of values

$$\mathsf{List}_1 : e, a, a^2, a^3, \dots, a^n.$$

Step 3: Compute $c = a^{-n}$, i.e., $c = (a^{-1})^n$, and compute a second list of values

$$\mathsf{List}_2 : b, bc, bc^2, bc^3, \dots, bc^n.$$

Step 4: Find a collision between the two lists, that is, find exponents i and j between 0 and n satisfying

$$a^i = bc^j.$$

Step 5: Compute the value $m = i + nj$.

Table 4.6: Shanks babystep-giantstep algorithm

4.29. Solve the following ECDLP's, either by naively computing multiples of P until you get to Q, or by the collision method described in Example 4.10 and Exercise 4.28.
  (a) $C : y^2 = x^3 + x^2 + x + 3, \quad p = 103, \quad P = (7, 14), \quad Q = (8, 22).$
  (b) $C : y^2 = x^3 - 2x^2 + 5x + 6, \quad p = 149, \quad P = (11, 16), \quad Q = (110, 46).$
  (c) $C : y^2 = x^3 + x^2 + x + 2, \quad p = 10037, \quad P = (8, 7358), \quad Q = (2057, 5437).$

Chapter 5

Integer Points on Cubic Curves

5.1 How Many Integer Points?

Let C be a non-singular cubic curve given by an equation

$$ax^3 + bx^2y + cxy^2 + dy^3 + ex^2 + fxy + gy^2 + hx + iy + j = 0$$

with integer coefficients. We have seen that if C has a rational point (possibly at infinity), then the set of all rational points on C forms a finitely generated abelian group. So we can get every rational point on C by starting from some finite set and adding points using the geometrically defined group law.

Another natural number theoretic problem is that of describing the solutions (x, y) to the cubic equation with x and y both *integers*. Since the cubic equation may have infinitely many rational points, we are asking which of those rational points have integer coordinates.

For a curve given by a Weierstrass equation

$$C : y^2 = x^3 + ax^2 + bx + c,$$

the Nagell–Lutz theorem tells us that points of finite order have integer coordinates. It is natural to ask if the converse is true. A little experimentation shows that it is not. We saw one example in Section 4.3, where we showed

© Springer International Publishing Switzerland 2015   167
J.H. Silverman, J.T. Tate, *Rational Points on Elliptic Curves*,
Undergraduate Texts in Mathematics, DOI 10.1007/978-3-319-18588-0_5

that the curve $y^2 = x^3 + 3$ has no points of finite order, but it clearly has the integer point $(1, 2)$. Similarly, it is easy to show that the curve $y^2 = x^3 + 17$ has no points of finite order, yet it has lots of integer points, including

$$(-2, \pm 3), \quad (-1, \pm 4), \quad (2, \pm 5), \quad (4, \pm 9), \quad (8, \pm 23),$$

and six other points that we leave as an exercise for you to discover.

Let's think a little bit about how many integer points we expect. If the rank of C is zero, then $C(\mathbb{Q})$ is finite, and the Nagell–Lutz theorem says that those finitely many points are integer points. This is the trivial case because if there are only finitely many rational points, then there are certainly only finitely many integer points.

The situation becomes much more interesting when the rank is positive. Suppose, for example, that the rank is 1 and that there are no non-trivial points of finite order. Then we can choose a generator P of $C(\mathbb{Q})$, and every point in $C(\mathbb{Q})$ has the form nP for some integer n. We look at the sequence of points $P, 2P, 3P, \ldots$. Writing $nP = (x_n, y_n)$ and using $nP = (n-1)P + P$, the explicit formula for the group law says that for $n \geq 3$ we have

$$x_n = \left(\frac{y_{n-1} - y_1}{x_{n-1} - x_1} \right)^2 - a - x_{n-1} - x_1.$$

So even if P and $(n-1)P$ have integer coordinates, there is no reason to expect that nP has integer coordinates. Indeed, looking at the formula, it seems quite unlikely that there will be very many nP's having integer coordinates.

This intuition turns out to be correct, although the proof is far from easy. Here is the general result, which was proven by Siegel [45, 46] in the 1920s.

Theorem 5.1 (Siegel's Theorem). *Let C be a non-singular cubic curve given by an equation $F(x, y) = 0$ with integer coefficients. Then C has only finitely many points with integer coordinates.*

One warning is in order. The curve C consists of the points satisfying $F(x, y) = 0$, together with one or more points at infinity. In order for the theorem to apply, the curve C must be non-singular at every point, including the points at infinity.

By way of contrast, we can compare Siegel's theorem to the situation for linear, quadratic, and singular cubic equations. If a linear equation

$$ax + by = c \quad \text{with } a, b, c \in \mathbb{Z}$$

has a solution (x_0, y_0) in integers, then it has infinitely many solutions given by the recipe

$$(x_0 + bn, y_0 - an) \quad \text{with } n \in \mathbb{Z}.$$

Similarly, quadratic equation can have infinitely many integer solutions. For example, consider the equation

$$x^2 - 2y^2 = 1.$$

This clearly has the solution $(3, 2)$. Further, it is easy to check that if (x, y) is a solution, then so is $(3x + 4y, 2x + 3y)$. So if we start with $(3, 2)$ and repeatedly apply this procedure, then we get infinitely many solutions

$$(3, 2), \quad (17, 12), \quad (99, 70), \quad (577, 408), \dots,$$

since the coordinates are clearly growing. A harder problem, which we shall not undertake, is to prove that up to sign, this gives every solution. This is a special case of Pell's equation

$$x^2 - Dy^2 = 1,$$

which you may have seen. If D is a positive square-free integer, then one can show that the solutions to Pell's equation form a group of the form $\mathbb{Z}/2\mathbb{Z} \times \mathbb{Z}$. More precisely, if (x_1, y_1) is the solution with smallest positive x-coordinate, then every solution has the form $(\pm x_n, \pm y_n)$ with x_n and y_n determined by the formula

$$x_n + y_n \sqrt{D} = \left(x_1 + y_1 \sqrt{D} \right)^n \quad \text{for } n \in \mathbb{Z}.$$

This, in turn, is a special case of Dirichlet's unit theorem, which says that the group of units in the ring of integers of a number field is finitely generated and gives a precise formula for the rank.

Finally, we mention that the singular cubic curves

$$C_1 : y^2 = x^3 \quad \text{and} \quad C_2 : y^2 = x^3 - x^2$$

have infinitely many integer points. This is clear for C_1, since $(t^2, t^3) \in C_1$ for all $t \in \mathbb{Z}$. Similarly one checks that $(t^2 + 1, t^3 + t) \in C_2$ for all $t \in \mathbb{Z}$. So the non-singularity of the cubic is essential in the statement of Siegel's theorem.

There are several different proofs of Siegel's theorem, none of them easy. In the next section we consider a special case where the proof is very easy and discuss some interesting questions that arise. The remainder of this chapter is devoted to a proof of a less trivial case of Siegel's theorem, due to Axel Thue in 1909. Thue's proof, which uses many of the tools needed to prove the general case, is quite complicated, but just as in the proof of Mordell's theorem, the proof can be broken down into several manageable steps.

The proofs of Siegel's and Thue's theorems have one other thing in common with the proof of Mordell's theorem. Recall that although Mordell's theorem tells us that the group of rational points is finitely generated, it does not provide a guaranteed method for finding generators. Similarly, Siegel's and Thue's theorems tell us that the set of points with integer coordinates is finite, but their proofs do not provide us with a method that is guaranteed to find all of the integer points. In the 1930s, Skolem [50] came up with a new proof of Siegel's theorem that, in practice, often allows one to find all solutions, but it, too, was not guaranteed to work. Finally, in 1966, Baker [2] gave an effective method for finding all solutions.

5.2 Taxicabs and Sums of Two Cubes

The title of this section may provoke some curiosity since it is the first time in the book that we have referred to methods of conveyance. The reference has to do with a famous mathematical story. When the brilliant Indian mathematician Ramanujan was in the hospital in London, his colleague G.H. Hardy came to visit. Hardy remarked that he had come in taxicab number 1729, and surely that was a rather dull number. Ramanujan instantly replied that, to the contrary, 1729 is a very interesting number. It is the smallest number expressible as a sum of two cubes in two different ways. Thus

$$1729 = 9^3 + 10^3 = 1^3 + 12^3.$$

So the taxicab number 1729 gives a cubic curve

$$x^3 + y^3 = 1729$$

that has two integer points. Of course, we can switch x and y, so we end up with four points,

$$(9, 10), \quad (10, 9), \quad (1, 12), \quad (12, 1).$$

We claim that there are no other integer points. This is a special case of Siegel's theorem (Theorem 5.1), but in this case the proof is easy because the cubic $x^3 + y^3$ factors.

So suppose that x and y are integers satisfying $x^3 + y^3 = 1729$. Then

$$(x + y)(x^2 - xy + y^2) = 1729 = 7 \cdot 13 \cdot 19.$$

So we have just to consider all possible factorizations $1729 = AB$ and solve the simultaneous equations

$$x + y = A \quad \text{and} \quad x^2 - xy + y^2 = B.$$

Substituting $y = A - x$ into the second equation, we find that

$$3x^2 - 3Ax + A^2 - B = 0,$$

so for each factorization $1729 = AB$, we need to check if

$$\frac{3A \pm \sqrt{12B - 3A^2}}{6}$$

is an integer. Doing this, we find that we get integer solutions only for the pairs $(A, B) = (13, 133)$ and $(A, B) = (91, 19)$, and these lead to the four known solutions to $x^3 + y^3 = 1729$.

More generally, most cubic equations that factor as

$$(ax + by + c)(dx^2 + exy + fy^2 + gx + hy + i) = j$$

with $j \neq 0$ have only finitely many solutions.[1] Merely look at all possible factorizations $j = AB$, solve the pair of equations

$$ax + by + c = A, \qquad dx^2 + exy + fy^2 + gx + hy + i = B,$$

and see which integer solutions arise. This might be called the trivial case of Siegel's theorem since it can be solved by an elementary argument.

But there are still many interesting questions that we can ask about the "taxicab equation" $x^3 + y^3 = m$ and other cubic equations for which Siegel's theorem is trivial. For example, we know that there are finitely many solutions, but can we bound how large they are? Well, yes, we can do that rather easily. We know that the solutions satisfy

$$x + y = A \quad \text{and} \quad x^2 - xy + y^2 = B$$

---

[1] But one has to be a little careful, since a silly equation such as $x^3 = 1$ has infinitely many solutions because y is arbitrary. Similarly, the equation $x(x^2 + xy - y) = 1$ has infinitely many solutions $(1, y)$.

for some factorization $m = AB$. Hence

$$m \geq |B| = |x^2 - xy + y^2| = \frac{3}{4}x^2 + \left(\frac{1}{2}x - y\right)^2 \geq \frac{3}{4}x^2.$$

Hence $|x| \leq 2\sqrt{m/3}$, and the same argument gives the same bound for $|y|$. This proves the following theorem for the "taxicab equation."

Proposition 5.2. *Let $m \geq 1$ be an integer. Then every solution to the equation*

$$x^3 + y^3 = m$$

in integers $x, y \in \mathbb{Z}$ satisfies

$$\max\{|x|, |y|\} \leq 2\sqrt{m/3}.$$

Another natural question is that of the number of solutions. Ramanujan's observation is that for every $1 \leq m \leq 1728$, the equation $x^3 + y^3 = m$ has at most one solution in positive integers, where we treat (x, y) and (y, x) as the same solution, but for $m = 1729$, there are two solutions. So we might ask whether there is a value of m for which there are three solutions, and four solutions, and so on. The answer is that for any $N \geq 1$ we can find an m so that the equation $x^3 + y^3 = m$ has at least N solutions.

To prove this, we first observe that there are equations

$$x^3 + y^3 = m$$

that have infinitely many *rational* solutions. For example, consider the curve

$$x^3 + y^3 = 9,$$

which has the solution $(2, 1)$. As we saw in Section 1.3, there is essentially a one-to-one correspondence between the rational points on $x^3 + y^3 = 9$ and the rational points on the curve $Y^2 = X^3 - 48$ given by the formulas

$$X = \frac{12}{x+y}, \qquad Y = 12\frac{x-y}{x+y}.$$

The point $(1, 2)$ on the curve $x^3 + y^3 = 9$ corresponds to the point $Q = (4, 4)$ on the curve $Y^2 = X^3 - 48$. We compute $2Q = (28, -148)$ and $3Q = (\frac{73}{9}, \frac{595}{27})$, which proves that Q has infinite order, because the Nagell–Lutz theorem (Section 2.4) says that points of finite order have integer coordinates. Hence both $Y^2 = X^3 - 48$ and $x^3 + y^3 = 9$ have infinitely many rational points.

Since there are infinitely many rational points on $x^3 + y^3 = 9$, we can certainly find N distinct points, say $P_1, \ldots, P_N$. If $P = \left(\frac{a}{b}, \frac{c}{d} \right)$ is any rational point written in lowest terms with positive denominators, then substituting into the equation and clearing denominators gives

$$a^3 d^3 + c^3 b^3 = 9b^3 d^3.$$

Thus b^3 divides $a^3 d^3$ and d^3 divides $c^3 b^3$. But $\gcd(a, b) = 1$ and $\gcd(c, d) = 1$, so $b^3 \mid d^3$ and $d^3 \mid b^3$, and hence $b = d$. This means that we can write the coordinates of $P_1, \ldots, P_N$ as

$$P_1 = \left(\frac{a_1}{d_1}, \frac{c_1}{d_1} \right), \ldots, P_N = \left(\frac{a_N}{d_N}, \frac{c_N}{d_N} \right).$$

Now for the main idea. We choose an m that, in essence, clears the denominators of the P_i's, thereby making them into integer points. The P_i are on the curve $x^3 + y^3 = 9$, so we let

$$D = d_1 d_2 \cdots d_N \quad \text{and take} \quad m = 9D^3.$$

Then the points

$$P_i' = \left(\frac{Da_i}{d_i}, \frac{Dc_i}{d_i} \right) \quad \text{for } i = 1, 2, \ldots, N$$

have integer coordinates and are on the curve

$$x^3 + y^3 = 9D^3.$$

This proves our assertion, which we restate as a formal proposition.

Proposition 5.3. *For every integer $N \geq 1$ there is an integer $m \geq 1$ such that the cubic curve*

$$x^3 + y^3 = m$$

has at least N points with integer coordinates.

Of course, this does not strictly generalize Ramanujan's example since he referred only to sums of positive cubes. However, it is not hard to prove that if $m > 0$ and if the curve $x^3 + y^3 = m$ has infinitely many rational solutions, then there are infinitely many rational solutions with x and y both positive. The idea is that the set of real points on this curve looks like the circle group, so the subgroup generated by a point of infinite order is dense in the set of real points. Since there are real points with $x, y > 0$ (see Figure 5.1), an open

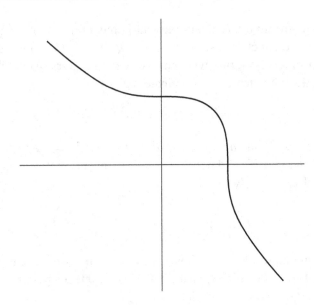

Figure 5.1: A taxicab curve

subset of such points contains infinitely many rational points with $x, y > 0$. So if you want, you can add the words "and with x and y both positive" on the end of Proposition 5.3.

This shows that if we take m large enough, then the equation $x^3 + y^3 = m$ can have an arbitrarily large number of positive integer solutions. But Ramanujan's observation was also that 1729 is the *smallest* m with two positive solutions. So what is the smallest m that has three positive solutions? The answer is

$$87539319 = 167^3 + 436^3 = 228^3 + 423^3 = 255^3 + 414^3.$$

Based on the Hardy–Ramanujan story, people have defined the *N'th Taxicab Number* to be

$$\mathsf{Taxi}(N) = \min \left\{ m \geq 1 : \begin{array}{l} x^3 + y^3 = m \text{ has at least } N \text{ integer} \\ \text{solutions with } x \geq y > 0 \end{array} \right\}.$$

So $\mathsf{Taxi}(2) = 1729$ and $\mathsf{Taxi}(3) = 87539319$. The proof that we gave of Proposition 5.3 can be turned into a (very poor) upper bound for $\mathsf{Taxi}(N)$, but in practice it is quite difficult to exactly determine $\mathsf{Taxi}(N)$ due to the difficulty of ruling out smaller m's that might work. Here is the current state of knowledge (as of 2015):

$$\text{Taxi}(1) = 2$$
$$\text{Taxi}(2) = 1729$$
$$\text{Taxi}(3) = 87539319$$
$$\text{Taxi}(4) = 6963472309248$$
$$\text{Taxi}(5) = 48988659276962496$$
$$\text{Taxi}(6) = 24153319581254312065344$$

Not surprisingly, taxicab numbers have lots of factors. For example,

$$\text{Taxi}(6) = 2^6 \cdot 3^3 \cdot 7^4 \cdot 13 \cdot 19 \cdot 43 \cdot 73 \cdot 79^3 \cdot 97 \cdot 157.$$

In some sense, Proposition 5.3 provides a satisfactory answer to our question of how many integer points can a cubic curve have. But it may leave you a bit uneasy because we haven't really found a lot of points that are intrinsically integral. Instead, we found lots of rational points and cleared their denominators. This leads to solutions (x, y) in which x and y tend to have a large common factor. If we disallow common factors, we are led to the following question.

> Given an integer N, is it possible to find an integer $m \geq 1$ so that the equation $x^3 + y^3 = m$ has at least N integer solutions with $x \geq y > 0$ and $\gcd(x, y) = 1$?

For $N = 2$, the answer is yes, since $1729 = 12^3 + 1^3 = 10^3 + 9^3$. For $N = 3$, the answer is also yes, as discovered by Paul Vojta in 1983 via a 3-day calculation on an early desktop computer. Vojta's number is

$$15170835645 = 2468^3 + 517^3 = 2456^3 + 709^3 = 2152^3 + 1733^3.$$

Two decades later Stuart Gascoigne and Duncan Moore (independently) found an example with four representations,

$$1801049058342701083 = 1216500^3 + 92227^3 = 1216102^3 + 136635^3$$
$$= 1207602^3 + 341995^3 = 1165884^3 + 600259^3.$$

And that's where the situation stands. No one knows whether the answer for $N = 5$ is yes or no.

We conclude this section by discussing an interesting relationship between the number of integer points and the rank of the group of rational points. Serge Lang made a general conjecture that has been proven for certain types of cubic curves, including the taxicab curves studied in this section.

Theorem 5.4. (Silverman [47]) *There is a constant $K > 1$ with the following property. For every integer $m \geq 1$, the number of relatively prime integer points on the cubic curve*

$$C_m : x^3 + y^3 = m$$

is bounded by the rank of the group of rational points via the estimate

$$\#\{(x,y) \in C_m(\mathbb{Q}) : x, y \in \mathbb{Z} \text{ and } \gcd(x,y) = 1\} \leq K^{1 + \operatorname{rank} C_m(\mathbb{Q})}.$$

Theorem 5.4 says that integer points with $\gcd(x,y) = 1$ tend to be somewhat linearly independent in the group of rational points. In particular, if one could find a sequence of m's so that the number of such integral points goes to infinity, then one could conclude that the ranks go to infinity. Conversely, if one could prove that the rank of $C_m(\mathbb{Q})$ is bounded independent of m, then the same would be true for the number of no-common-factor integer points.

5.3 Thue's Theorem and Diophantine Approximation

In the last section we saw how easy it is to find all integer solutions to equations of the form $x^3 + y^3 = m$. The reason why it is easy is because the polynomial $x^3 + y^3$ factors as $(x + y)(x^2 - xy + y^2)$, and by considering the finitely many factorizations of m, we end up with finitely many pairs of equations for the two unknowns x and y.

Suppose instead that we take a polynomial that does not factor, for example,

$$x^3 + 2y^3 = m.$$

It is not clear whether an equation of this sort may have infinitely many integer solutions. For equations of degree two, we observe that $x^2 - y^2 = 1$ has finitely many solutions, while $x^2 - 2y^2 = 1$ has infinitely many solutions. So that fact that $x^3 + y^3 = m$ has finitely many solutions is not a strong argument for or against the same being true of $x^3 + 2y^3 = m$.

More generally, consider a cubic equation of the form

$$ax^3 + by^3 = c$$

with $abc \neq 0$. It turns out that such an equation has only finitely many solutions in integers, regardless of whether it factors. In this section we explain how to reduce this problem to a question of approximating certain irrational number by rational numbers. We also give a rough outline of the proof of the approximation theorem that we need. The remainder of Chapter 5 is then devoted to giving the details of the proof of the approximation theorem.

Theorem 5.5. (Thue [54]) *Let* a, b, c *be non-zero integers. Then the equation*

$$ax^3 + by^3 = c$$

has only finitely many solutions in integers x, y.

One trivial observation is that if (x, y) is a solution to $ax^3 + by^3 = c$, then (ax, y) is a solution to

$$X^3 + a^2bY^3 = a^2c,$$

so it suffices to prove Thue's theorem for equations with $a = 1$. A second observation is that replacing y by $-y$ and/or b by $-b$ if necessary, it is enough to consider equations of the form

$$x^3 - by^3 = c \quad \text{with } b, c \in \mathbb{Z}, b > 0, \text{ and } c > 0.$$

This is the equation that we will prove has only finitely many integer solutions.

The factorization method that we used in the last section worked extremely well, so let's try to use it again. Of course, if b is not a perfect cube, then we cannot factor $x^3 - by^3$ over the rational numbers. We need to use a cube root of b. So we let

$$\beta = \sqrt[3]{b},$$

and then we can factor

$$x^3 - by^3 = (x - \beta y)(x^2 + \beta xy + \beta^2 y^2).$$

It is important to note that this is *not* a factorization of integers, so we cannot factor c and get two equations for x and y.

However, what we observe is that if (x, y) is a solution to $x^3 - by^3 = c$ with x and y large, then the difference $|x - \beta y|$ must be quite small. This is true because

$$x^2 + \beta xy + \beta^2 y^2 = \left(x + \frac{1}{2}\beta\right)^2 + \frac{3}{4}\beta^2 y^2 \geq \frac{3}{4}\beta^2 y^2,$$

which in turn implies that

$$|c| = |x^3 - by^3| = |x - \beta y| \cdot |x^2 + \beta xy + \beta^2 y^2|. \geq |x - \beta y| \cdot \frac{3}{4}\beta^2 y^2.$$

Dividing by $\frac{3}{4}\beta^2 y^2$, we obtain the important inequality

$$\left|\frac{x}{y} - \beta\right| \le \frac{4|c|}{3\beta^2} \cdot \frac{1}{|y|^3}.$$

This inequality says that if (x, y) is an integer solution to the equation $x^3 - by^3 = c$ with $|y|$ large, then the rational number x/y is extremely close to the irrational number $\beta = \sqrt[3]{b}$. Hence in order to prove that there are finitely many solutions, it suffices to show that there are only finitely many rational numbers with this approximation property. The study of rational approximations to irrational quantities is called the *Theory of Diophantine Approximation*. Our goal is to the prove the following theorem.

Theorem 5.6 (Diophantine Approximation Theorem). (Thue [54]) *Let b be a positive integer that is not a perfect cube, and let $\beta = \sqrt[3]{b}$. Let C be any fixed positive constant. Then there are only finitely many pairs of integers (p, q) with $q > 0$ that satisfy the inequality*

$$\left|\frac{p}{q} - \beta\right| \le \frac{C}{q^3}. \tag{$*$}$$

Assuming the truth of the Diophantine approximation theorem, how can we finish the proof that $x^3 - by^3 = c$ has only finitely many solutions? If b is a perfect cube, then $x^3 - by^3$ factors, so the elementary argument given in Section 5.2 works. Next, if $y = 0$, then $x^3 = c$, so there is at most one solution with $y = 0$. Finally, suppose that b is not a cube and that (x, y) is a solution with $y \ne 0$. Then we showed earlier that

$$\left|\frac{x}{y} - \beta\right| \le \frac{C}{|y|^3} \quad \text{with} \quad C = \frac{4|c|}{3\beta^2},$$

and the Diophantine approximation theorem tells us that there are only finitely many pairs (x, y) with $y > 0$. To deal with solutions having $y < 0$, we rewrite the inequality as

$$\left|\frac{-x}{-y} - \beta\right| \le \frac{C}{|y|^3},$$

so applying the Diophantine approximation theorem again shows that there are only finitely many pairs of integer solutions (x, y).

So "all" that remains to do is to prove the Diophantine approximation theorem. To motivate the argument used in the actual proof, we first describe an idea for the proof that almost, but not quite, works.

As observed earlier, we may consider the factorization

$$x^3 - by^3 = (x - \beta y)(x^2 + \beta xy + \beta^2 y^2).$$

Suppose that p/q satisfies the estimate $(*)$ in the Diophantine approximation theorem. Substituting $x = p$ and $y = q$ into our identity and dividing by q^3 yields

$$\frac{p^3 - bq^3}{q^3} = \left(\frac{p}{q} - \beta\right)\left(\frac{p^2}{q^2} + \beta\frac{p}{q} + \beta^2\right). \qquad (\dagger_1)$$

We make two observations concerning this last equation. First, since b is not a perfect cube, the integer $p^3 - bq^3$ is not zero, and hence

$$\left|\frac{p^3 - bq^3}{q^3}\right| \geq \frac{1}{q^3}. \qquad (\dagger_2)$$

Second, from $(*)$ we have

$$\left|\frac{p}{q}\right| \leq \beta + \frac{C}{q^3} \leq \beta + C,$$

so

$$\left|\frac{p^2}{q^2} + \beta\frac{p}{q} + \beta^2\right| \leq (\beta + C)^2 + \beta(\beta + C) + \beta^2 \leq C', \qquad (\dagger_3)$$

where we have written C' for the constant $3\beta^2 + 3\beta C + C^2$. The crucial fact is that C' depends only on β and C; it is the same for every choice of p/q.

Substituting the two inequalities $(\dagger_2)$ and $(\dagger_3)$ into the equation $(\dagger_1)$, we have shown that there is a constant C' so that for every rational number $\frac{p}{q}$,

$$\left|\frac{p}{q} - \beta\right| = \frac{\left|\dfrac{p^3 - bq^3}{q^3}\right|}{\left|\dfrac{p^2}{q^2} + \beta\dfrac{p}{q} + \beta^2\right|} \geq \frac{1}{C'q^3}. \qquad (**)$$

Recall that we are trying to prove that for every constant C, there are only finitely many rational numbers p/q satisfying the inequality

$$\left|\frac{p}{q} - \beta\right| \leq \frac{C}{q^3}. \qquad (*)$$

Comparing $(*)$ and $(**)$, we do not seem to have learned anything, other than the fact that $C' \geq 1/C$, which is not helpful since we already know that C' is fairly large. The problem is that the bounds in both $(*)$ and $(**)$ involve a multiple of $1/q^3$.

There is nothing that we can do about (∗), that's what we're trying to prove. But suppose that we could prove a stronger version of (∗∗) with some exponent smaller than 3. For the sake of illustration, suppose that we could prove that

$$\left| \frac{p}{q} - \beta \right| \geq \frac{1}{C' q^{2.9}} \quad \text{for all} \quad \frac{p}{q}. \tag{∗∗′}$$

Then combining (∗) and (∗∗′) gives

$$\frac{1}{C' q^{2.9}} \leq \left| \frac{p}{q} - \beta \right| \leq \frac{C}{q^3},$$

and so

$$q \leq (CC')^{10}.$$

Thus every solution p/q to (∗) has its denominator bounded by the number $(CC')^{10}$, which depends only on C and b. Then (∗) implies that the numerator is also bounded, so we could conclude that (∗) has only finitely many solutions.

How might we improve on (∗∗)? Let's summarize how we proved (∗∗). We took the polynomial $f(X) = X^3 - b$ that has integer coefficients and β as a root. Evaluating $f(X)$ at p/q, we noted that $\left| f(p/q) \right|$ is no smaller than $1/q^3$, since its numerator is an integer and its denominator divides q^3. On the other hand, factoring $f(X)$, we saw that $\left| f(p/q) \right|$ equals $|p/q - \beta|$ times something that is bounded. Comparing upper and lower bounds for $f(p, q)$ yields (∗∗).

One way to improve (∗∗) might be to use some other polynomial in place of $X^3 - b$. More precisely, suppose that we find a polynomial $F(X)$ with integer coefficients that is divisible by $(X^3 - b)^n$ for some (presumably large) integer n. Then $F(X)$ factors as

$$F(X) = (X - \beta)^n G(X)$$

for some polynomial $G(X) \in \mathbb{R}[X]$, and just as before we can show that

$$\left| F\left(\frac{p}{q} \right) \right| \leq C'' \left| \frac{p}{q} - \beta \right|^n.$$

Here C'' depends on C and the polynomial $F(X)$, but it is the same for all p/q's.

On the other hand, if $F(p/q) \neq 0$, then we immediately derive the lower bound

$$\left| F\left(\frac{p}{q}\right) \right| = \frac{|\text{non-zero integer}|}{q^d} \geq \frac{1}{q^d},$$

where d is the degree of F. Comparing the upper and lower bounds and taking n'th roots, we find that

$$\left| \frac{p}{q} - \beta \right| \geq \frac{1}{\sqrt[n]{C''}} \cdot \frac{1}{q^{d/n}}.$$

So if $d < 3n$ (strict inequality), then we are done.

Unfortunately, it turns out that $d \geq 3n$. To see why, we note that $F(X)$ is divisible by $(X - \beta)^n$, where $\beta = \sqrt[3]{b}$. Further, $F(X)$ has integer coefficients. Hence $F(X)$ is divisible by the n'th power of the minimal polynomial of β, which is $X^3 - b$. And clearly if $(X^3 - b)^n$ divides $F(X)$, then $\deg F(X) \geq 3n$. So this attempt to prove (∗∗) meets with failure.

Thue's brilliant idea, which enabled him to improve (∗∗), was to instead use a two-variable polynomial $F(X, Y) \in \mathbb{Z}[X, Y]$. He chose a polynomial that vanishes to high order at the point (β, β), and he then compared upper and lower bounds for the value $|F(p_1/q_1, p_2/q_2)|$, where p_1/q_1 and p_2/q_2 are solutions to (∗). Thue's proof naturally divides into three parts:

(1) Find a suitable polynomial $F(X, Y)$.

(2) Compute a good upper bound for $|F(p_1/q_1, p_2/q_2)|$ in terms of the quantities $|p_1/q_1 - \beta|$ and $|p_2/q_2 - \beta|$.

(3) Derive a lower bound for $|F(p_1/q_1, p_2/q_2)|$, and in particular, show that this value is not zero. This is the technically hardest part of the proof.

This description of the proof is certainly very sketchy. We now describe each of the steps in more detail, leaving the proofs to subsequent sections of this chapter. But it is important to understand the outline of the proof before proceeding, since otherwise it is easy to become bogged down in the numerous details.

Step I: Construction of an Auxiliary Polynomial

We begin by constructing a polynomial $F(X, Y)$ with integer coefficients so that $F(X, Y)$ vanishes to very high order at the point (β, β). We will need to find an F whose coefficients are not too large.

$\boxed{\text{Step II: The Auxiliary Polynomial Is Small}}$

We assume that there are infinitely many pairs of integers (p, q) that satisfy the Diophantine inequality $(*)$ and aim to derive a contradiction. Under this assumption, we can find a rational number p_1/q_1 satisfying $(*)$ and with q_1 quite large. Then we can find a second rational number p_2/q_2 satisfying $(*)$ with q_2 much larger than q_1. Having done this, we consider the value of the polynomial $F(X, Y)$ at the point $(p_1/q_1, p_2/q_2)$. Since $F(X, Y)$ vanishes to high order at (β, β) and since $(*)$ says that each p_i/q_i is close to β, we find that $F(p_1/q_1, p_2/q_2)$ is quite small.

$\boxed{\text{Step III: The Auxiliary Polynomial Does Not Vanish}}$

This is the subtlest part of the proof. We want to show that $F(p_1/q_1, p_2/q_2)$ is not zero. Then, by writing

$$F\left(\frac{p_1}{q_1}, \frac{p_2}{q_2}\right) = \frac{\text{non-zero integer}}{q_1^d q_2^e},$$

we get a lower bound

$$\left| F\left(\frac{p_1}{q_1}, \frac{p_2}{q_2}\right) \right| \geq \frac{1}{q_1^d q_2^e}.$$

The hope is that this lower bound contradicts the upper bound in Step II, thereby completing the proof of the theorem.

Unfortunately, there is one additional complication to the proof. In Step III, we will not actually be able to show that $F(p_1/q_1, p_2/q_2)$ is not zero. Instead, we will show that some derivative of F does not vanish at $(p_1/q_1, p_2/q_2)$. This means that in Step II we need to give an upper bound for the values of the derivatives of F. It is not hard to do this, so we hope that you will not be deterred by the small notational inconveniences that this entails.

5.4 Construction of an Auxiliary Polynomial

In this section we are going to construct a polynomial $F(X, Y)$ with reasonably small integer coefficients and the property that F vanishes to high order at (β, β). The way that we will build F is by solving a system of linear equations with integer coefficients. Results describing integer solutions of systems of linear equations are often named after Siegel because he was the first to formalize this procedure.

Lemma 5.7 (Siegel's Lemma). *Let $N > M$ be positive integers and let*

$$a_{11}T_1 + \cdots + a_{1N}T_N = 0$$
$$\vdots \qquad \ddots \qquad \vdots \qquad \vdots$$
$$a_{M1}T_1 + \cdots + a_{MN}T_N = 0$$

be a non-trivial system of linear equations with integer coefficients. Then there is a solution $(t_1, \ldots, t_N)$ to this system with $t_1, \ldots, t_N$ integers, not all zero, and satisfying

$$\max_{1 \le i \le N} |t_i| < 2 \left(4N \max_{\substack{1 \le i \le M \\ 1 \le j \le N}} |a_{ij}| \right)^{\frac{M}{N-M}}.$$

The statement of Siegel's lemma looks complicated, but it is really saying something very easy. The system of homogeneous equations has more variables than equations, so we know that it has non-trivial solutions. Since the coefficients are integers, there are solutions in rational numbers, and clearing denominators, we can create integer solutions. So it is obvious that there are non-zero integer solutions. The last part of the lemma says that we can find a solution whose coordinates are not too large. More precisely, we can find a solution whose coordinates are bounded explicitly in terms of the number of equations M, the number of variables N, and the size of the coefficients a_{ij}. This, too, is not surprising, so the real content of Siegel's lemma is the precise form of the bound.

Proof of Siegel's lemma. For any vector $t = (t_1, \ldots, t_N)$ with integer coordinates, we let

$$\|t\| = \max_{1 \le i \le N} |t_i|$$

be the largest of the absolute values of its coordinates. Similarly, we let A be the matrix

$$A = \begin{pmatrix} a_{11} & \cdots & a_{1N} \\ \vdots & \ddots & \vdots \\ a_{M1} & \cdots & a_{MN} \end{pmatrix} \quad \text{and} \quad \|A\| = \max_{\substack{1 \le i \le M \\ 1 \le j \le N}} |a_{ij}|.$$

Siegel's lemma asserts that the equation $At = 0$ has a vector solution $t \ne 0$ satisfying

$$\|t\| < 2 \left(4N\|A\| \right)^{M/(N-M)}.$$

If $t = (t_1, \ldots, t_N)$ is any vector, we can estimate the size of the vector

$$At = \left(\sum_{j=1}^{N} a_{1j} t_j, \ldots, \sum_{j=1}^{N} a_{Mj} t_j \right)$$

by estimating the size of the i'th coordinate of At. Thus

$$\left| \sum_{j=1}^{N} a_{ij} t_j \right| \leq \sum_{j=1}^{N} |a_{ij} t_j| \quad \text{by the triangle inequality,}$$

$$\leq N \left(\max_{1 \leq j \leq N} |a_{ij}| \right) \left(\max_{1 \leq j \leq N} |t_j| \right)$$

$$\leq N \|A\| \cdot \|t\|.$$

Taking the maximum over j gives

$$\|At\| \leq N \|A\| \cdot \|t\|.$$

Thus if t is a vector with size $\|t\| \leq H$, then its image At has size $\|At\| \leq N\|A\|H$. In particular, multiplication by the matrix A maps the set of integer vectors

$$T_H = \left\{ t = (t_1, \ldots, t_N) : t_i \in \mathbb{Z}, \ \|t\| \leq H \right\}$$

into the set of integer vectors

$$U_H = \left\{ u = (u_1, \ldots, u_M) : u_i \in \mathbb{Z}, \ \|u\| \leq N\|A\|H \right\}.$$

We claim that if H is large enough, then T_H has more elements than U_H, so there will be two vectors in T_H with the same image in U_H. This last statement is an application of the famous *pigeonhole principle*, where T_H is our set of pigeons, U_H is our set of pigeonholes, and multiplication by the matrix A assigns each pigeon to a pigeonhole.

How many vectors are in T_H and U_H? Each vector in T_H has N coordinates, and each coordinate is an integer satisfying $-H \leq t_i \leq H$, so

$$\#T_H = \left(2\lfloor H \rfloor + 1 \right)^N,$$

where $\lfloor H \rfloor$ denotes the greatest integer that is less than or equal to H. Similarly

$$\#U_H = \left(2\lfloor N\|A\|H \rfloor + 1 \right)^M.$$

Since $N > M$, we see that $\#T_H$ will be larger than $\#U_H$ provided that H is large enough, but we need to be more precise. We assume that $H \geq 1$. Then

$$\#T_H \geq \big(2(H-1)+1\big)^N = (2H-1)^N \geq H^N,$$

and similarly,

$$\#U_H \leq \big(2N\|A\|H+1\big)^M \leq (3N\|A\|H)^M.$$

Combining these two estimates, we find that

$$\#T_H > \#U_H \quad \text{for all } H \text{ satisfying} \quad H > \big(3N\|A\|\big)^{M/(N-M)}.$$

Now we can finish the proof of Siegel's lemma. Let

$$H = \big(4N\|A\|\big)^{M/(N-M)}.$$

Then T_H contains more vectors than U_H, and since we showed that multiplication by A sends T_H to U_H, it follows that there must be distinct vectors $t', t'' \in T_H$ with the same image $At' = At''$. Then

$$t = t' - t'' \neq 0 \quad \text{satisfies} \quad At = 0,$$

and the coordinates of t satisfy

$$\|t\| = \|t' - t''\| \leq \|t'\| + \|t''\| \leq 2H = 2\big(4N\|A\|\big)^{M/(N-M)}.$$

This shows that the vector t has all of the properties specified in the statement of Siegel's lemma. $\qquad\qquad\qquad\qquad\qquad\qquad\qquad\qquad\qquad\qquad$ $\square$

Now we are ready to construct our auxiliary polynomial. We recall that $b > 0$ is a fixed integer and that $\beta = \sqrt[3]{b}$. We let n be a large positive integer that we will specify later. (For those who are curious, the value that we eventually choose for n will depend on b and on the two rational numbers p_1/q_1 and p_2/q_2 that are close to β.) Then we let m be the integer satisfying

$$m \leq \frac{2}{3}n < m + 1,$$

that is, m is the greatest integer in $\frac{2}{3}n$. We are going to construct a non-zero polynomial

$$F(X, Y) = P(X) + YQ(X)$$

with integer coefficients so that $P(X)$ and $Q(X)$ have degree at most $m + n$ and so that $F(X, \beta)$ is divisible by $(X - \beta)^n$. We will also need to keep track of the size of the coefficients of F.

It is convenient to use superscripts to denote differentiation with respect to X. However, we also want to cancel common factors from the integer coefficients of the derivatives of a polynomial. For example, consider the polynomial $f(x) = x^n$. Its k'th derivative is

$$\frac{d^k(x^n)}{dx^k} = n(n-1)\cdots(n-k+1)x^{n-k} = \frac{n!}{(n-k)!}x^{n-k}.$$

You probably already noticed that the ratio $n!/(n-k)!$ is always divisible by $k!$, since the quantity $n!/(n-k)!k!$ is the binomial coefficient $\binom{n}{k}$, which is an integer. Hence in the k'th derivative of any polynomial, every coefficient is a multiple of $k!$. This suggests that we define a modified k'th derivative by

$$F^{(k)}(X, Y) = \frac{1}{k!}\frac{\partial^k}{\partial X^k}F(X, Y) = \frac{1}{k!}\left(\frac{d^k P(X)}{dX^k} + Y\frac{d^k Q(X)}{dX^k}\right).$$

Then $F^{(k)}(X, Y)$ has integer coefficients if $F(X, Y)$ does.

The condition that $F(X, \beta)$ be divisible by $(X - \beta)^n$ is equivalent to its first $n - 1$ derivatives vanishing at $X = \beta$, so we want to choose coefficients for $F(X, Y)$ so as to force

$$F(\beta, \beta) = F^{(1)}(\beta, \beta) = \cdots = F^{(n-1)}(\beta, \beta) = 0.$$

We write

$$P(X) = \sum_{i=0}^{m+n} u_i X^i \quad \text{and} \quad Q(X) = \sum_{i=0}^{m+n} v_i X^i.$$

Then

$$F^{(k)}(X, Y) = \sum_{i=k}^{m+n} \binom{i}{k}(u_i X^{i-k} + v_i X^{i-k}Y),$$

so

$$F^{(k)}(\beta, \beta) = \sum_{i=k}^{m+n} \binom{i}{k} (u_i \beta^{i-k} + v_i \beta^{i-k+1})$$

$$= \sum_{i=0}^{m+n-k} \binom{i+k}{k} \beta^i u_{i+k} + \sum_{i=1}^{m+n-k+1} \binom{i+k-1}{k} \beta^i v_{i+k-1}$$

$$= \sum_{i=0}^{m+n-k+1} \left\{ \binom{i+k}{k} \beta^i u_{i+k} + \binom{i+k-1}{k} \beta^i v_{i+k-1} \right\},$$

where for the last equation we make the convention that $u_i = v_i = 0$ if either $i < 0$ or $i > m + n$.

Our goal is to choose the u_i's and v_i's so that this last quantity vanishes for all $0 \le k < n$. We can simplify matters a bit by recalling that $\beta^3 = b$, so every power β^i is an integer times one of 1, β, or β^2. Writing $i = 3j + \ell$, we break the last sum into a double sum over j and ℓ. Thus

$$F^{(k)}(\beta, \beta) = \sum_{\ell=0}^{2} \left\{ \sum_j \left(\binom{3j+\ell+k}{k} b^j u_{3j+\ell+k} \right. \right.$$

$$\left. \left. + \binom{3j+\ell+k-1}{k} b^j v_{3j+\ell+k-1} \right) \right\} \beta^\ell.$$

The quantity in braces is an integer. On the other hand, 1, β, and β^2 are linearly independent over $\mathbb{Q}$, i.e., if $A + B\beta + C\beta^2 = 0$ with $A, B, C \in \mathbb{Q}$, then necessarily $A = B = C = 0$. So we are forced to choose the u_i's and the v_i's so that they satisfy

$$\sum_j \left(\binom{3j+\ell+k}{k} b^j u_{3j+\ell+k} + \binom{3j+\ell+k-1}{k} b^j v_{3j+\ell+k-1} \right) = 0$$

for every $\ell \in \{0, 1, 2\}$ and every $k \in \{0, 1, \ldots, n-1\}$.

Although our equations are rather messy, the astute reader will see that we are in exactly the right situation to apply Siegel's lemma. We have $3n$ homogeneous equations, one for each pair (ℓ, n) with $0 \le \ell \le 2$ and $0 \le k < n$, and we have $2(m+n+1)$ variables $\{u_0, \ldots, u_{m+n}, v_0, \ldots, v_{m+n}\}$. Further, these equations have integer coefficients. So Siegel's lemma (Lemma 5.7) tells us that there is a non-zero solution in integers satisfying

$$\max_{0 \le i \le m+n} \{|u_i|, |v_i|\} \le 2\big(4 \cdot 2(m+n+1) \cdot \mu\big)^{\frac{3n}{2(m+n+1)-3n}}.$$

Here we let μ denote the largest coefficient in the equations, which we now need to estimate.

First we observe that

$$\binom{N}{M} \le (1+1)^N = 2^N \quad \text{for all integers } N, M \ge 0.$$

Hence

$$\max_{\substack{j,\ell,k \\ 0 \le 3j+\ell \le m+n \\ 0 \le k \le n}} \binom{3j+\ell+k}{k} b^j \le \max_{\substack{0 \le i \le m+n \\ 0 \le k < n}} 2^{i+k} b^{i/3}$$

$$= 2^{m+2n-1} b^{(m+n)/3}$$

$$< (4b)^{m+n}.$$

For the other part of our upper bound for $\max\{|u_i|, |v_i|\}$, we can use the coarse estimate

$$4 \cdot 2(m+n+1) \le 2^{m+n+3} \le 4^{m+n}.$$

(We assume that $m \ge 3$.) Putting this together gives

$$\max_{0 \le i \le m+n} \{|u_i|, |v_i|\} \le 2 \cdot \big((16b)^{m+n}\big)^{\frac{3n}{2(m+n+1)-3n}}.$$

We can also simplify the exponent. Since m satisfies $m+1 > \frac{2}{3}n$, we find that

$$\frac{3n}{2(m+n+1)-3n} = \frac{3}{2\frac{m+1}{n}-1} \le 9.$$

Using this estimate gives a bound for $\max\{|u_i|, |v_i|\}$, thereby proving the following result, which was the main goal of this section.

Theorem 5.8 (Auxiliary Polynomial Theorem). *Let b be an integer, let $\beta = \sqrt[3]{b}$, and let m and n be integers satisfying*

$$m+1 > \frac{2}{3}n \ge m \ge 3.$$

Then there is a non-zero polynomial

$$F(X, Y) = P(X) + Q(X)Y = \sum_{i=0}^{m+n} (u_i X + v_i X^i Y)$$

having the following properties:

$$F^{(k)}(\beta, \beta) = 0 \text{ for all } 0 \le k < n. \tag{i}$$

$$\max_{0 \le i \le m+n} \{|u_i|, |v_i|\} \le 2 \cdot (16b)^{9(m+n)}. \tag{ii}$$

Example 5.9. Although the computations in this section have been somewhat complicated, they are not hard to carry out in practice. For example, suppose that we take

$$n = 5, \quad m = 3, \quad b = 2, \quad \beta = \sqrt[3]{2}.$$

So we are looking for a polynomial

$$F(X, Y) = \sum_{i=0}^{8} (u_i X^i + v_i X^i Y)$$

satisfying $F^{(k)}(\beta, \beta) = 0$ for all $0 \le k \le 4$. Writing this out explicitly leads to 15 homogeneous linear equations in 18 variables $\{u_0, \ldots, u_8, v_0, \ldots, v_8\}$. Solving for the first 15 variables in terms of the last 3, we can substitute small integer values for v_6, v_7, v_8 to find non-zero integer solutions.

For example, $v_6 = v_7 = 0$ and $v_8 = 1$ gives the polynomial

$$F(X, Y) = -8 - 64X^3 - 20X^6 + 40X^2Y + 32X^5Y + X^8Y.$$

It's an easy exercise to check that $F^{(k)}(\beta, \beta) = 0$ for $0 \le k \le 4$. We observe that the largest coefficient of this F has magnitude 64, while Theorem 5.8 only guarantees a polynomial whose coefficients are no larger than

$$2 \cdot (16b)^{9(m+n)} = 2 \cdot 32^{72} \approx 4.7 \cdot 10^{108}.$$

It is superfluous to point out that the estimate provided by Theorem 5.8 is far from optimal!

We now use F to illustrate a further point. The rational numbers

$$\frac{29}{23} = 1.2608\ldots \quad \text{and} \quad \frac{635}{504} = 1.2599206\ldots$$

are quite close to

$$\sqrt[3]{2} = 1.2599210\ldots.$$

So we expect that F evaluated at these rational numbers should be quite small, and indeed we find that

$$F\left(\frac{29}{23}, \frac{635}{504}\right) = \frac{2816387629}{23^8 \cdot 504} = -0.0000714\ldots.$$

This serves to illustrate the Smallness Theorem, which we prove in the next section.

5.5   The Auxiliary Polynomial Is Small

The auxiliary polynomial $F(X, Y)$ that we constructed in the last section vanishes to high order at the point (β, β). So if p_1/q_1 and p_2/q_2 are rational numbers that are close to β, then we expect $F(p_1/q_1, p_2/q_2)$ to be small. This is indeed true, as we now prove.

Theorem 5.10 (Smallness Theorem). *Let $F(X, Y)$ be a polynomial as described in the Auxiliary Polynomial Theorem (Theorem 5.8). Then there is a constant $c_1 > 0$, depending only on b, so that for any real numbers x and y with $|x - \beta| \le 1$ and for any integer $0 \le t \le n$, we have*

$$\left| F^{(t)}(x, y) \right| \le c_1^n \left(|x - \beta|^{n-t} + |y - \beta| \right).$$

N.B. It is essential that c_1 depends only on b and does not depend on n or t or F or x or y.

Proof. We know that many of the partial derivatives of $F(X, Y)$ vanish at (β, β). We exploit this fact by using the Taylor series expansion of F around the point (β, β). Since Y appears only to the first power in $F(X, Y) = P(X) + Q(X)Y$, we find that

$$F(X, Y) = \sum_{k,j} \frac{1}{k!j!} \cdot \frac{\partial^{k+j} F}{\partial X^k \partial Y^j}(\beta, \beta) \cdot (X - \beta)^k (Y - \beta)^j$$

$$= \sum_{k=0}^{m+n} F^{(k)}(\beta, \beta)(X - \beta)^k + \sum_{k=0}^{m+n} Q^{(k)}(\beta)(X - \beta)^k (Y - \beta).$$

We know that $F^{(k)}(\beta, \beta) = 0$ for all $0 \leq k < n$, so the first sum starts with the $k = n$ term. Thus,

$$F(X, Y) = \sum_{k=n}^{m+n} F^{(k)}(\beta, \beta)(X - \beta)^k + \sum_{k=0}^{m+n} Q^{(k)}(\beta)(X - \beta)^k(Y - \beta).$$

But we really want to estimate $F^{(t)}(x, y)$, so we differentiate t times with respect to X and divide by $t!$. This yields

$$F^{(t)}(X, Y) = \sum_{k=n}^{m+n} F^{(k)}(\beta, \beta)\binom{k}{t}(X - \beta)^{k-t}$$

$$+ \sum_{k=0}^{m+n} Q^{(k)}(\beta)\binom{k}{t}(X - \beta)^{k-t}(Y - \beta)$$

$$= \left(\sum_{k=n}^{m+n} F^{(k)}(\beta, \beta)\binom{k}{t}(X - \beta)^{k-n}\right) \cdot (X - \beta)^{n-t}$$

$$+ \left(\sum_{k=0}^{m+n} Q^{(k)}(\beta)\binom{k}{t}(X - \beta)^{k-t}\right) \cdot (Y - \beta).$$

This last formula reveals the reason that we've done this computation. If we substitute values for X and Y that are close to β, then the last expression will be small due to the presence of the factors $(X - \beta)^{n-t}$ and $Y - \beta$.

So now we put $X = x$ and $Y = y$, take the absolute value of both sides, and use the triangle inequality. We find that

$$\left|F^{(t)}(x, y)\right| \leq \left\{\sum_{k=n}^{m+n} \left|F^{(k)}(\beta, \beta)\right|\binom{k}{t}|x - \beta|^{k-n}\right\} \cdot |x - \beta|^{n-t}$$

$$+ \left\{\sum_{k=0}^{m+n} \left|Q^{(k)}(\beta)\right|\binom{k}{t}|x - \beta|^{k-t}\right\} \cdot |y - \beta|. \quad (**)$$

Compare this estimate with the estimate that we are trying to prove. All that remains is to show that the quantities in braces are bounded by c_1^n for some constant c_1 that depends only on b.

We first observe that for any integer $k \leq m + n$ and any exponent $e \geq 0$,

$$\binom{k}{t}|x - \beta|^e \leq 2^{m+n}, \quad \text{since } |x - \beta| \leq 1.$$

We next write $F(X, Y) = P(X) + Q(X)Y = \sum u_i X^i + v_i X^i Y$ as usual, and use this to estimate

$$\left| F^{(k)}(\beta, \beta) \right| = \left| \sum_{i=k}^{m+n} \binom{i}{k} (u_i \beta^{i-k} + v_i \beta^{i-k+1}) \right|$$

$$\leq (m + n + 1) \max_{0 \leq i \leq m+n} \binom{i}{k} \cdot 2 \max_{0 \leq i \leq m+n} \{|u_i|, |v_i|\} \cdot \beta^{m+n}$$

$$\leq 2^{m+n} \cdot 2^{m+n} \cdot 2 \cdot 2(16b)^{9(m+n)} \cdot b^{(m+n)/3}$$

$$= 4(2^{38} b^{28/3})^{m+n}.$$

Notice that we have made use of the upper bound for the coefficients of F provided by the Auxiliary Polynomial Theorem (Theorem 5.8).

This allows us to bound the first sum in braces in $(**)$ by

$$\sum_{k=n}^{m+n} \left| F^{(k)}(\beta, \beta) \right| \binom{k}{t} |x - \beta|^{k-n} \leq (m + 1) \cdot 4(2^{38} b^{38/3}) \cdot 2^{m+n}$$

$$\leq (2^{42} b^{28/3})^{m+n}$$

$$\leq (2^{70} b^{140/9})^n \quad \text{since } m \leq \frac{2}{3} n.$$

A similar calculation gives a bound for $Q^{(k)}(\beta)$,

$$\left| Q^{(k)}(\beta) \right| \leq \left| \sum_{i=k}^{m+n} \binom{i}{k} v_i \beta^{i-k} \right|$$

$$\leq (m + n + 1) \cdot 2^{m+n} \max_{0 \leq i \leq m+n} |v_i| \cdot \beta^{m+n}$$

$$\leq 2^{2(m+n)} \cdot 2(16b)^{9(m+n)} \cdot b^{(m+n)/3}$$

$$= 2(2^{38} b^{28/3})^{m+n}.$$

And then the second sum in braces in $(**)$ is bounded by

$$\sum_{k=0}^{m+n} \left| Q^{(k)}(\beta) \right| \binom{k}{t} |x - \beta|^{k-t} \leq (m + n + 1) \cdot 2(2^{38} b^{28/3})^{m+n} \cdot 2^{m+n}$$

$$\leq (2^{41} b^{28/3})^{m+n}$$

$$\leq (2^{205/3} b^{140/9})^n.$$

We now have upper bounds for both of the bracketed expressions in $(**)$. Substituting these bounds into $(**)$ gives

$$\left|F^{(t)}(x,y)\right| \le (2^{70}b^{140/9})^n|x-\beta|^{n-t} + (2^{205/3}b^{140/9})^n|y-\beta|$$
$$\le c_1^n\left(|x-\beta|^{n-t} + |y-\beta|\right),$$

where we may take $c_1 = 2^{70}b^{140/9}$. This is precisely the estimate that we have been aiming to prove. $\qquad\square$

5.6 The Auxiliary Polynomial Does Not Vanish

In the last section we showed that an auxiliary polynomial $F(X,Y)$ is small if it is evaluated at a point that is close to (β,β). In this section we would like to show that if x and y are rational numbers, then $F(x,y)$ is not zero. Unfortunately, we are not able to prove such a strong result. Instead, we will show that some derivative $F^{(t)}(X,Y)$, with t not too large, does not vanish.

Theorem 5.11 (Non-Vanishing Theorem). *Let $F(X,Y)$ be an auxiliary polynomial as described in the Auxiliary Polynomial Theorem (Theorem 5.8). Let p_1/q_1 and p_2/q_2 be rational numbers in lowest terms. Then there is a constant c_2, depending only on b, and an integer t satisfying*

$$0 \le t \le 1 + \frac{c_2 n}{\log q_1}$$

so that

$$F^{(t)}\left(\frac{p_1}{q_1}, \frac{p_2}{q_2}\right) \neq 0.$$

N.B. As always, it is crucial that the constant c_2 depend only on b.

Proof. We write $F(X,Y) = P(X) + YQ(X)$ as usual. We are going to look at the *Wronskian polynomial* $W(X)$ defined by

$$W(X) = \det\begin{pmatrix} P(X) & Q(X) \\ P'(X) & Q'(X) \end{pmatrix} = P(X)Q'(X) - Q(X)P'(X).$$

Why is the Wronskian a natural object to look at?

We are searching for some derivative of $F(X, Y)$ that does not vanish as $(p_1/q_1, p_2/q_2)$. Suppose, for example, that we are unlucky and that both F and its first derivative vanish,

$$F\left(\frac{p_1}{q_1}, \frac{p_2}{q_2}\right) = 0 \quad \text{and} \quad F^{(1)}\left(\frac{p_1}{q_1}, \frac{p_2}{q_2}\right) = 0.$$

This means that

$$P\left(\frac{p_1}{q_1}\right) + Q\left(\frac{p_1}{q_1}\right)\frac{p_2}{q_2} = 0,$$

$$P'\left(\frac{p_1}{q_1}\right) + Q'\left(\frac{p_1}{q_1}\right)\frac{p_2}{q_2} = 0.$$

Eliminating p_2/q_2 from these two equations, we find that

$$W\left(\frac{p_1}{q_1}\right) = P\left(\frac{p_1}{q_1}\right)Q'\left(\frac{p_1}{q_1}\right) - Q\left(\frac{p_1}{q_1}\right)P'\left(\frac{p_1}{q_1}\right) = 0.$$

So rather than looking at a two variable polynomial $F(X, Y)$ with certain vanishing properties at $(p_1/q_1, p_2/q_2)$, we can instead study the vanishing properties of the one variable polynomial $W(X)$ at p_1/q_1.

We now work more generally. Let T be the largest integer such that

$$F^{(t)}\left(\frac{p_1}{q_1}, \frac{p_2}{q_2}\right) = P^{(t)}\left(\frac{p_1}{q_1}\right) + Q^{(t)}\left(\frac{p_1}{q_1}\right)\frac{p_2}{q_2} = 0 \quad \text{for all } 0 \le t < T.$$

Our goal is to show that T cannot be too large.

If we take pairs of these equations and eliminate p_2/q_2 from them, we get relations

$$P^{(t)}\left(\frac{p_1}{q_1}\right)Q^{(s)}\left(\frac{p_1}{q_1}\right) - Q^{(t)}\left(\frac{p_1}{q_1}\right)P^{(s)}\left(\frac{p_1}{q_1}\right) = 0 \quad \text{for all } 0 \le s, t \le T.$$

We can relate this to the Wronskian by differentiating $W(X)$. Thus

$$W^{(r)}(X) = \sum_{i+j=r} \frac{i!(j+1)!}{r!}\left(P^{(i)}(X)Q^{(j+1)}(X) - Q^{(i)}(X)P^{(j+1)}(X)\right).$$

Taking any $r < T - 1$ and substituting $X = p_1/q_1$, we find that every term in the sum vanishes, so

$$W^{(r)}\left(\frac{p_1}{q_1}\right) = 0 \quad \text{for all } 0 \le r < T - 1.$$

This means that p_1/q_1 is a $(T-1)$-fold root of $W(X)$, so

$$\left(X - \frac{p_1}{q_1}\right)^{T-1} \bigg| \, W(X).$$

But $W(X)$ has integer coefficients, so Gauss' lemma says that $W(X)$ is divisible by $(q_1 X - p_1)^{T-1}$ in the polynomial ring $\mathbb{Z}[X]$. (Recall that Gauss' lemma says that if a polynomial with integer coefficients factors in $\mathbb{Q}[X]$, then it factors in $\mathbb{Z}[X]$.) In other words, there is a polynomial $V(X)$ with integer coefficients such that

$$W(X) = (q_1 X - p_1)^{T-1} V(X).$$

In order to exploit this factorization, we need to estimate the size of the coefficients of $W(X)$. This is not difficult because the Auxiliary Polynomial Theorem gives us a bound for the coefficients of $P(X)$ and $Q(X)$. We write as usual $P(X) = \sum u_i X^i$ and $Q(X) = \sum v_i X^i$, and then

$$W(X) = P(X)Q'(X) - Q(X)P'(X) = \sum_{i,j} j(u_i v_j - v_i u_j)X^{i+j-1}.$$

So the largest coefficient of $W(X)$ is bounded by

$$\max_{i,j \leq m+n} |j(u_i v_j - v_i u_j)| \leq 2(m+n)\left(\max_{i \leq m+n}\{u_i, v_i\}\right)^2$$

$$\leq 2(m+n)\left(2(16b)^{9(m+n)}\right)^2$$

$$\leq c_3^n,$$

where c_3 is a constant depending only on b. (Note that we always assume that $m \leq \frac{2}{3}n$, as specified in the Auxiliary Polynomial Theorem.)

On the other hand, since $V(X)$ has integer coefficients, the leading coefficient[2] of the product $(q_1 X - p_1)^{T-1}V(X)$ is at least q_1^{T-1}. Thus $W(X)$ has a coefficient that is at least as large as q_1^{T-1}. So we have shown that

$$q_1^{T-1} \leq \left(\text{largest coefficient of } W(X)\right) \leq c_3^n.$$

---

[2]Actually, we also need to check that $W(X)$ is not the zero polynomial. We will verify this at the end of the proof.

Taking logarithms and defining a new constant $c_2 = \log c_3$, we find that

$$T \leq 1 + \frac{c_2 n}{\log q_1}.$$

It only remains to recall that we chose T as the largest integer for which the derivatives $F^{(t)}(p_1/q_1, p_2/q_2)$ vanish for all $0 \leq t < T$. We have just found an upper bound for T. It follows that there is some integer

$$0 \leq t \leq +\frac{c_2 n}{\log q_1} \quad \text{such that} \quad F^{(t)}\left(\frac{p_1}{q_1}, \frac{p_2}{q_2}\right) \neq 0.$$

This (almost) concludes the proof of the Non-Vanishing Theorem.

What is left is that we must show that the Wronskian polynomial $W(X)$ is not identically zero. Suppose to the contrary that $W(X) = 0$. This means that $P'(X)Q(X) = Q'(X)P(X)$, so by the quotient rule we have

$$\frac{d}{dX}\left(\frac{P(X)}{Q(X)}\right) = 0.$$

Thus the ratio $P(X)/Q(X)$ is constant, say $P(X) = aQ(X)$. Note that $a \in \mathbb{Q}$.

Now we have

$$F(X,Y) = P(X) + YQ(X) = (a + Y)Q(X).$$

From the Auxiliary Polynomial Theorem (Theorem 5.8) we know that

$$0 = F^{(k)}(\beta, \beta) = (a + \beta)Q^{(k)}(\beta) \quad \text{for all } 0 \leq k < n.$$

The fact that a is rational means that $a + \beta \neq 0$, so β is an n-fold root of $Q(X)$. Hence

$$(X - \beta)^n \mid Q(X).$$

But $\beta = \sqrt[3]{b}$ and $Q(X)$ has rational coefficients, so $Q(X)$ must be divisible by the n'th power of the minimal polynomial of β,

$$(X^3 - b)^n \mid Q(X).$$

In particular, the degree of $Q(X)$ must be at least $3n$. But we know that the degree of $Q(X)$ is at most $m + n$, and m satisfies $m \leq \frac{2}{3}n$, so the degree of $Q(X)$ is at most $\frac{5}{3}n$. This contradiction shows that $W(X)$ is not the zero polynomial, which completes the proof of the Non-Vanishing Theorem. $\square$

5.7 Proof of the Diophantine Approximation Theorem

We have now assembled all of the tools needed to prove the Diophantine Approximation Theorem.

Theorem 5.12 (Diophantine Approximation Theorem (Thue)). *Let b be a positive integer that is not a perfect cube, and let $\beta = \sqrt[3]{b}$. Let C be a fixed positive constant. Then there are only finitely many pairs of integers (p, q) with $q > 0$ that satisfy the inequality*

$$\left| \frac{p}{q} - \beta \right| \le \frac{C}{q^3}. \tag{$*$}$$

Proof. We give a proof by contradiction. So we suppose that there are infinitely many pairs (p, q) satisfying the inequality $(*)$. Let c_1 and c_2 be the constants appearing in the Smallness Theorem and the Non-Vanishing Theorem (Theorems 5.10 and 5.11), respectively. We emphasize again that these constants depend only on the integer b, which is fixed throughout our discussion.

The inequality $(*)$ implies in particular (since $q \ge 1$) that

$$|p - \beta q| \le C.$$

We are assuming that $(*)$ has infinitely many solutions (p, q), so we see that the q values must tend toward infinity, since otherwise both p and q would be bounded, which would mean that there are only finitely many pairs. Hence we can find a solution (p_1, q_1) to $(*)$ whose second coordinate satisfies

$$q_1 > e^{9c_2} \quad \text{and} \quad q_1 > (2c_1 C)^{18}. \tag{5.1}$$

Then, since our assumption says that there are infinitely many more solutions, we can find another solution (p_2, q_2) whose second coordinate is even larger, say satisfying[3]

$$q_2 > q_1^{65}. \tag{5.2}$$

---

[3]How do we know to choose exponents 9 and 18 and 65 in (5.1) and (5.2)? The answer is that initially we did not know. What we did was to write down the proof leaving the exponents as unknowns. Then, at the end, we could see which values would work. But there is nothing magical about 9, 18, and 65. Any larger values will also work, and if you redo the calculations with more care, you'll find that there are smaller values that work, too.

Next we let n be the integer satisfying

$$n \le \frac{9}{8} \cdot \frac{\log q_2}{\log q_1} < n+1.$$

Exponentiating, this becomes

$$q_1^{\frac{8}{9}n} \le q_2 < q_1^{\frac{8}{9}(n+1)}. \qquad (5.3)$$

Notice that (5.2) implies that $\dfrac{\log q_2}{\log q_1} > 65$, so

$$n \ge \frac{9}{8} \cdot 65 - 1 > 72. \qquad (5.4)$$

Now we start to make use of our theorems. We use the Auxiliary Polynomial Theorem (Theorem 5.8) and our chosen value of n to find a polynomial $F(X, Y)$. Then we apply the Non-Vanishing Theorem (Theorem 5.11) to find an integer t such that

$$t \le 1 + \frac{c_2 n}{\log q_1} \quad \text{and} \quad F^{(t)}\left(\frac{p_1}{q_1}, \frac{p_2}{q_2}\right) \ne 0. \qquad (5.5)$$

Notice that from (5.1) we get the estimate

$$t \le 1 + \frac{1}{9}n. \qquad (5.6)$$

We are going to take the rational number $F^{(t)}(p_1/q_1, p_2/q_2)$ and derive contradictory upper and lower bounds for its size, which will finish the proof of the theorem. We begin with the lower bound.

The auxiliary polynomial $F^{(t)}(X, Y)$ has integer coefficients, degree at most $m + n$ in X, and degree 1 in Y. So putting everything over a common denominator, we find that

$$F^{(t)}\left(\frac{p_1}{q_1}, \frac{p_2}{q_2}\right) = \frac{\text{integer}}{q_1^{m+n} q_2}.$$

Further, we know from (5.5) that the integer in the numerator is not zero. There being no integers strictly between 0 and 1, we deduce that the absolute value of the numerator is at least 1. Hence

$$\left| F^{(t)}\left(\frac{p_1}{q_1}, \frac{p_2}{q_2}\right) \right| \ge \frac{1}{q_1^{m+n} q_2}.$$

The Auxiliary Polynomial Theorem tells us that $m \leq \frac{2}{3}n$, while (5.3) says that $q_2 < q_1^{(8/9)(n+1)}$, so we obtain the fundamental lower bound

$$\left| F^{(t)}\left(\frac{p_1}{q_1}, \frac{p_2}{q_2}\right) \right| \geq \frac{1}{q_1^{\frac{23}{9}n+\frac{8}{9}}}. \tag{5.7}$$

To find a complementary upper bound, we turn to the Smallness Theorem. Thus

$$
\left| F^{(t)}\left(\frac{p_1}{q_1}, \frac{p_2}{q_2}\right) \right| \leq c_1^n \left(\left| \frac{p_1}{q_1} - \beta \right|^{n-t} + \left| \frac{p_2}{q_2} - \beta \right| \right) \quad \text{Smallness Theorem,}
$$

$$
\leq c_1^n \left(\left(\frac{C}{q_1^3}\right)^{n-t} + \frac{C}{q_2^3} \right) \quad \text{from } (*),
$$

$$
\leq c_1^n \left(\left(\frac{C}{q_1^3}\right)^{\frac{8}{9}n-1} + \frac{C}{q_1^{\frac{8}{3}n}} \right) \quad \text{from (5.6) and (5.3),}
$$

$$
\leq \frac{(2c_1 C)^n}{q_1^{\frac{8}{3}n-3}}
$$

$$
\leq \frac{1}{q_1^{\frac{47}{18}n-3}} \quad \text{from (5.1).} \tag{5.8}
$$

Combining our lower bound (5.7) and our upper bound (5.8), we find that

$$\frac{1}{q_1^{\frac{23}{9}n+\frac{8}{9}}} \leq \left| F^{(t)}\left(\frac{p_1}{q_1}, \frac{p_2}{q_2}\right) \right| \leq \frac{1}{q_1^{\frac{47}{18}n-3}},$$

so

$$q_1^{\frac{1}{18}n-\frac{35}{9}} \leq 1.$$

On the other hand, (5.4) says that $n \geq 72$, so we find that

$$q_1^{\frac{1}{9}} \leq 1.$$

This is an obvious absurdity, since the integer q_1 is certainly larger than 2, e.g., from (5.1). We have arrived at the desired contradiction, which completes the proof that there are only finitely many pairs of integers (p, q) with $q > 0$ satisfying the inequality $(*)$. $\qquad \square$

5.8   Further Developments

In this chapter we have proven that an equation of the form

$$ax^3 + by^3 = c$$

has only finitely many solutions in integers x, y. The proof depends on a Diophantine Approximation Theorem which says, roughly, that it is not possible to use rational numbers p/q to very closely approximate a cube root $\sqrt[3]{b}$. With small modifications, the proof that we gave can be adapted to prove the following stronger result.

Theorem 5.13. (Thue 1909 [54]) *Let $\beta \in \mathbb{R}$ be the root of an irreducible polynomial $f(X) \in \mathbb{Q}[X]$ with $d = \deg(f) \geq 3$. Let $\epsilon > 0$ and $C > 0$ be positive numbers. Then there are only finitely many pairs of integers (p, q) with $q > 0$ that satisfy the inequality*

$$\left| \frac{p}{q} - \beta \right| \leq \frac{C}{q^{\frac{1}{2}d+1+\epsilon}}.$$

We proved this theorem for the polynomial $f(X) = X^3 - b$ with $d = 3$ and $\epsilon = \frac{1}{2}$. A number of mathematicians have strengthened Thue's result. Notice that one way to make it stronger is to decrease the exponent of q appearing on the right-hand side. So we might ask for what value of $\tau(d)$ is it true that there are only finitely many rational numbers satisfying

$$\left| \frac{p}{q} - \beta \right| \leq \frac{C}{q^{\tau(d)+\epsilon}}.$$

Thue's result says that we may take $\tau(d) = \frac{1}{2}d + 1$. The following list illustrates the history of this problem.

Liouville	1851	$\tau(d) = d$
Thue	1909	$\tau(d) = \frac{1}{2}d + 1$
Siegel	1921	$\tau(d) = 2\sqrt{d}$
Gelfond, Dyson	1947	$\tau(d) = \sqrt{2d}$
Roth	1955	$\tau(d) = 2$

Roth's theorem, which is somewhat surprising, says that for every degree d we may take $\tau(d) = 2$. It is the strongest theorem of this form in the sense that if we take any $\tau(d) < 2$, then the theorem would not be true. However, Roth's theorem is not the end of the story. There are higher dimensional generalizations (both proven and conjectural) due to Schmidt [39, 40], Vojta [55, 56], and Faltings [15].

The proof that we gave for our special case of Thue's theorem contains all of the ingredients that appear in general. One constructs an auxiliary polynomial, evaluates it at some rational numbers, shows that it (or a small derivative) does not vanish, and derives a contradiction by giving upper and lower bounds for its magnitude. Siegel, Gelfond, and Dyson obtain their stronger results by using a general polynomial $F(X, Y)$, rather than a polynomial of the form $P(X) + Y Q(X)$ as used by Thue. Roth improves this by using an auxiliary polynomial $F(X_1, \ldots, X_r)$ of many variables. However, when working with a multi-variable polynomial, it is quite difficult to prove the analogue of what we called the Non-Vanishing Theorem. The major new technique developed by Roth was an intricate inductive procedure designed to show that some fairly small partial derivative of his auxiliary polynomial $F(X_1, \ldots, X_r)$ does not vanish when evaluated at $(p_1/q_1, \ldots, p_r/q_r)$.

In our concentration on proving the Diophantine Approximation Theorem, we ignored the problem of *effectivity*. That is, we proved that there are only finitely many pairs of integers (p, q) satisfying the inequality

$$\left| \frac{p}{q} - \sqrt[3]{b} \right| \leq \frac{1}{q^3}. \tag{$*$}$$

But for any particular value of b, for example $b = 2$, does our proof give us a method for finding all such pairs?

The answer is NO! If you look at the proof, you will find that it says the following. If we can find a solution (p_1, q_1) to $(*)$ with q_1 very large (how large depends on b), then we can bound the coordinates of every other solution in terms of b and q_1. So if we can find that first large solution, then we can find all of them. But suppose that there are no large solutions? "Ah," you might say, "then we just take the small solutions and we're done." However, nothing in our proof gives us a way of verifying that there are no large solutions. So if we find one large solution, we can find all solutions, but if we cannot find a large solution, then we have no way of proving that the set of solutions that we already have is complete. (This is somewhat subtle. You should stop and think about it for a minute.)

This is not a good state of affairs. In 1966, Baker devised a new method to prove a version of the Diophantine Approximation Theorem that is effective. Although Baker's theorem is not even as strong as Thue's result, it is strong enough to deduce effective bounds for the integer solutions to cubic equations. The bounds tend to be quite large, as illustrated by the following result.

Theorem 5.14. (Baker [2, page 45]) *Let* $a, b, c \in \mathbb{Z}$ *be integers, and let*

$$H = \max\{|a|, |b|, |c|\}.$$

Then every point (x, y) *on the elliptic curve*

$$y^2 = x^3 + ax^2 + bx + c$$

with integer coordinates $x, y \in \mathbb{Z}$ *satisfies*

$$\max\{|x|, |y|\} \le \exp\left((10^6 H)^{10^6}\right).$$

For special curves such as the ones that we have considered in this chapter, Baker's method yields somewhat better estimates. For example, Baker [1] gives the estimate

$$\left|\frac{p}{q} - \sqrt[3]{2}\right| \le \frac{10^{-6}}{q^{2.9955}},$$

valid for all rational numbers p/q. In Section 5.3 we showed that any solution to

$$x^3 - 2y^3 = c$$

in integers $x, y \in \mathbb{Z}$ satisfies

$$\left|\frac{x}{y} - \sqrt[3]{2}\right| \le \frac{4|c|}{3\sqrt[3]{4}} \cdot \frac{1}{y^3}.$$

Combining this inequality with Baker's result, we find that

$$|y| \le 10^{1317} \cdot |c|^{2000/9}.$$

So again the bound is large, but at least it grows only like a power of $|c|$, rather than an exponential of a power of $|c|$.

Exercises

5.1. Define a sequence of pairs of integers by the following rule:

$$(x_0, y_0) = (1, 0)$$
$$(x_{i+1}, y_{i+1}) = (3x_i + 4y_i, 2x_i + 3y_i) \quad \text{for } i \ge 0.$$

(a) Prove that every (x_i, y_i) is a solution to the equation

$$x^2 - 2y^2 = 1.$$

(b) Suppose that (x, y) is a solution to the equation in (a) and that x and y are positive integers. Prove that there is some index $i \geq 0$ such that $(x, y) = (x_i, y_i)$. *Hint.* Using the assumption that $y > 0$, prove that there are positive integers u and v satisfying

$$(x, y) = (3u + 4v, 2u + 3v) \quad \text{and} \quad u^2 - 2v^2 = 1 \quad \text{and} \quad v < y.$$

5.2. Let a, b, c be non-zero integers, and suppose that (x, y) is a solution in integers to the equation

$$ax^3 + bxy^2 = c.$$

Prove that

$$\max\{|ax^2|, |by^2|\} \leq 1 + \max\{|a|, |b|\}c^2.$$

5.3. Find all integer solutions to the following equations.
(a) $x^2 y + xy^2 = 240$.
(b) $(x - 2y + 1)(79x^2 + 4xy - 34y^2) = 98$.
(c) $(x - 2y + 1)(403x^2 - 388xy + 394y^2 + 1412x - 1612y) = 1218$.

5.4. Let $m \geq 2$ be an integer, and let $d(m)$ denote the number of distinct positive divisors of m. Prove that the equation $x^3 + y^3 = m$ has no more than $d(m)$ solutions in pairs of integers (x, y) with $x \leq y$.

5.5. Let p be an odd prime.
(a) Prove that the equation $x^3 + y^3 = p$ has a solution in integers if and only if $p = 3u^2 + 3u + 1$ for some integer u.
(b) Find all primes $p < 300$ for which the equation in (a) has a solution in integers.

5.6. For this exercise we look at the curves

$$C_d : y^2 = x^3 + d.$$

We let $C_d(\mathbb{Z})$ denote the set of integer points,

$$C_d(\mathbb{Z}) = \{(x, y) : x, y \in \mathbb{Z} \text{ and } y^2 = x^3 + d\}.$$

(a) Prove that for every integer $N \geq 1$ there is an integer $d \geq 1$ so that $C_d(\mathbb{Z})$ contains at least N points.
(b) More precisely, prove that there is a constant $\kappa > 0$ and a sequence of integers $1 \leq d_1 < d_2 < d_3 < \cdots$ such that

$$\#C_{d_i}(\mathbb{Z}) \geq \kappa \log \log d_i.$$

(*Hint.* Take a rational point P of infinite order on some C_d, look at the rational points $P, 2P, 4P, 8P, \ldots$, and clear denominators. Use the height formula $h(2P) \leq 4h(P) + \kappa$ to keep track of the size of the denominators.)

(c) * Same as (b), but prove the better lower bound

$$\#C_d(\mathbb{Z}) \geq \kappa(\log d_i)^{1/3}.$$

(*Hint.* Same as (c), but use all of the multiples $P, 2P, 3P, \ldots$.)

(d) Show that C_{17} has at least 16 integer points. How many integer points can you find on C_{2089}?

(e) ** Call an integer point *primitive* if $\gcd(x, y) = 1$. Either prove that the number of primitive integer points in $C_d(\mathbb{Z})$ is bounded independently of d, or else find a sequence of d's so that the number of primitive integer points in $C_d(\mathbb{Z})$ goes to infinity.

5.7. Let $\beta \in \mathbb{R}$ be a real number.

(a) Prove that for every integer $q \geq 1$ there is an integer p so that the rational number p/q satisfies

$$\left| \frac{p}{q} - \beta \right| \leq \frac{1}{2q}.$$

(b) * Assuming that $\beta \notin \mathbb{Q}$, prove that there are infinitely many rational numbers p/q satisfying

$$\left| \frac{p}{q} - \beta \right| \leq \frac{1}{q^2}.$$

This result is due to Dirichlet. It shows that the exponent 2 in Roth's theorem cannot be decreased.

5.8. Let $\beta \in \mathbb{R}$ be a real number. In this exercise we consider solutions to the inequality

$$\left| \frac{p}{q} - \beta \right| \leq \frac{1}{q^3}.$$

(a) Suppose that p/q and p'/q' are distinct solutions with $q' \geq q$. Prove that $q' \geq \frac{1}{2}q^2$.

(b) Suppose that $p_0/q_0, p_1/q_1, \ldots, p_r/q_r$ is a list of distinct solutions with $4 \leq q_0 \leq q_1 \leq \cdots \leq q_r$. Prove that

$$q_r \geq 2^{2^r}.$$

This shows that the solutions are very widely spaced. It is an example of what is known as a *gap principle*.

5.9. Let $d \geq 3$ be an integer, and let b be an integer that is not a perfect d'th power.

(a) Let C be a constant. Prove that there are only finitely many rational numbers p/q satisfying the inequality

$$\left| \frac{p}{q} - \sqrt[d]{b} \right| \leq \frac{1}{q^{(d+3)/2}}.$$

(b) Let a, b, c be non-zero integers. Prove that the equation

$$ax^d + by^d = c$$

has only finitely many solutions $x, y \in \mathbb{Z}$.

5.10. (a) * Prove the general version of Thue's Approximation Theorem (Theorem 5.13) stated in Section 5.8.

(b) * Let $a_0 t^d + a_1 t^{d-1} + \cdots + a_d$ be a polynomial of degree at least 3 with integer coefficients, and assume that it is irreducible in $\mathbb{Q}[t]$. Prove that for any non-zero integer c, the equation

$$a_0 x^d + a_1 x^{d-1} y + a_2 x^{d-2} y^2 \cdots + a_{d-1} xy^{d-1} + a_d y^d = c$$

has only finitely many solutions in integers $x, y \in \mathbb{Z}$.

5.11. Let C be a non-singular cubic curve given by a Weierstrass equation

$$y^2 = x^3 + ax^2 + bx + c$$

with integer coefficients. Let $P \in C(\mathbb{Q})$, and suppose that there is an integer $n \geq 1$ such that nP has integer coordinates. Prove that P has integer coordinates. (*Hint.* Consider the subgroups $C(p)$ defined in Section 2.4.)

5.12. Let C be a non-singular cubic curve given by a Weierstrass equation

$$y^2 = x^3 + ax^2 + bx + c$$

with integer coefficients, and let $P \in C(\mathbb{Q})$ be a point of infinite order.

(a) For each $n \geq 1$, prove that the coordinates of nP can be written in the form

$$x(nP) = \left(\frac{a_n}{d_n^2}, \frac{b_n}{d_n^3} \right) \quad \text{with } \gcd(a_n, d_n) = \gcd(b_n, d_n) = 1.$$

Changing the sign of b_n if necessary, we may assume that $d_n \geq 1$.

(b) Prove that if m and n are integers with m dividing n, then d_m divides d_n. The sequence $(d_n)_{n \geq 1}$ is called an *elliptic divisibility sequence*. You are probably familiar with other sequences having this property, for example the sequence $2^n - 1$ and the Fibonacci sequence.

5.13. Let $a, b, c \in \mathbb{Z}$ be non-zero integers with $\gcd(a, b) = 1$.

(a) Prove that the linear equation

$$ax + by = c$$

has a solution (x_0, y_0) in integers.

(b) Prove that the complete set of integer solutions is then given by

$$\{(x_0 + bn, y_0 - an) : n \in \mathbb{Z}\}.$$

(c) Suppose that $\gcd(a, b) > 1$. Formulate and prove appropriate versions of (a) and (b) for this situation.

Chapter 6

Complex Multiplication

6.1 Abelian Extensions of $\mathbb{Q}$

In this chapter we describe how points of finite order on certain elliptic curves can be used to generate interesting extension fields of $\mathbb{Q}$. Here we mean points of finite order with arbitrary complex coordinates, not just the ones with rational coordinates that we studied in Chapter 2. So we will need to use some basic theorems about extension fields and Galois groups, but nothing very fancy. We start by reminding you of most of the facts that we need, and you can look in any basic algebra text such as [14, 23, 26] for the proofs and additional background material.

We are interested in subfields of the complex numbers $\mathbb{Q} \subset K \subset \mathbb{C}$. We may view K as a $\mathbb{Q}$-vector space, and the *degree of K over $\mathbb{Q}$* is defined to be

$$[K : \mathbb{Q}] = \text{dimension of } K \text{ as a } \mathbb{Q}\text{-vector space}.$$

If $[K : \mathbb{Q}]$ is finite, then we call K a *number field*.

An important technique for studying number fields is to look at the set of field homomorphisms

$$\sigma : K \hookrightarrow \mathbb{C}.$$

We recall that a homomorphism of fields is always one-to-one because a field has no non-trivial ideals. Also, since by definition $\sigma(1) = 1$, we see

© Springer International Publishing Switzerland 2015
J.H. Silverman, J.T. Tate, *Rational Points on Elliptic Curves*,
Undergraduate Texts in Mathematics, DOI 10.1007/978-3-319-18588-0_6

that $\sigma(a) = a$ for every $a \in \mathbb{Q}$. It is a theorem that the number of homomorphisms $K \hookrightarrow \mathbb{C}$ is exactly equal to the degree $[K : \mathbb{Q}]$.

It sometimes happens that the image $\sigma(K)$ is equal to the original field K. Then σ is an isomorphism from K to itself, in which case we call σ an *automorphism of K*. Note that this does not mean that $\sigma(\alpha) = \alpha$ for every $\alpha \in K$, but merely that $\sigma(\alpha) \in K$. If this is true for every σ, then we say that K is a *Galois extension of $\mathbb{Q}$*. More generally, let

$$\mathrm{Aut}(K) = \{\text{automorphisms } \sigma : K \to K\}.$$

We make $\mathrm{Aut}(K)$ into a group in the usual way. If $\sigma, \tau \in \mathrm{Aut}(K)$, then we define $\sigma\tau$ to be the composition, $(\sigma\tau)(\alpha) = \sigma(\tau(\alpha))$. A number field K is a Galois extension of $\mathbb{Q}$ if and only if

$$\# \mathrm{Aut}(K) = [K : \mathbb{Q}].$$

In this case, we write $\mathrm{Gal}(K/\mathbb{Q})$ instead of $\mathrm{Aut}(K)$, and we call $\mathrm{Gal}(K/\mathbb{Q})$ the *Galois group of $K/\mathbb{Q}$*.

This is all somewhat abstract. How does one actually find number fields that are Galois over $\mathbb{Q}$? The answer is simple. Take any polynomial with rational coefficients $f(X) \in \mathbb{Q}[X]$. Factor $f(X)$ over the complex numbers,

$$f(X) = a(X - \alpha_1)(X - \alpha_2) \cdots (X - \alpha_n),$$

and let

$$K = \mathbb{Q}(\alpha_1, \alpha_2, \ldots, \alpha_n)$$

be the smallest subfield of $\mathbb{C}$ containing all of the α_i's. Then any homomorphism $\sigma : K \to \mathbb{C}$ is determined by the values of $\sigma(\alpha_1), \ldots \sigma(\alpha_n)$, and each $\sigma(\alpha_i)$ has to be a root of $f(X)$, so equals some α_j. In particular, $\sigma(\alpha_i) \in K$, so $\sigma(K) = K$. (The inclusion $\sigma(K) \subset K$ is clear, and then equality follows by comparing the degrees of K and $\sigma(K)$ over $\mathbb{Q}$.) The field K is called the *splitting field of $f(X)$ over $\mathbb{Q}$*, and we have just seen that such a splitting field is a Galois extension. Conversely, one can prove that if a number field K is a Galois extension of $\mathbb{Q}$, then it is the splitting field of some polynomial $f(X) \in \mathbb{Q}[X]$.

This fact helps to explain why Galois extensions are both useful and important. The study of roots of polynomials lies at the classical base of much of algebra and number theory. In order to study those roots, one might instead study the fields that the roots generate. And if one takes the field generated by all of the roots, then one gets a Galois extension, which has attached to it a certain finite group. So by using basic facts from group theory, one can often make interesting deductions about the roots of the original polynomial. Schematically, one might imagine the process as follows:

Roots of Polynomials	*Field Theory* $\longleftrightarrow$	Extension Fields	*Galois Theory* $\longleftrightarrow$	Group Theory

The easiest sorts of groups are abelian groups, so it is natural to begin by looking at Galois extensions $K/\mathbb{Q}$ whose Galois groups are abelian. One way that such extensions arise is in the study of Fermat's equation

$$x^n + y^n = 1.$$

If we try to apply the factorization techniques used throughout this book, we might move the x^n to the other side of the equation and factor

$$y^n = 1 - x^n = (1 - x)(1 - \zeta x)(1 - \zeta^2 x) \cdots (1 - \zeta^{n-1}x).$$

Here $\zeta \in \mathbb{C}$ is a *primitive n'th root of unity*, that is, a complex number satisfying $\zeta^n = 1$ and $\zeta^j \neq 1$ for all $1 \leq j < n$. For example, we could take $\zeta = e^{2\pi i/n}$. In order to study Fermat's equation, we are led, following Kummer, to look at the field $\mathbb{Q}(\zeta)$.

A field generated by roots of unity, such at the field $\mathbb{Q}(\zeta)$, is called a *cyclotomic field*. The name comes from the Greek word *kyklos* ($\kappa \upsilon \kappa \lambda o \zeta$) for cycle, because roots of unity lie cyclically around the unit circle $|z| = 1$ in the unit plane. Note that $\mathbb{Q}(\zeta)$ contains all of the powers of ζ, so it is the splitting field over $\mathbb{Q}$ of the polynomial $X^n - 1$. Thus $\mathbb{Q}(\zeta)$ is a Galois extension of $\mathbb{Q}$, and as we now explain, it is possible to give a very explicit and concrete description of its Galois group.

An automorphism $\sigma : \mathbb{Q}(\zeta) \to \mathbb{Q}(\zeta)$ is determined by the value of $\sigma(\zeta)$, and that value will also be a primitive n'th root of unity, since σ preserves the order of an element. Every primitive n'th root of unity is a power of ζ, and more precisely, has the form ζ^t for some integer t that is relatively prime to n. Thus we obtain a one-to-one map of sets

$$t : \mathrm{Gal}\big(\mathbb{Q}(\zeta)/\mathbb{Q}\big) \longrightarrow (\mathbb{Z}/n\mathbb{Z})^*$$

that is completely determined by the property

$$\sigma(\zeta) = \zeta^{t(\sigma)} \quad \text{for } \sigma \in \mathrm{Gal}\big(\mathbb{Q}(\zeta)/\mathbb{Q}\big).$$

Here $(\mathbb{Z}/n\mathbb{Z})^*$ is the group of units in $\mathbb{Z}/n\mathbb{Z}$,

$$(\mathbb{Z}/n\mathbb{Z})^* = \big\{a \bmod n : \gcd(a, n) = 1\big\}.$$

We claim that the map t is a homomorphism of groups. The proof is easy. If $\sigma, \tau \in \mathrm{Gal}\big(\mathbb{Q}(\zeta)/\mathbb{Q}\big)$, then

$$\zeta^{t(\sigma\tau)} = (\sigma\tau)(\zeta) = \sigma\big(\tau(\zeta)\big) = \sigma(\zeta^{t(\tau)})$$
$$= \big(\sigma(\zeta)\big)^{t(\tau)} = (\zeta^{t(\sigma)})^{t(\tau)} = \zeta^{t(\sigma)t(\tau)}.$$

Hence

$$t(\sigma\tau) \equiv t(\sigma)t(\tau) \pmod{n},$$

which proves our assertion.

We have proven that there is a one-to-one homomorphism

$$t : \mathrm{Gal}\big(\mathbb{Q}(\zeta)/\mathbb{Q}\big) \longrightarrow (\mathbb{Z}/n\mathbb{Z})^*.$$

Since $(\mathbb{Z}/n\mathbb{Z})^*$ is an abelian group, the same is true of $\mathrm{Gal}\big(\mathbb{Q}(\zeta)/\mathbb{Q}\big)$. This completes the proof of the following proposition.

Proposition 6.1. *The Galois group of a cyclotomic extension is abelian. More precisely, if ζ is a primitive n'th root of unity, then there is a one-to-one homomorphism*

$$t : \mathrm{Gal}\big(\mathbb{Q}(\zeta)/\mathbb{Q}\big) \longrightarrow (\mathbb{Z}/n\mathbb{Z})^*$$

determined by the property that $\sigma(\zeta) = \zeta^{t(\sigma)}$.

In fact, the map t is an isomorphism, but the proof is not easy except in the case that $n = p$ is prime, in which case it can be proven by checking that $[\mathbb{Q}(\zeta) : \mathbb{Q}] = p - 1$.

We now want to talk more generally about field extensions $F \subset K$ with F not necessarily equal to $\mathbb{Q}$. For such an extension of fields, we let

$$\mathrm{Aut}_F(K) = \left\{ \begin{array}{c} \text{automorphisms } \sigma : K \to K \text{ such} \\ \text{that } \sigma(a) = a \text{ for all } a \in F \end{array} \right\}.$$

If $[K : F] = \#\,\mathrm{Aut}_F(K)$, then we say that K/F is a Galois extension, and we write $\mathrm{Gal}(K/F)$ instead of $\mathrm{Aut}_F(K)$.

Now suppose that we have a subextension of a cyclotomic field,

$$\mathbb{Q} \subset F \subset \mathbb{Q}(\zeta).$$

The Fundamental Theorem of Galois Theory tells us that $F/\mathbb{Q}$ is a Galois extension if and only if $\mathrm{Gal}\big(\mathbb{Q}(\zeta)/F\big)$ is a normal subgroup of $\mathrm{Gal}\big(\mathbb{Q}(\zeta)/\mathbb{Q}\big)$. But we just saw that $\mathrm{Gal}\big(\mathbb{Q}(\zeta)/\mathbb{Q}\big)$ is an abelian group, so all of its subgroups are normal. Hence $F/\mathbb{Q}$ is Galois, and Galois theory says that there is an isomorphism

$$\frac{\mathrm{Gal}\big(\mathbb{Q}(\zeta)/\mathbb{Q}\big)}{\mathrm{Gal}\big(\mathbb{Q}(\zeta)/F\big)} \xrightarrow{\sim} \mathrm{Gal}(F/\mathbb{Q}).$$

Hence every subfield of a cyclotomic field is a Galois extension of $\mathbb{Q}$ with abelian Galois group. Amazingly, the converse is also true.

Theorem 6.2. (Kronecker–Weber Theorem) *Let F be a number field that is Galois over $\mathbb{Q}$, and suppose that $\mathrm{Gal}(F/\mathbb{Q})$ is abelian. Then there exists a cyclotomic extension $\mathbb{Q}(\zeta)/\mathbb{Q}$ such that*

$$F \subset \mathbb{Q}(\zeta).$$

Hence the Galois extensions of $\mathbb{Q}$ with abelian Galois groups are precisely the subfields of cyclotomic fields.

The proof of the Kronecker–Weber theorem is quite difficult, although nowadays people might say that it is an immediate corollary of class field theory (which is too complicated for us to even describe). But we can prove a special case for you.

Suppose that $F = \mathbb{Q}(\sqrt{p})$ is a quadratic extension of $\mathbb{Q}$, where p is a prime. Then $F/\mathbb{Q}$ is a Galois extension whose Galois group is a cyclic group of order two. In particular, the Galois group is abelian, so the Kronecker–Weber theorem says that F should be contained in come cyclotomic extension.

To prove this for odd p, we let $\zeta \in \mathbb{C}$ be a primitive p'th root of unity, and we let γ be the quadratic Gauss sum

$$\gamma = \sum_{a=0}^{p-1} \zeta^{a^2}.$$

Then one can check that

$$\gamma^2 = \begin{cases} p & \text{if } p \equiv 1 \ (\mathrm{mod}\ 4), \\ -p & \text{if } p \equiv -1 \ (\mathrm{mod}\ 4). \end{cases}$$

(See Exercise 4.4, where γ is $2\alpha+1$.) Hence if $p \equiv 1 \ (\mathrm{mod}\ 4)$, then $\mathbb{Q}(\sqrt{p}) = \mathbb{Q}(\gamma) \subset \mathbb{Q}(\zeta)$. On the other hand, if $p \equiv -1 \ (\mathrm{mod}\ 4)$, then we let $\zeta' = i\zeta$, so ζ' is a primitive $4p$'th root of unity, and we have inclusions

$$\mathbb{Q}(\sqrt{p}) \subset \mathbb{Q}(i, \sqrt{p}) = \mathbb{Q}(i, \gamma) \subset \mathbb{Q}(i, \zeta) = \mathbb{Q}(\zeta').$$

This proves the Kronecker–Weber theorem for the quadratic extension $\mathbb{Q}(\sqrt{p})$ when p is an odd prime. And for $p = 2$, we leave it to the reader to check that

$$\sqrt{2} = \zeta + \zeta^{-1} \quad \text{with} \quad \zeta = e^{2\pi i/8}.$$

If we use a little complex analysis, the Kronecker–Weber theorem becomes even more remarkable. To calculate an n'th root of unity, we can use the Taylor series for the exponential function

$$f(z) = e^{2\pi i z} = \sum_{k=0}^{\infty} \frac{(2\pi i z)^k}{k!}.$$

This is an entire, i.e., everywhere holomorphic, function on $\mathbb{C}$. If we evaluate this function at a rational number $1/n$, we get a complex number

$$f\left(\frac{1}{n}\right) = \sum_{k=0}^{\infty} \frac{(2\pi i)^k}{n^k k!}$$

given by a convergent power series. We now have three amazing facts:
(i) The series converges to a number that is a root of a polynomial having rational coefficients, viz., it is a root of $X^n - 1$.
(ii) The field extension of $\mathbb{Q}$ generated by $f(1/n)$ is a Galois extension of $\mathbb{Q}$ with abelian Galois group.
(iii) Every Galois extension of $\mathbb{Q}$ with abelian Galois group is contained in one of these extensions.
So the abelian extensions of $\mathbb{Q}$ may be described in terms of certain specific values of the holomorphic function $f(z) = e^{2\pi i z}$. Further, recall our homomorphism (really an isomorphism)

$$t : \mathrm{Gal}\big(\mathbb{Q}(\zeta)/\mathbb{Q}\big) \longrightarrow (\mathbb{Z}/n\mathbb{Z})^*,$$

where we can take $\zeta = f(1/n) = e^{2\pi i/n}$. Then we can describe the action of an element $\sigma \in \mathrm{Gal}\big(\mathbb{Q}(\zeta)/\mathbb{Q}\big)$ on ζ very easily in terms of f and t,

$$\sigma\left(f\left(\frac{1}{n}\right)\right) = f\left(\frac{t(\sigma)}{n}\right).$$

The question now arises whether a similar theory exists for other fields. Kronecker's Jugendtraum ("Dream of Youth") was to construct a similar theory for extensions of other fields F. Kronecker's hope was to find a holomorphic (or meromorphic) function $f(z)$ with the property that for every Galois extension K/F with abelian Galois group, there are special values $f(a_1), \ldots, f(a_n)$ of $f(z)$ so that the field $F\big(f(a_1), \ldots, f(a_n)\big)$ generated by these values is Galois, has abelian Galois group, and so that

$$K \subset F\big(f(a_1), \ldots, f(a_n)\big).$$

Further, he desired that for $\sigma \in \mathrm{Gal}\big(F\big(f(a_1), \ldots, f(a_n)\big)/F\big)$, the value of $\sigma\big(f(a_i)\big)$ could be described in terms of the value of $f(z)$ at some value of z obtained by applying a simple operation to a_i. We have seen that Kronecker's Jugendtraum is true for $F = \mathbb{Q}$ by taking $f(z) = e^{2\pi i z}$ and special values $f(j/n)$ with $j \in (\mathbb{Z}/n\mathbb{Z})^*$. The action of σ on $f(j/n)$ is given by evaluating f at tj/n, where $t = t(\sigma) \in (\mathbb{Z}/n\mathbb{Z})^*$.

Kronecker and his contemporaries were largely able to construct such a theory for imaginary quadratic fields, that is, for quadratic extensions F of $\mathbb{Q}$ such that F is not contained in $\mathbb{R}$. Their construction is intimately tied up with the theory of elliptic curves. This is the material that we plan to discuss in the remainder of this chapter.

More generally, if one starts with any number field F, one can ask for a description of all Galois extensions K/F with abelian Galois group. The class field theory alluded to above gives such a description, but it does so in a somewhat indirect manner. Except in certain special cases, the extension of Kronecker's Jugendtraum to number fields is still very much an open problem.

6.2 Algebraic Points on Cubic Curves

As usual, let C be an elliptic curve given by a Weierstrass equation

$$C : y^2 = x^3 + ax^2 + bx + c$$

with rational coefficients $a, b, c \in \mathbb{Q}$. Up to now we have been mainly concerned with points on such curves having either rational or integer coordinates, although we have also talked about real points $C(\mathbb{R})$ and complex points $C(\mathbb{C})$. More generally, if $K \subset \mathbb{C}$ is any subfield of the complex numbers, then we can look at the set of K-rational points,

$$C(K) = \big\{ (x, y) : x, y \in K \text{ and } y^2 = x^3 + ax^2 + bx + c \big\} \cup \{\mathcal{O}\}.$$

It is clear from the formulas for the addition law on C that $C(K)$ is closed under addition, so it is a subgroup of $C(\mathbb{C})$.

For example, consider the curve

$$y^2 = x^3 - 4x^2 + 16.$$

The discriminant of the cubic polynomial is $D = 45056 = 2^{12} \cdot 11$, and one can easily check, for example using the Nagell–Lutz theorem, that the rational points of finite order on C form a group of order five,

$$\{\mathcal{O}, (0, \pm 4), (4, \pm 4)\}.$$

With somewhat more effort, it is possible to prove that $C(\mathbb{Q})$ consists of only these five points. There are no points of infinite order in $C(\mathbb{Q})$.

However, if we replace $\mathbb{Q}$ by an extension field, matters may drastically change. For example, if we take the field $\mathbb{Q}(\sqrt{-2})$, then C contains the point

$$P = (8 + 4\sqrt{-2}, 12 + 16\sqrt{-2}) \in C\big(\mathbb{Q}(\sqrt{-2})\big).$$

We can use the duplication formula to compute $2P$, thus

$$2P = \left(\frac{-124 + 56\sqrt{-2}}{(3 + 4\sqrt{-2})^2}, \frac{-276 - 448\sqrt{-2}}{(3 + 4\sqrt{-2})^3} \right).$$

The point P has infinite order, so $C\big(\mathbb{Q}(\sqrt{-2})\big)$ contains infinitely many points.

Suppose now that K is a Galois extension of $\mathbb{Q}$. Then for any point $P = (x, y) \in C(K)$ and any element $\sigma \in \text{Gal}(K/\mathbb{Q})$, we define a new point

$$\sigma(P) = \big(\sigma(x), \sigma(y)\big).$$

We also set $\sigma(\mathcal{O}) = \mathcal{O}.$[1] We now check that $\sigma(P)$ is a point in $C(K)$ and that the map $P \to \sigma(P)$ interacts nicely with the group law on C. This, and more, is contained in the following elementary proposition.

Proposition 6.3. *Let C be an elliptic curve defined by an equation with coefficients in $\mathbb{Q}$, and let K be a Galois extension of $\mathbb{Q}$.*
(a) *The set $C(K)$ of points with coordinates in K is a subgroup of $C(\mathbb{C})$.*

---

[1]This makes sense because $\mathcal{O} = [0, 1, 0]$ in homogeneous coordinates, so $\sigma(\mathcal{O}) = [\sigma(0), \sigma(1), \sigma(0)] = [0, 1, 0] = \mathcal{O}$.

(b) *For $P \in C(K)$ and $\sigma \in \mathrm{Gal}(K/\mathbb{Q})$, define*

$$\sigma(P) = \begin{cases} \big(\sigma(x), \sigma(y)\big) & \text{if } P = (x, y), \\ \mathcal{O} & \text{if } P = \mathcal{O}. \end{cases}$$

Then $\sigma(P) \in C(K)$.

(c) *For all $P \in C(K)$ and all $\sigma, \tau \in \mathrm{Gal}(K/\mathbb{Q})$,*

$$(\sigma\tau)(P) = \sigma\big(\tau(P)\big).$$

Further the identity element $e \in \mathrm{Gal}(K/\mathbb{Q})$ acts trivially, $e(P) = P$.

(d) *For all $P, Q \in C(K)$ and all $\sigma \in \mathrm{Gal}(K/\mathbb{Q})$,*

$$\sigma(P + Q) = \sigma(P) + \sigma(Q) \quad \text{and} \quad \sigma(-P) = -\sigma(P).$$

In particular, $\sigma(nP) = n\big(\sigma(P)\big)$ for all integers n.

(e) *Let $P \in C(K)$ be a point of order n and let $\sigma \in \mathrm{Gal}(K/\mathbb{Q})$. Then $\sigma(P)$ also has order n.*

Proof. (a) If P_1 and P_2 are in $C(K)$, then their x and y-coordinates are in K, so it is clear from the explicit formulas for the addition law on C that $P_1 \pm P_2$ have coordinates in K. Hence $C(K)$ is closed under addition and subtraction, so it is a subgroup of $C(\mathbb{C})$.

(b) Let $P = (x, y) \in C(K)$. The coordinates of $\sigma(P)$ are in K, so we just need to check that the point $\sigma(P)$ is on the curve C. We know that P is on C and that $\sigma : K \to K$ is a field homomorphism that fixes $\mathbb{Q}$, so we find that

$$P = (x, y) \in C(K)$$
$$\implies y^2 - x^3 - ax^2 - bx - c = 0$$
$$\implies \sigma(y^2 - x^3 - ax^2 - bx - c) = 0$$
$$\implies \sigma(y)^2 - \sigma(x)^3 - \sigma(a)\sigma(x)^2 - \sigma(b)\sigma(x) - \sigma(c) = 0$$
$$\qquad\qquad \text{because } \sigma \text{ is a field homomorphism,}$$
$$\implies \sigma(y)^2 - \sigma(x)^3 - a\sigma(x)^2 - b\sigma(x) - c = 0$$
$$\qquad\qquad \text{because } \sigma \text{ fixes } \mathbb{Q} \text{ and } a, b, c \in \mathbb{Q},$$
$$\implies \sigma(P) = \big(\sigma(x), \sigma(y)\big) \in C(K).$$

(c) We leave this as an exercise.

(d) As in (b), this part follows from the fact that the addition law is given by rational functions with coefficients in $\mathbb{Q}$. There are several cases to check. We will do one and leave the others to you.

Write

$$P = (x_1, y_1), \quad Q = (x_2, y_2), \quad \text{and} \quad P + Q = (x_3, y_3).$$

Assuming that $P \neq \pm Q$, the formulas in Section 1.4 say that

$$x_3 = \left(\frac{y_2 - y_1}{x_2 - x_1}\right)^2 - a - x_1 - x_2, \quad y_3 = \frac{y_2 - y_1}{x_2 - x_1}(x_1 - x_3) - y_1.$$

Using the fact that σ is a field homomorphism that fixes $\mathbb{Q}$, we find that

$$\sigma(x_3) = \left(\frac{\sigma(y_2) - \sigma(y_1)}{\sigma(x_2) - \sigma(x_1)}\right)^2 - a - \sigma(x_1) - \sigma(x_2),$$

$$\sigma(y_3) = \frac{\sigma(y_2) - \sigma(y_1)}{\sigma(x_2) - \sigma(x_1)}(\sigma(x_1) - \sigma(x_3)) - \sigma(y_1).$$

Hence

$$\begin{aligned}
\sigma(P + Q) &= \big(\sigma(x_3), \sigma(y_3)\big) \\
&= \big(\sigma(x_1), \sigma(y_1)\big) + \big(\sigma(x_2), \sigma(y_2)\big) \\
&= \sigma(P) + \sigma(Q).
\end{aligned}$$

The fact that $\sigma(-P) = -\sigma(P)$ is even easier, since if $P = (x, y)$, then

$$\sigma(-P) = \sigma(x, -y) = \big(\sigma(x), \sigma(-y)\big) = \big(\sigma(x), -\sigma(y)\big) = -\sigma(P).$$

Finally, by repeatedly applying the formula $\sigma(P+Q) = \sigma(P)+\sigma(Q)$, we easily find that $\sigma(nP) = n\sigma(P)$ for all $n \geq 0$, and then $\sigma(-P) = -\sigma(P)$ shows that it is also true for $n < 0$.

(e) Let $P \in C(K)$ have order n, and let m be the order of $\sigma(P)$. Using (d), we find that

$$n\sigma(P) = \sigma(nP) = \sigma(\mathcal{O}) = \mathcal{O},$$

so m divides n. Conversely, using the fact that $\mathcal{O} = m\sigma(P) = \sigma(mP)$ and applying σ^{-1} to both sides, we find that

$$\mathcal{O} = \sigma^{-1}(\mathcal{O}) = \sigma^{-1}\big(\sigma(mP)\big) = (\sigma^{-1}\sigma)(mP) = mP.$$

Hence n divides m, which completes the proof that $m = n$. $\square$

In the last section we defined a cyclotomic field as the splitting field over $\mathbb{Q}$ of a polynomial $X^n - 1$. To clarify the analogy with elliptic curves, we want to reformulate this as follows.

Consider the group $\mathbb{C}^*$ of non-zero complex numbers with the group law being multiplication. For any integer n, raising to the n'th power gives a group homomorphism from $\mathbb{C}^*$ to itself,

$$\lambda_n : \mathbb{C}^* \longrightarrow \mathbb{C}^*, \quad \lambda_n(z) = z^n.$$

The kernel of the homomorphism λ_n consists of precisely the set of n'th roots of unity. So a cyclotomic field is a field generated over $\mathbb{Q}$ by the elements in the kernel of some n'th power homomorphism $\lambda_n : \mathbb{C}^* \to \mathbb{C}^*$.

We now do the same thing with the group $\mathbb{C}^*$ replaced by the elliptic curve $C(\mathbb{C})$ and the n'th-power homomorphism replaced by the *multiplication-by-n map*

$$\lambda_n : C(\mathbb{C}) \longrightarrow C(\mathbb{C}), \quad \lambda_n(P) = nP.$$

The kernel of λ_n is a subgroup of $C(\mathbb{C})$, which we denote by

$$C[n] = \ker(\lambda_n) = \{P \in C(\mathbb{C}) : nP = \mathcal{O}\}.$$

It is easy to describe $C[n]$ as an abstract group, at least if you believe the analytic description of $C(\mathbb{C})$ that we discussed in Section 2.2. (See the exercises for an algebraic proof.)

Proposition 6.4. *As an abstract group,*

$$C[n] \cong (\mathbb{Z}/n\mathbb{Z}) \oplus (\mathbb{Z}/n\mathbb{Z}).$$

In other words, $C[n]$ is the direct sum of two cyclic groups of order n.

Proof. Recall from Section 2.2 that $C(\mathbb{C})$ is isomorphic, as a group, to $\mathbb{C}/L$, where

$$L = \mathbb{Z}\omega_1 + \mathbb{Z}\omega_2 = \{m_1\omega_1 + m_2\omega_2 : m_1, m_2 \in \mathbb{Z}\}$$

is a lattice in $\mathbb{C}$. With this description of the group $C(\mathbb{C})$, we see that a point $z \in \mathbb{C}/L$ is in $C[n]$ if and only if $nz \in L$. This gives us an explicit isomorphism,

$$(\mathbb{Z}/n\mathbb{Z}) \oplus (\mathbb{Z}/n\mathbb{Z}) \longrightarrow C[n] \subset \mathbb{C}/L,$$

$$(a_1, a_2) \longmapsto \frac{a_1}{n}\omega_1 + \frac{a_2}{n}\omega_2,$$

which completes the proof of the proposition. $\qquad\square$

As we have seen, cyclotomic extensions are generated by the elements in the kernel of the n'th power map $\mathbb{C}^* \to \mathbb{C}^*$. In a similar manner, we want to look at the field extensions generated by the points in $C[n]$. A point $P = (x, y) \in C[n]$ has two coordinates, so we might consider the field generated by all of the coordinates of all of the points in $C[n]$. The next proposition suggests that this is an interesting field.

Proposition 6.5. *Let C be an elliptic curve given by a Weierstrass equation*

$$C : y^2 = x^3 + ax^2 + bx + c$$

with rational coefficients $a, b, c \in \mathbb{Q}$.

(a) *Let $P = (x_1, y_1) \in C[n]$ be a point of order dividing n. Then x_1 and y_1 are algebraic over $\mathbb{Q}$, i.e., x_1 and y_1 are roots of polynomials with rational coefficients.*

(b) *Let*

$$C[n] = \big\{ (x_1, y_1), \ldots, (x_m, y_m), \mathcal{O} \big\}$$

be the complete set of points of $C(\mathbb{C})$ of order dividing n, where Proposition 6.4 tells us that $m = n^2 - 1$. Let

$$K = \mathbb{Q}(x_1, y_1, \ldots, x_m, y_m)$$

be the field generated by the coordinates of all of the points in $C[n]$. Then K is a Galois extension of $\mathbb{Q}$. N.B. In general, $\mathrm{Gal}(K/\mathbb{Q})$ will not *be abelian.*

Proof. (a) We give a computational proof, although in truth it is not difficult to adapt the proof of (b) so as to simultaneously prove (a).

If we are given a point $P = (x, y)$ and an integer $n \geq 2$, how can we tell whether $nP = \mathcal{O}$? For $n = 2$ we have seen that

$$2P = \mathcal{O} \quad \Longleftrightarrow \quad x^3 + ax^2 + bx + c = 0,$$

so the x-coordinate of a point of order two is clearly algebraic. In general, if we repeatedly use the addition formula, we can find a multiplication-by-n formula that is similar to the duplication formula. For large values of n, the formula will be very complicated, but the fact that the addition law is given by rational functions means that if $P = (x, y)$, then

$$(x\text{-coordinate of } nP) = \frac{\text{polynomial in } x \text{ and } y}{\text{polynomial in } x \text{ and } y}.$$

In fact, since the x-coordinates of nP and $-nP = (x, -y)$ are the same, it is not hard to see, for example by induction, that we can choose polynomials that depend only on x. In other words,

$$(x\text{-coordinate of } nP) = \frac{\phi_n(x)}{\psi_n(x)},$$

where $\phi_n(x)$ and $\psi_n(x)$ are relatively prime polynomials in $\mathbb{Q}[x]$. Then a point $P = (x_1, y_1)$ has order dividing n if and only if $\psi_n(x_1) = 0$.

This proves that the x-coordinate of a point of order n is algebraic, since it is a root of the polynomial $\psi_n(x)$. And then the y-coordinate is also algebraic, since it satisfies $y^2 = x^3 + ax^2 + bx + c$.

(b) Let $\sigma : K \to \mathbb{C}$ be a field homomorphism. In order to prove that K is Galois over $\mathbb{Q}$, we must verify that $\sigma(K) = K$.

The map σ is completely determined by where it sends the x_i's and the y_i's. What are the allowable possibilities? By assumption, each point P_i is in $C[n]$, so Proposition 6.3(e) tells us that $\sigma(P_i)$ is also in $C[n]$. This means that $\sigma(P_i)$ is one of the P_j's, with $i = j$ being allowed. This is true for every $1 \leq i \leq m$, which proves that $\sigma(K) \subset K$. This completes the proof that K is a Galois extension of $\mathbb{Q}$.

Addendum: Here is the alternative, albeit fancier, proof of (a) that we mentioned. We have just seen that every field homomorphism $\sigma : K \to \mathbb{C}$ is determined by specifying some permutation of the points $P_1, \ldots, P_m$. In particular, this means that there are only finitely many such homomorphisms. But if some x_i or y_i were not algebraic over $\mathbb{Q}$, then the field K would have infinite degree over $\mathbb{Q}$, so there would be infinitely many distinct homomorphisms $K \to \mathbb{C}$. Therefore all of the x_i's and y_i's are algebraic over $\mathbb{Q}$. $\square$

Example 6.6. Let's see how Proposition 6.5 works in practice. We consider the elliptic curve

$$C : y^2 = x^3 + x.$$

Let $P = (x, y)$ be a point on C. Then it is easy to compute $2P$,

$$2P = \left(\frac{x^4 - 2x^2 + 1}{4y^2}, \frac{x^6 + 5x^4 - 5x^2 - 1}{8y^3} \right).$$

We first look at points of order three. We observe that

$$P = (x, y) \text{ has order } 3 \iff \begin{pmatrix} \text{the } x\text{-coordinate of } 2P \text{ equals} \\ \text{the } x\text{-coordinate of } P \end{pmatrix}$$

$$\iff \frac{x^4 - 2x^2 + 1}{4y^2} = x$$

$$\iff 3x^4 + 6x^2 - 1 = 0,$$

where for the last line we used the fact that $y^2 = x^3 + x$. So the points of order three in $C(\mathbb{C})$ are the points whose x-coordinates satisfy the polynomial equation

$$3x^4 + 6x^2 - 1 = 0.$$

In particular, the coordinates of the points of order three on C are algebraic numbers.

Each x gives two possible values for y, since the points with $y = 0$ have order two, not order three. This gives eight points of order three, and together with $\mathcal{O}$ they form the group $C[3] \cong \mathbb{Z}/3\mathbb{Z} \oplus \mathbb{Z}/3\mathbb{Z}$.

Since our equation is so simple, we can solve it explicitly. Thus

$$\alpha = \sqrt{\frac{2\sqrt{3} - 3}{3}} \quad \text{satisfies} \quad 3\alpha^4 + 6\alpha^2 - 1 = 0,$$

and the other three roots are $-\alpha$, $(i\sqrt{3}\alpha)^{-1}$, and $-(i\sqrt{3}\alpha)^{-1}$. Substituting into $y^2 = x^3 + x$, we then find the y-coordinates. Thus if we let

$$\beta = \sqrt[4]{\frac{8\sqrt{3} - 12}{9}} = \sqrt{\frac{2\alpha}{\sqrt{3}}},$$

then the nine points in $C[3]$ are

$$C[3] = \left\{ \mathcal{O}, (\alpha, \pm\beta), (-\alpha, \pm i\beta), \left(\frac{i}{\sqrt{3}\alpha}, \pm\frac{2\sqrt{-i}}{\sqrt[4]{27}\beta} \right), \left(\frac{-i}{\sqrt{3}\alpha}, \pm\frac{2\sqrt{i}}{\sqrt[4]{27}\beta} \right) \right\}.$$

It is a nice exercise to check that the field generated by the coordinates of these points is $\mathbb{Q}(\beta, i)$, and that $\mathrm{Gal}(\mathbb{Q}(\beta, i)/\mathbb{Q})$ is a non-abelian group of order 16. Recall that we never claimed that elliptic curves would give abelian Galois groups over $\mathbb{Q}$. Instead, we said that in certain cases they would give abelian extensions of imaginary quadratic fields. For this elliptic curve, we will prove in Section 6.5, as a special case of our main theorem, that $\text{Gal}\big(\mathbb{Q}(\beta, i)/\mathbb{Q}(i)\big)$ is an abelian group. You might try to prove this directly, without any reference to elliptic curves.

Next we look at points of order four on C. Since a point has order two if and only if its y-coordinate is zero, we find that

$$P = (x, y) \text{ has order four} \iff 2P \text{ has order two}$$
$$\iff \text{the } y\text{-coordinate of } 2P \text{ is } 0$$
$$\iff x^6 + 5x^4 - 5x^2 - 1 = 0.$$

So the points of order four in $C(\mathbb{C})$ are the 12 points whose x-coordinates satisfy the polynomial equation

$$x^6 + 5x^4 - 5x^2 - 1 = 0$$

Of course, there are also three points of order two, and one point of order one, which altogether gives the 16 points in $C[4]$.

The sextic polynomial giving the points of order four factors as

$$x^6 + 5x^4 - 5x^2 - 1 = (x - 1)(x + 1)(x^4 + 6x^2 + 1).$$

Further, if we let $\alpha = (\sqrt{2} - 1)i$, then

$$x^4 + 6x^2 + 1 = (x - \alpha)(x + \alpha)(x - \alpha^{-1})(x + \alpha^{-1}).$$

And letting $\beta = (1 + i)(\sqrt{2} - 1)$, we find that $\beta^2 = \alpha^3 + \alpha$, and then a little algebra gives us a complete description of the points of order four,

$$C[4] = \Big\{ (1, \pm\sqrt{2}), (-1, \pm i\sqrt{3}), (\alpha, \pm\beta), (-\alpha, \pm i\beta),$$
$$(\alpha^{-1}, \pm\alpha^{-2}\beta), (-\alpha^{-1}, \pm i\alpha^{-2}\beta) \Big\}.$$

Hence the points of order four generate the field $\mathbb{Q}(i, \sqrt{2})$.

6.3 A Galois Representation

In the last section we considered the field

$$\mathbb{Q}(x_1, y_1, \ldots, x_m, y_m),$$

where $\{\mathcal{O}, (x_1, y_1), \ldots, (x_m, y_m)\}$ is the set $C[n]$ of points having order dividing n. This field will be our primary object of study for the remainder of this chapter, so it is convenient to give it a name. We call it the *field of definition of $C[n]$ over* $\mathbb{Q}$ and denote it by

$$\mathbb{Q}(C[n]) = \begin{pmatrix} \text{field generated over } \mathbb{Q} \text{ by the } x \text{ and} \\ y\text{-coordinates of all points in } C[n] \end{pmatrix}.$$

Later, if we need to replace $\mathbb{Q}$ by some other field F, we write $F(C[n])$.

We proved in Section 6.2 that $\mathbb{Q}(C[n])$ is a Galois extension of $\mathbb{Q}$. We now begin to describe its Galois group. For $\sigma \in \mathrm{Gal}(\mathbb{Q}(C[n])/\mathbb{Q})$ and $P \in C[n]$, we know from Section 6.2 that $\sigma(P) \in C[n]$. Thus each σ induces a permutation of the set $C[n]$. This permutation is not completely arbitrary, because for example we showed in Section 6.2 that

$$\sigma(P + Q) = \sigma(P) + \sigma(Q), \quad \sigma(-P) = -\sigma(P), \quad \text{and} \quad \sigma(\mathcal{O}) = \mathcal{O}.$$

In other words, if we view $C[n]$ as being an abelian group, then each $\sigma \in \mathrm{Gal}(\mathbb{Q}(C[n])/\mathbb{Q})$ gives a group homomorphism from $C[n]$ to itself,

$$C[n] \longrightarrow C[n], \quad P \longmapsto \sigma(P).$$

Further, this homomorphism has an inverse, namely the homomorphism corresponding to σ^{-1}. Thus each $\sigma \in \mathrm{Gal}(\mathbb{Q}(C[n])/\mathbb{Q})$ gives a group isomorphism from $C[n]$ to itself.

Using the description of $C[n]$ proven in Proposition 6.4, we can describe these isomorphisms quite explicitly. Recall that we proved that $C[n]$ is a direct sum of two cyclic groups of order n,

$$C[n] \cong (\mathbb{Z}/n\mathbb{Z}) \oplus (\mathbb{Z}/n\mathbb{Z}).$$

So $C[n]$ is generated by two "basis" elements,[2] say P_1 and P_2, and the n^2 elements of $C[n]$ are exactly described by

$$C[n] = \{a_1 P_1 + a_2 P_2 : a_1, a_2 \in \mathbb{Z}/n\mathbb{Z}\}.$$

In other words, every element of $C[n]$ may be written as $a_1 P_1 + a_2 P_2$ for a *unique* pair of elements $a_1, a_2 \in \mathbb{Z}/n\mathbb{Z}$.

---

[2]There are many possible choices for P_1 and P_2, just as a vector space has many different bases. It will not matter which basis we choose.

Now suppose that $h : C[n] \to C[n]$ is any homomorphism from $C[n]$ to itself. Then
$$h(a_1 P_1 + a_2 P_2) = a_1 h(P_1) + a_2 h(P_2),$$
so h is completely determined once we know the values of $h(P_1)$ and $h(P_2)$. Conversely, if we take any two points $Q_1, Q_2 \in C[n]$, then we can define a homomorphism from $C[n]$ to itself by the rule

$$a_1 P_1 + a_2 P_2 \longmapsto a_1 Q_1 + a_2 Q_2.$$

Notice the analogy with linear algebra. A linear map between vector spaces can be given by specifying the image of each element in a basis. So we are really just doing linear algebra, except that the scalars of our "vector space" are in the ring $\mathbb{Z}/n\mathbb{Z}$, rather than in a field. A vector space with scalars in a ring R is called an R-*module*. Not every R-module has a basis, but luckily for us, $C[n]$ does.

Thus a homomorphism $h : C[n] \to C[n]$ is determined by the values of $h(P_1)$ and $h(P_2)$. Each of $h(P_1)$ and $h(P_2)$ is itself a linear combination of P_1 and P_2, say

$$h(P_1) = \alpha_h P_1 + \gamma_h P_2,$$
$$h(P_2) = \beta_h P_1 + \delta_h P_2.$$

Here $\alpha_h, \beta_h, \gamma_h, \delta_h$ are elements of $\mathbb{Z}/n\mathbb{Z}$ that are uniquely determined by h. It is suggestive to write these equations using matrix notation,

$$\big(h(P_1), h(P_2)\big) = (P_1, P_2) \begin{pmatrix} \alpha_h & \beta_h \\ \gamma_h & \delta_h \end{pmatrix}.$$

Then, if $g : C[n] \to C[n]$ is another homomorphism, it is easy to check that the composition $g \circ h$ is given by the usual matrix product

$$\begin{pmatrix} \alpha_{g \circ h} & \beta_{g \circ h} \\ \gamma_{g \circ h} & \delta_{g \circ h} \end{pmatrix} = \begin{pmatrix} \alpha_g & \beta_g \\ \gamma_g & \delta_g \end{pmatrix} \begin{pmatrix} \alpha_h & \beta_h \\ \gamma_h & \delta_h \end{pmatrix}.$$

We illustrate by checking the first column. Thus

$$\begin{aligned}
\alpha_{g \circ h} P_1 + \gamma_{g \circ h} P_2 &= (g \circ h)(P_1) \\
&= g(\alpha_h P_1 + \gamma_h P_2) \\
&= \alpha_h g(P_1) + \gamma_h g(P_2) \\
&= \alpha_h(\alpha_g P_1 + \gamma_g P_2) + \gamma_h(\beta_g P_1 + \delta_g P_2) \\
&= (\alpha_h \alpha_g + \gamma_h \beta_g) P_1 + (\alpha_h \gamma_g + \gamma_h \delta_g) P_2.
\end{aligned}$$

The homomorphisms $C[n] \to C[n]$ defined by $P \mapsto \sigma(P)$ that we studied earlier are actually isomorphisms, that is, they have inverses. How is the existence of an inverse to a homomorphism $h : C[n] \to C[n]$ reflected in the matrix for h? If we take $g = h^{-1}$, then the matrix for $g \circ h$ is the identity matrix, so we find that

$$\begin{pmatrix} 1 & 0 \\ 0 & 1 \end{pmatrix} = \begin{pmatrix} \alpha_{h-1} & \beta_{h-1} \\ \gamma_{h-1} & \delta_{h-1} \end{pmatrix} \begin{pmatrix} \alpha_h & \beta_h \\ \gamma_h & \delta_h \end{pmatrix}.$$

Thus the matrix associated to an isomorphism is invertible. And conversely, any invertible matrix can be used to define an isomorphism of $C[n]$ to itself.

This suggests that we should look at the set (actually group) of invertible 2×2 matrices with coefficients in $\mathbb{Z}/n\mathbb{Z}$. More generally, we can look at square matrices of any size with coefficients in any commutative ring R. The resulting group is called the *general linear group* and is denoted

$$\mathrm{GL}_r(R) = \left\{ \begin{array}{l} r \times r \text{ matrices } A \text{ with coefficients} \\ \text{in } R \text{ and satisfying } \det(A) \in R^* \end{array} \right\}.$$

The condition that the determinant be a unit is equivalent to requiring that A^{-1} exist, where we emphasize that A^{-1} is required to have coefficients in the ring R. The proof of this fact for general rings, which we leave as an exercise, is the same as the proof that you saw in linear algebra when R is a field. However, for 2×2 matrices, we can just write everything out explicitly. So let A be a 2×2 matrix with coefficients in R and with determinant a unit in R,

$$A = \begin{pmatrix} \alpha & \beta \\ \gamma & \delta \end{pmatrix} \quad \text{with} \quad \Delta = \alpha\delta - \beta\gamma \in R^*.$$

Then the inverse of A is the matrix

$$\begin{pmatrix} \alpha & \beta \\ \gamma & \delta \end{pmatrix}^{-1} = \begin{pmatrix} \delta/\Delta & -\beta/\Delta \\ -\gamma/\Delta & \alpha/\Delta \end{pmatrix}.$$

Conversely, if A has an inverse with coefficients in R, then

$$1 = \det(I) = \det(AA^{-1}) = \det(A)\det(A^{-1}),$$

so $\det(A)$ is a unit in R.

Let's look at an example, say the group of 2×2 matrices with coefficients in $\mathbb{Z}/2\mathbb{Z}$. It is easy to list all such matrices with non-zero determinant. There are six of them:

$$\begin{pmatrix} 1 & 0 \\ 0 & 1 \end{pmatrix}, \begin{pmatrix} 1 & 0 \\ 1 & 1 \end{pmatrix}, \begin{pmatrix} 1 & 1 \\ 0 & 1 \end{pmatrix}, \begin{pmatrix} 1 & 1 \\ 1 & 0 \end{pmatrix}, \begin{pmatrix} 0 & 1 \\ 1 & 0 \end{pmatrix}, \begin{pmatrix} 0 & 1 \\ 1 & 1 \end{pmatrix}.$$

The group $\mathrm{GL}_2(\mathbb{Z}/2\mathbb{Z})$ is isomorphic to the symmetric group on three letters. A quick way to get an isomorphism is to look at the way that the matrices permute the three non-zero vectors in the vector space $(\mathbb{Z}/2\mathbb{Z})^2$.

Let us briefly recapitulate. To each element $\sigma \in \mathrm{Gal}\big(\mathbb{Q}(C[n])/\mathbb{Q}\big)$ we have associated an isomorphism from $C[n]$ to itself. And to each such isomorphism we have associated a matrix in $\mathrm{GL}_2(\mathbb{Z}/n\mathbb{Z})$. So we get a map

$$\rho_n : \mathrm{Gal}\big(\mathbb{Q}(C[n])/\mathbb{Q}\big) \longrightarrow \mathrm{GL}_2(\mathbb{Z}/n\mathbb{Z}), \qquad \rho_n(\sigma) = \begin{pmatrix} \alpha_\sigma & \beta_\sigma \\ \gamma_\sigma & \delta_\sigma \end{pmatrix},$$

where $\alpha_\sigma, \beta_\sigma, \gamma_\sigma, \delta_\sigma$ are determined by the formulas

$$\sigma(P_1) = \alpha_\sigma P_1 + \gamma_\sigma P_2,$$
$$\sigma(P_2) = \beta_\sigma P_1 + \delta_\sigma P_2.$$

Further, the matrix computation that we did earlier shows that

$$\rho_n(\sigma\tau) = \rho_n(\sigma)\rho_n(\tau) \quad \text{for all } \sigma, \tau \in \mathrm{Gal}\big(\mathbb{Q}(C[n])/\mathbb{Q}\big),$$

so ρ_n is a group homomorphism. We have thus constructed a homomorphism from the complicated group $\mathrm{Gal}\big(\mathbb{Q}(C[n])/\mathbb{Q}\big)$ that we are trying to study into the group of matrices $\mathrm{GL}_2(\mathbb{Z}/n\mathbb{Z})$. Such a homomorphism is called a *representation*.[3] Since $\mathrm{Gal}\big(\mathbb{Q}(C[n])/\mathbb{Q}\big)$ is a Galois group, the representation ρ_n is called a *Galois representation*.

We have now proven a lot of important facts, which we record in the following theorem.

Theorem 6.7 (Galois Representation Theorem). *Let C be an elliptic curve given by a Weierstrass equation with rational coefficients, and let $n \geq 2$ be an integer. Fix generators P_1 and P_2 for $C[n]$. Then the map*

$$\rho_n : \mathrm{Gal}\big(\mathbb{Q}(C[n])/\mathbb{Q}\big) \longrightarrow \mathrm{GL}_2(\mathbb{Z}/n\mathbb{Z})$$

described in this section is a one-to-one group homomorphism.

---

[3]The theory of group representations is an extremely powerful tool for studying groups, and it is used extensively in mathematics, physics, and chemistry. We do not need the general theory, but for those who are interested, a very nice introduction to the representation theory of finite groups is given in Serre [43].

Proof. We have proven everything except that ρ_n is one-to-one. Suppose that $\sigma \in \mathrm{Gal}\big(\mathbb{Q}(C[n])/\mathbb{Q}\big)$ is in the kernel of ρ_n, so $\rho_n(\sigma) = \left(\begin{smallmatrix} 1 & 0 \\ 0 & 1 \end{smallmatrix}\right)$. This means that $\sigma(P_1) = P_1$ and $\sigma(P_2) = P_2$, from which it follows that $\sigma(P) = P$ for every $P \in C[n]$. Since by definition $\sigma(x,y) = \big(\sigma(x), \sigma(y)\big)$, this means that σ fixes the x and y-coordinates of every point in $C[n]$. Now recall that $\mathbb{Q}(C[n])$ is generated over $\mathbb{Q}$ by the x and y-coordinates of the points in $C[n]$. Hence σ fixes the generators of $\mathbb{Q}(C[n])$, so it fixes the entire field $\mathbb{Q}(C[n])$. This means that σ is the identity element of $\mathrm{Gal}\big(\mathbb{Q}(C[n])/\mathbb{Q}\big)$, which proves that the kernel of ρ_n consists of only the identity element. Therefore ρ_n is one-to-one.   □

Notice the analogy with the cyclotomic extensions studied in Section 6.1. If we choose a generator $\zeta \in \mathbb{C}^*$ for the group of n'th roots of unity, then we get a homomorphism

$$t : \mathrm{Gal}\big(\mathbb{Q}(\zeta)/\mathbb{Q}\big) \longrightarrow \mathrm{GL}_1(\mathbb{Z}/n\mathbb{Z}) = (\mathbb{Z}/n\mathbb{Z})^*$$

determined by the rule $\sigma(\zeta) = \zeta^{t(\sigma)}$. The homomorphism t is called the n'th cyclotomic representation of $\mathbb{Q}$. As we mentioned but did not prove in Section 6.1, the cyclotomic representation is not only one-to-one, it is also onto, so it is an isomorphism. Hence $\mathrm{Gal}\big(\mathbb{Q}(\zeta)/\mathbb{Q}\big)$ is isomorphic to the unit group $(\mathbb{Z}/n\mathbb{Z})^*$ of the ring $\mathbb{Z}/n\mathbb{Z}$.

We have now done a lot of abstract theory, so this might be a good time to look at some particular elliptic curves and explicitly determine the representation ρ_n for some small values of n, such as $n = 2$.

Example 6.8. Consider the elliptic curve given by the equation

$$C : y^2 = x(x - 1)(x - 2).$$

Then

$$C[2] = \big\{\mathcal{O}, (0,0), (1,0), (2,0)\big\}$$

consists entirely of rational points, so $\mathbb{Q}(C[2]) = \mathbb{Q}$. It follows that the Galois group $\mathrm{Gal}\big(\mathbb{Q}(C[2])/\mathbb{Q}\big)$ is the trivial group $\{\sigma_0\}$. The representation

$$\rho_2 : \mathrm{Gal}\big(\mathbb{Q}(C[2])/\mathbb{Q}\big) \longrightarrow \mathrm{GL}_2(\mathbb{Z}/2\mathbb{Z})$$

is given by $\rho_2(\sigma_0) = \left(\begin{smallmatrix} 1 & 0 \\ 0 & 1 \end{smallmatrix}\right)$. In particular, the image of ρ_2 is definitely not all of $\mathrm{GL}_2(\mathbb{Z}/2\mathbb{Z})$, so in contrast to the case of the cyclotomic representation, the Galois representations associated to elliptic curves need not be isomorphisms.

Example 6.9. Next we look at the elliptic curve

$$C : y^2 = x^3 + x.$$

The points of order two are not all rational, but they are easy to describe:

$$C[2] = \{\mathcal{O}, (0,0), (i,0), (-i,0)\},$$

where as usual we let $i = \sqrt{-1}$. Thus $\mathbb{Q}(C[2]) = \mathbb{Q}(i)$, and the Galois group $\mathrm{Gal}(\mathbb{Q}(C[2])/\mathbb{Q}) = \{\sigma_0, \sigma_1\}$ contains two elements, the identity element σ_0 and complex conjugation σ_1.

To describe the representation ρ_2, we need to choose generators for $C[2]$, say we take $P_1 = (0,0)$ and $P_2 = (i,0)$. Then

$$\sigma_1(P_1) = \sigma_1(0,0) = (0,0) = P_1.$$
$$\sigma_1(P_2) = \sigma_2(i,0) = (-i,0) = P_1 + P_2.$$

So the matrix associated to σ_1 is $\left(\begin{smallmatrix} 1 & 1 \\ 0 & 1 \end{smallmatrix}\right)$, and the representation ρ_2 is given explicitly by

$$\rho_2(\sigma_0) = \begin{pmatrix} 1 & 0 \\ 0 & 1 \end{pmatrix} \quad \text{and} \quad \rho_2(\sigma_1) = \begin{pmatrix} 1 & 1 \\ 0 & 1 \end{pmatrix}.$$

Notice that if we had instead used $P_1 = (i,0)$ and $P_2 = (-i,0)$ as our basis, then $\sigma_1(P_1) = P_2$ and $\sigma_1(P_2) = P_1$, so for this basis the value of the representation ρ_2 at σ_1 is the matrix $\left(\begin{smallmatrix} 0 & 1 \\ 1 & 0 \end{smallmatrix}\right)$; see Exercise 6.12. This illustrates how the choice of basis for $C[n]$ affects the values of ρ_n. See Exercise 6.22 for further details.

Example 6.10. Finally, we examine the elliptic curve

$$C : y^2 = x^3 - 2.$$

We let

$$\zeta = e^{2\pi i/3} = \frac{-1 + \sqrt{-3}}{2} \quad \text{and} \quad \beta = \sqrt[3]{2},$$

so ζ is a primitive cube root of unity and β is the positive cube root of 2. Then the points of order two on C are

$$C[2] = \{\mathcal{O}, (\beta,0), (\zeta\beta,0), (\zeta^2\beta,0)\} = \{\mathcal{O}, P_1, P_2, P_3\},$$

so the field generated by the points of order two is

$$\mathbb{Q}(C[2]) = \mathbb{Q}(\zeta, \beta) = \mathbb{Q}(\sqrt{-3}, \sqrt[3]{2}).$$

The Galois group $\mathrm{Gal}\big(\mathbb{Q}(C[2])/\mathbb{Q}\big)$ has order six. It is the full symmetric group on the set consisting of the three non-zero points in $C[2]$.

We write this Galois group as

$$\mathrm{Gal}\big(\mathbb{Q}(C[2])/\mathbb{Q}\big) = \{e, \sigma, \sigma^2, \tau, \sigma\tau, \sigma^2\tau\},$$

where σ and τ are the automorphisms determined by the formulas

$$\sigma(\sqrt[3]{2}) = \zeta\sqrt[3]{2}, \qquad\qquad \tau(\sqrt[3]{2}) = \sqrt[3]{2},$$
$$\sigma(\sqrt{-3}) = \sqrt{-3}, \qquad\qquad \tau(\sqrt{-3}) = -\sqrt{-3},$$

or equivalently, by the formulas

$$\sigma(\beta) = \zeta\beta, \quad \sigma(\zeta) = \zeta, \quad \tau(\beta) = \beta, \quad \tau(\zeta) = \zeta^2.$$

Then one easily checks that σ and τ satisfy the relations

$$\sigma^3 = \tau^2 = e \quad\text{and}\quad \sigma\tau = \tau\sigma^2.$$

Next, for generators of $C[2]$ we take the points

$$P_1 = (\beta, 0) \quad\text{and}\quad P_2 = (\zeta\beta, 0).$$

Then the action of σ and τ on P_1 and P_2 is given by

$$\sigma(P_1) = \sigma(\beta, 0) = (\sigma(\beta), 0) = (\zeta\beta, 0) = P_2,$$
$$\sigma(P_2) = \sigma(\zeta\beta, 0) = (\sigma(\zeta)\sigma(\beta), 0) = (\zeta^2\beta, 0) = P_3 = P_1 + P_2,$$
$$\tau(P_1) = \tau(\beta, 0) = (\tau(\beta), 0) = (\beta, 0) = P_1,$$
$$\tau(P_2) = \tau(\zeta\beta, 0) = (\tau(\zeta)\tau(\beta), 0) = (\zeta^2\beta, 0) = P_3 = P_1 + P_2.$$

So the matrices for σ and τ are, respectively,

$$\rho_2(\sigma) = \begin{pmatrix} 0 & 1 \\ 1 & 1 \end{pmatrix} \quad\text{and}\quad \rho_2(\tau) = \begin{pmatrix} 1 & 1 \\ 0 & 1 \end{pmatrix}.$$

Since the representation ρ_2 is a homomorphism, and since σ and τ generate $\mathrm{Gal}\big(\mathbb{Q}(C[2])/\mathbb{Q}\big)$, we can use the values of $\rho_2(\sigma)$ and $\rho_2(\tau)$ to compute ρ_2 for any element of $\mathrm{Gal}\big(\mathbb{Q}(C[2])/\mathbb{Q}\big)$. For example,

$$\rho_2(\sigma^2\tau) = \rho_2(\sigma)^2\rho_2(\tau) = \begin{pmatrix} 0 & 1 \\ 1 & 1 \end{pmatrix}^2 \begin{pmatrix} 1 & 1 \\ 0 & 1 \end{pmatrix} = \begin{pmatrix} 1 & 0 \\ 1 & 1 \end{pmatrix}.$$

Of course, one can also compute directly that

$$(\sigma^2\tau)(P_1) = P_1 + P_2 \quad \text{and} \quad (\sigma^2\tau)(P_2) = P_2.$$

Recall that one of our goals in this chapter is to construct field extensions with abelian Galois groups. Naturally we plan to use the fields $\mathbb{Q}(C[n])$ that we have been studying. We have proven that there is a one-to-one homomorphism

$$\rho_n : \text{Gal}(\mathbb{Q}(C[n])/\mathbb{Q}) \longrightarrow \text{GL}_2(\mathbb{Z}/n\mathbb{Z}).$$

We have also seen that ρ_n need not be onto, which is good, since the group $\text{GL}_2(\mathbb{Z}/n\mathbb{Z})$ with $n \geq 2$ is never an abelian group. For example, the matrices $\left(\begin{smallmatrix} 1 & 0 \\ 1 & 1 \end{smallmatrix}\right)$ and $\left(\begin{smallmatrix} 0 & 1 \\ 1 & 0 \end{smallmatrix}\right)$ never commute. (You should check this.)

It turns out that for most elliptic curves and most values of n, the representation ρ_n is "almost" onto. It is only for a very special class of elliptic curves, called *elliptic curves with complex multiplication*, that we get abelian Galois groups. We save the precise definition of complex multiplication for the next section, but to finish our general discussion of representations coming from elliptic curves, we quote a beautiful and difficult theorem of Serre that explains in some sense what it means to say that the ρ_n are "almost" onto.

Theorem 6.11. (Serre [41, 42]) *Let C be an elliptic curve given by a Weierstrass equation with rational coefficients. Assume that C does not have complex multiplication.*

(a) *There is an integer $M \geq 1$, depending only on the curve C, so that for all n, the index of $\rho_n(\text{Gal}(\mathbb{Q}(C[n])/\mathbb{Q}))$ inside $\text{GL}_2(\mathbb{Z}/n\mathbb{Z})$ is smaller than M.*

(b) *There is an integer $N \geq 1$, depending only on the curve C, so that for all integers n satisfying $\gcd(n, N) = 1$, the Galois representation*

$$\rho_n : \text{Gal}(\mathbb{Q}(C[n])/\mathbb{Q}) \longrightarrow \text{GL}_2(\mathbb{Z}/n\mathbb{Z})$$

is an isomorphism.

Conjecture 6.12. *The integer M in Theorem 6.11(a) may be chosen independently of the curve C. In other words, there is a single integer M so that for all rational elliptic curves C that don't have complex multiplication and all $n \geq 1$, we have*

$$\Big(\text{GL}_2(\mathbb{Z}/n\mathbb{Z}) : \rho_n(\text{Gal}(\mathbb{Q}(C[n])/\mathbb{Q}))\Big) \leq M.$$

6.4   Complex Multiplication

The complex points on an elliptic curve $C(\mathbb{C})$ form an abelian group, and for any abelian group and any integer n, there is a *multiplication-by-n homomorphism*,

$$C(\mathbb{C}) \xrightarrow[\text{by } n]{\text{multiplication}} C(\mathbb{C}), \quad P \longmapsto nP.$$

The kernel of this homomorphism is precisely $C[n]$, the set of points of order dividing n.

The multiplication-by-n homomorphism on $C(\mathbb{C})$ has the special property that it is defined by rational functions, that is, the x and y-coordinates of nP are rational functions of the x and y-coordinates of P. For example, if $P = (x, y)$ is a point on the elliptic curve

$$C : y^2 = x^3 + ax^2 + bx + c,$$

then after some computation we find that

$$2P = \left(\frac{x^4 - 2bx^2 - 8cx + b^2 - 4ac}{y^2}, \frac{\begin{array}{c} x^6 + 2ax^5 + 5bx^4 + 20cx^3 + 5(4ac - b^2)x^2 \\ + 2(4a^2c - ab^2 + 2bc)x + 4abc - b^3 - 8c^2 \end{array}}{y^3} \right).$$

In general, a non-trivial homomorphism $\phi : C(\mathbb{C}) \to C(\mathbb{C})$ that is defined by rational functions is called an *isogeny*. That is, an isogeny is a homomorphism $\phi : C(\mathbb{C}) \to C(\mathbb{C})$ that has the form

$$\phi(x, y) = \left(\frac{\text{polynomial in } x \text{ and } y}{\text{polynomial in } x \text{ and } y}, \frac{\text{polynomial in } x \text{ and } y}{\text{polynomial in } x \text{ and } y} \right).$$

More generally, one can look at isogenies $\phi : C(\mathbb{C}) \to \overline{C}(\mathbb{C})$ between two possibly different elliptic curves. For example, consider the two elliptic curves

$$C : y^2 = x^3 + ax^2 + bx \quad \text{and} \quad \overline{C} : y^2 = x^3 + \overline{a}x^2 + \overline{b}x$$

that we studied in Chapter 3, where $\overline{a} = -2a$ and $\overline{b} = a^2 - 4b$. We showed in Chapter 3 that the function

$$\phi : C(\mathbb{C}) \longrightarrow \overline{C}(\mathbb{C}), \quad \phi(x, y) = \left(\frac{y^2}{x^2}, \frac{y(x^2 - b)}{x^2} \right),$$

is a homomorphism. Thus ϕ is an isogeny from the elliptic curve C to the elliptic curve $\overline{C}$.

We are particularly interested in isogenies from an elliptic curve to itself. Such isogenies are called *endomorphisms*, or sometimes *algebraic endomorphisms* to emphasize the fact that they are defined by rational functions. We have just seen that every elliptic curve has a multiplication-by-n endomorphism for each integer n. For most elliptic curves, that's the whole story, there are no other endomorphisms. However, there are some elliptic curves with additional endomorphisms. We will focus our attention on these special elliptic curves, which provides some justification for giving them a name.

Definition. Let C be an elliptic curve. We say that C has *complex multiplication*, or CM for short, if there is an endomorphism $\phi : C \to C$ that is not a multiplication-by-n map.

It might be helpful at this point to give a few examples of elliptic curves having complex multiplication.

Example 6.13. The elliptic curve

$$C : y^2 = x^3 + x$$

has the complex multiplication

$$\phi(x, y) = (-x, iy),$$

since if $(x, y) \in C$, then the computation

$$(iy)^2 = -y^2 = -(x^3 + x) = (-x)^3 + (-x)$$

shows that $(-x, iy) \in C$. We leave you to check that $\phi \circ \phi(P) = -P$.

Example 6.14. Let $\zeta = e^{2\pi i/3} = \frac{-1+\sqrt{-3}}{2}$ be a primitive cube root of unity. Then the elliptic curve

$$C : y^2 = x^3 + 1$$

has the complex multiplication

$$\phi(x, y) = (\zeta x, -y).$$

We leave you to check that $(\zeta x, -y) \in C$ and that $\phi^3(P) = -P$, so $\phi^6(P) = P$. (Here we write ϕ^n for the n'iterate of ϕ, that is, $\phi^n = \phi \circ \phi \circ \cdots \circ \phi$ iterated n times.)

Example 6.15. We recalled earlier that there is an isogeny ϕ between two different curves

$$C : y^2 = x^3 + ax^2 + bx \quad \text{and} \quad \overline{C} : y^2 = x^3 + \overline{a}x^2 + \overline{b}x.$$

Suppose that we choose a and b so that C and $\overline{C}$ are isomorphic. Then composing the isogeny $\phi : C \to \overline{C}$ with the isomorphism $\overline{C} \xrightarrow{\sim} C$ gives an endomorphism of C. For example, if we take $a = 0$, then the curves $C : y^2 = x^3 + bx$ and $\overline{C} : \overline{y}^2 = \overline{x}^3 - 4b\overline{x}$ are isomorphic via the map

$$\overline{C} \longrightarrow C, \quad (\overline{x}, \overline{y}) \longmapsto \left(\frac{i}{2}\overline{x}, \frac{i-1}{4}\overline{y} \right).$$

Composing this with the isogeny $\phi : C \to \overline{C}$ gives the endomorphism

$$\psi : C \longrightarrow C, \quad (x,y) \longmapsto \left(\frac{iy^2}{2x^2}, \frac{(i-1)y(x^2 - b)}{4x^2} \right).$$

This endomorphism may look mysterious, but it really isn't. Notice that the curve $C; y^2 = x^3 + bx$ is essentially the same as the curve from Example 6.13. In particular, it has the obvious endomorphism defined by $(x, y) \mapsto (-x, -iy)$. Then it is not hard to check that the complicated map ψ is given by

$$\psi(x, y) = (x, y) + (-x, -iy).$$

N.B. The plus sign means addition on the elliptic curve C.

More generally, if ϕ_1 and ϕ_2 are endomorphisms of C, then we can define a new endomorphism $\phi_1 + \phi_2$ by

$$(\phi_1 + \phi_2) : C \longrightarrow C, \quad (\phi_1 + \phi_2)(P) = \phi_1(P) + \phi_2(P).$$

We also get a new endomorphism by taking the composition,

$$(\phi_1 \phi_2) : C \longrightarrow C, \quad (\phi_1 \phi_2)(P) = \phi_1\big(\phi_2(P)\big).$$

With this "addition" and "multiplication," the set of endomorphisms of C becomes a ring. If C does not have complex multiplication, then this ring is isomorphic to $\mathbb{Z}$, the ordinary ring of integers. But if C has complex multiplication, then the endomorphism ring of C is strictly larger than $\mathbb{Z}$. It is an interesting question, which we answer in part in Exercises 6.15 and 6.16, as to what sort of ring it can be.

You may have noticed that we did not completely verify that the maps in Examples 6.13, 6.14, and 6.15 are endomorphisms. We did show that they are maps from C to C given by rational functions, but we did not check that they are homomorphisms. Using the explicit formulas for the group law, it is tedious, but not difficult, to check this. However, as the following rigidity theorem shows, there is actually no need to do the work. Unfortunately, the proof is too complicated for us to give here, but you can find a proof in [49, III.4.8].

Theorem 6.16. *Let C and $\overline{C}$ be elliptic curves, and let $\phi : C(\mathbb{C}) \to \overline{C}(\mathbb{C})$ be a map given by rational functions and satisfying $\phi(\mathcal{O}) = \overline{\mathcal{O}}$. Then ϕ is automatically a homomorphism.*

Why is an elliptic curve with an extra endomorphism said to have "complex multiplication"? Recall from Section 2.2 that the complex points on an elliptic curve look like $\mathbb{C}/L$, where

$$L = \{a_1\omega_1 + a_2\omega_2 : a_1, a_2 \in \mathbb{Z}\}$$

is a lattice in $\mathbb{C}$. So an endomorphism $\phi : C(\mathbb{C}) \to C(\mathbb{C})$ gives a holomorphic map

$$f : \mathbb{C}/L \longrightarrow \mathbb{C}/L.$$

This means that in a neighborhood of 0, the map f is given by a convergent power series

$$f(z) = c_0 + c_1 z + c_2 z^2 + c_3 z^3 + \cdots .$$

We also know that f is a homomorphism, so

$$f(z_1 + z_2) = f(z_1) + f(z_2)$$

for all z_1 and z_2 in a neighborhood of 0. Of course, this equality is taking place in the quotient $\mathbb{C}/L$, so we should really say that

$$f(z_1 + z_2) - f(z_1) - f(z_2) \in L \quad \text{for all } z_1 \text{ and } z_2 \text{ close to } 0.$$

But L consists of a discrete set of points in $\mathbb{C}$, and therefore contains no non-empty open set. Since the image of an open set by a non-constant holomorphic function is open, it follows that the difference $f(z_1 + z_2) - f(z_1) - f(z_2)$ must be constant. Putting $z_1 = z_2 = 0$, we see that the constant is $-c_0$, so the power series for f satisfies

$$f(z_1 + z_2) + c_0 = f(z_1) + f(z_2) \quad \text{for all } z_1 \text{ and } z_2 \text{ close to } 0.$$

Since $f(0) = 0$ in $\mathbb{C}/L$, we see that $c_0 \in L$, so the maps $z \to f(z)$ and $z \to f(z) - c_0$ give the same endomorphism of $\mathbb{C}/L$, so we may as well take the latter in place of the former. This means that we may assume that $c_0 = 0$, so the power series for f satisfies

$$f(z_1 + z_2) = f(z_1) + f(z_2) \quad \text{for all } z_1 \text{ and } z_2 \text{ close to } 0.$$

As you may suspect, there are very few power series with this property.

Proposition 6.17. *Let $f(z)$ be a function that is holomorphic in a neighborhood of 0 and has the property that*

$$f(z_1 + z_2) = f(z_1) + f(z_2)$$

for all z_1 and z_2 in a neighborhood of 0. Then $f(z) = cz$ for some $c \in \mathbb{C}$.

Proof. Putting $z_1 = z_2 = 0$, we find that $f(0) = 2f(0)$, so $f(0) = 0$. Next we compute $f'(z)$ directly from the definition of derivative. Thus

$$
\begin{aligned}
f'(z) &= \lim_{h \to 0} \frac{f(z + h) - f(z)}{h} \\
&= \lim_{h \to 0} \frac{(f(z) + f(h)) - f(z)}{h} \quad \text{from the given property of } f, \\
&= \lim_{h \to 0} \frac{f(h) - f(0)}{h} \quad \text{since } f(0) = 0, \\
&= f'(0).
\end{aligned}
$$

In other words, the derivative of $f(z)$ is constant, which means that f is linear, say $f(z) = c_0 + c_1 z$. Then $0 = f(0) = c_0$, and so $f(z) = c_1 z$. $\qquad\square$

Now let $\phi : C(\mathbb{C}) \to C(\mathbb{C})$ be an endomorphism. From Proposition 6.17, there is some $c \in \mathbb{C}$ so that ϕ is given by a function of the form

$$f : \mathbb{C}/L \longrightarrow \mathbb{C}/L, \quad f(z) = cz \bmod L.$$

But c is not completely arbitrary, because f is a function on the quotient group $\mathbb{C}/L$. Thus suppose that $z_1, z_2 \in \mathbb{C}$ differ by an element of L, so they represent the same element of $\mathbb{C}/L$. Then we must have $f(z_1) = f(z_2)$. In terms of c, we find that

$$
\begin{aligned}
z_1 - z_2 \in L \quad &\Longrightarrow \quad f(z_1) = f(z_2) \\
&\Longrightarrow \quad cz_1 = cz_2 \quad \text{in } \mathbb{C}/L \\
&\Longrightarrow \quad c(z_1 - z_2) \in L.
\end{aligned}
$$

Hence c must satisfy the condition $cL \subset L$, and conversely, if $cL \subset L$, then $f(z) = cz$ gives an endomorphism of $\mathbb{C}/L\mathbb{C}$. (Here we are writing $cL = \{c\omega : \omega \in L\}$.)

So now we ask: "What are the possible values for c?" Since L is an abelian group, we certainly have $cL \subset L$ if c is an integer. These are just the multiplication-by-c maps on the elliptic curve. If the elliptic curve has complex multiplication, then by definition there is at least one more value of $c \in \mathbb{C}$ such that $cL \subset L$. We are going to prove that in this case, the number c is complex, i.e., it is not a real number. So it is natural to say that the lattice L has *complex multiplication*, since there is a complex (non-real) number c such that $cL \subset L$. This is the origin of the appellation "complex multiplication" for elliptic curves with an extra endomorphism. For additional information about the complex number c, see Exercise 6.15.

Proposition 6.18. *Let* $\mathbb{C}/L$ *be an elliptic curve with a complex multiplication*

$$f : \mathbb{C}/L \longrightarrow \mathbb{C}/L, \quad f(z) = cz \bmod L,$$

i.e., with $c \notin \mathbb{Z}$*. Then* c *is not a real number.*

Proof. Choose generators for L, say

$$L = \mathbb{Z}\omega_1 + \mathbb{Z}\omega_2 = \{a_1\omega_1 + a_1\omega_2 : a_1, a_2 \in \mathbb{Z}\}.$$

Note that ω_1 and ω_2 are linearly independent over $\mathbb{R}$, since otherwise L would lie on a line, so it could not be a lattice. In other words, if $r_1, r_2 \in \mathbb{R}$ and $r_1\omega_1 + r_2\omega_2 = 0$, then we must have $r_1 = r_2 = 0$.

We know that $cL \subset L$. In particular, we know that $c\omega_1 \in L$, so we can find integers A and B so that

$$c\omega_1 = A\omega_1 + B\omega_2.$$

Thus

$$(c - A)\omega_1 - B\omega_2 = 0.$$

If c were real, we could conclude that $c - A = B = 0$, so $c = A$. This contradicts our assumption that $c \notin \mathbb{Z}$. Therefore c is not real. $\qquad\square$

6.5   Abelian Extensions of $\mathbb{Q}(i)$

In this section we look at the elliptic curve

$$C : y^2 = x^3 + x$$

and the fields generated by its points of finite order. We have seen in Example 6.13 that C has a complex multiplication,

$$\phi : C \longrightarrow C, \quad \phi(x, y) = (-x, iy).$$

Since the endomorphism ϕ involves $i = \sqrt{-1}$, it is not surprising that we will look at extensions of the field $\mathbb{Q}(i)$. But there is a more intrinsic reason why $\mathbb{Q}(i)$ is the "right" field to study.

Let $K/\mathbb{Q}$ be any Galois extension with $i \in K$, and let $\sigma \in \mathrm{Gal}(K/\mathbb{Q})$. Then for any point $P \in C(K)$, we have two ways to get a new point in $C(K)$, namely we can apply the endomorphism ϕ to P or we can apply the Galois element σ to P. We ask whether these actions of σ and ϕ commute. In other words, is it true that

$$\sigma\big(\phi(P)\big) = \phi\big(\sigma(P)\big) \quad \text{for every } P \in C(K)?$$

Using the definitions, we see that

$$\sigma\big(\phi(P)\big) = \sigma(-x, iy) = \big(\sigma(-x), \sigma(iy)\big) = \big(-\sigma(x), \sigma(i)\sigma(y)\big),$$
$$\phi\big(\sigma(P)\big) = \phi\big(\sigma(x), \sigma(y)\big) = \big(-\sigma(x), i\sigma(y)\big).$$

So the actions of σ and ϕ on $C(K)$ commute provided that $\sigma(i) = i$. In other words, they commute if $\sigma \in \mathrm{Gal}\big(K/\mathbb{Q}(i)\big)$. So if we plan to use the map ϕ to study Galois groups, it makes sense to look at Galois extensions of $\mathbb{Q}(i)$ rather than of $\mathbb{Q}$.

Our main theorem says that the points of finite order on C generate abelian extensions of $\mathbb{Q}(i)$.

Theorem 6.19. *Let C be the elliptic curve*

$$C : y^2 = x^3 + x.$$

For each integer $n \geq 1$, let

$$K_n = \mathbb{Q}(i)(C[n])$$

be the field generated by i and the coordinates of the points in $C[n]$. Then K_n is a Galois extension of $\mathbb{Q}(i)$, and its Galois group is abelian.

Proof. We proved in Section 6.2 that $\mathbb{Q}(C[n])$ is Galois over $\mathbb{Q}$, and it is clear that $\mathbb{Q}(i)$ is Galois over $\mathbb{Q}$, so their compositum K_n is Galois over $\mathbb{Q}$. Hence K_n is certainly Galois over $\mathbb{Q}(i)$.

Now comes the interesting part of the theorem, namely the fact that the Galois group $\text{Gal}\big(K_n/\mathbb{Q}(i)\big)$ is abelian. We use the representation theory developed in Section 6.3. We fix generators $P_1, P_2 \in C[n]$, and then we obtain a one-to-one homomorphism

$$\rho_n : \text{Gal}\big(K_n/\mathbb{Q}(i)\big) \longrightarrow \text{GL}_2(\mathbb{Z}/n\mathbb{Z}), \qquad \rho_n(\sigma) = \begin{pmatrix} \alpha_\sigma & \beta_\sigma \\ \gamma_\sigma & \delta_\sigma \end{pmatrix},$$

where $\alpha_\sigma, \beta_\sigma, \gamma_\sigma, \delta_\sigma$ are determined by the formulas

$$\sigma(P_1) = \alpha_\sigma P_1 + \gamma_\sigma P_2,$$
$$\sigma(P_2) = \beta_\sigma P_1 + \delta_\sigma P_2.$$

In a similar manner, the endomorphism $\phi : C \to C$ gives a homomorphism $\phi : C[n] \to C[n]$, since if $P \in C[n]$, then

$$n\phi(P) = \phi(nP) = \phi(\mathcal{O}) = \mathcal{O},$$

so $\phi(P) \in C[n]$. There are thus numbers $a, b, c, d \in \mathbb{Z}/n\mathbb{Z}$ such that

$$\phi(P_1) = aP_1 + cP_2,$$
$$\phi(P_2) = bP_1 + dP_2.$$

In other words, the homomorphism $\phi : C[n] \to C[n]$ corresponds to the matrix $\begin{pmatrix} a & b \\ c & d \end{pmatrix}$.

Further, and this is one of the crucial steps in the proof, we saw earlier that for all $\sigma \in \text{Gal}\big(K_n/\mathbb{Q}(i)\big)$ and all $P \in C(K_n)$, we have

$$\sigma\big(\phi(P)\big) = \phi\big(\sigma(P)\big).$$

If we apply this with $P = P_1$ and $P = P_2$, we see that the matrices for σ and ϕ commute,

$$\begin{pmatrix} \alpha_\sigma & \beta_\sigma \\ \gamma_\sigma & \delta_\sigma \end{pmatrix} \begin{pmatrix} a & b \\ c & d \end{pmatrix} = \begin{pmatrix} a & b \\ c & d \end{pmatrix} \begin{pmatrix} \alpha_\sigma & \beta_\sigma \\ \gamma_\sigma & \delta_\sigma \end{pmatrix}.$$

There are two more steps required to complete the proof of Theorem 6.19. First we show that the matrix ϕ is not a *scalar matrix*, i.e., it is not a multiple of the identity matrix. Second, we use a little linear algebra to show that if a 2×2 matrix A is not a scalar matrix, then any two matrices that commute with A must also commute with one another. From this we will conclude that the image of ρ_n is an abelian subgroup of $\text{GL}_2(\mathbb{Z}/n\mathbb{Z})$, and then, since ρ_n is one-to-one, that $\text{Gal}\big(K_n/\mathbb{Q}(i)\big)$ is also abelian.

Lemma 6.20. *Let $A = \begin{pmatrix} a & b \\ c & d \end{pmatrix}$ be the matrix corresponding to ϕ.*

(a) $A \in \mathrm{GL}_2(\mathbb{Z}/n\mathbb{Z})$.

(b) *Let ℓ be a prime dividing n. When we reduce the matrix A modulo ℓ, it is not a scalar matrix. Equivalently, for every such prime ℓ, at least one of the following three conditions is true:*
  (i) $b \not\equiv 0 \pmod{\ell}$,
  (ii) $c \not\equiv 0 \pmod{\ell}$,
  (iii) $a \not\equiv d \pmod{\ell}$.

Proof. (a) We need to show that $\det(A)$ is a unit in $\mathbb{Z}/n\mathbb{Z}$. If we compose ϕ with itself, we find that

$$\phi\big(\phi(P)\big) = \phi\big(\phi(x,y)\big) = \phi(-x,iy) = (x,-y) = -P.$$

So the matrix A corresponding to ϕ satisfies $A^2 = \begin{pmatrix} -1 & 0 \\ 0 & -1 \end{pmatrix}$, and hence

$$1 = \det(A^2) = \det(A)^2.$$

This proves that $\det(A)$ is a unit in $\mathbb{Z}/n\mathbb{Z}$, so $A \in \mathrm{GL}_2(\mathbb{Z}/n\mathbb{Z})$.

(b) Suppose to the contrary that there is some prime ℓ dividing n and some integer m such that

$$A = \begin{pmatrix} a & b \\ c & d \end{pmatrix} \equiv \begin{pmatrix} m & 0 \\ 0 & m \end{pmatrix} \pmod{\ell}.$$

This means that $\phi : C[\ell] \to C[\ell]$ is the same as the multiplication-by-m map, i.e.,

$$\phi(P) = mP \quad \text{for all } P \in C[\ell].$$

Let $\tau : \mathbb{C} \to \mathbb{C}$ be complex conjugation. We fix an inclusion $K_n \subset \mathbb{C}$, and then we may view τ as being an element of $\mathrm{Gal}(K_n/\mathbb{Q})$. From Section 6.2 we know that $\tau(mP) = m\tau(P)$. On the other hand, since $\tau(i) = -i$, we find that

$$\tau\big(\phi(P)\big) = \tau(-x,iy) = \big(\tau(-x),\tau(iy)\big)$$
$$= \big(-\tau(x),-i\tau(y)\big) = -\phi\big(\tau(P)\big).$$

This is true for all points in $C(K_n)$, so in particular, it is true for points in $C[\ell]$. We thus find that for every $P \in C[\ell]$,

$$
\begin{aligned}
m\tau(P) &= \tau(mP) \\
&= \tau\big(\phi(P)\big) \\
&= -\phi\big(\tau(P)\big) \\
&= -m\tau(P) \quad \text{since } \tau(P) \text{ is also in } C[\ell].
\end{aligned}
$$

Hence $2m\tau(P) = \mathcal{O}$ for every $P \in C[\ell]$.

But τ just permutes the elements in $C[\ell]$, and thus $2mP = \mathcal{O}$ for every $P \in C[\ell]$. There are two possibilities. Either $\ell = 2$ or ℓ divides m. (Note that ℓ is prime.) But if $\ell \mid m$, then $\phi(P) = \mathcal{O}$ for every $P \in C[\ell]$, which is absurd, since for example $\phi\big(\phi(P)\big) = -P$. So we must have $\ell = 2$.

But for $\ell = 2$ we can explicitly compute the matrix ϕ. We take $P_1 = (0,0)$ and $P_2 = (i,0)$ as generators for $C[2]$, and then

$$
\phi(P_1) = (0,0) = P_1 \quad \text{and} \quad \phi(P_2) = (-i,0) = P_1 + P_2,
$$

so the matrix for $\phi : C[2] \to C[2]$ is $\left(\begin{smallmatrix} 1 & 1 \\ 0 & 1 \end{smallmatrix}\right)$. This matrix is not a scalar matrix modulo any prime. This eliminates $\ell = 2$ as a possibility, which completes the proof of Lemma 6.20. $\qquad\square$

Lemma 6.21. *Let $A \in \mathrm{GL}_2(\mathbb{Z}/n\mathbb{Z})$ be a matrix that is not a scalar matrix modulo ℓ for all primes ℓ dividing n. Then*

$$
\big\{ B \in \mathrm{GL}_2(\mathbb{Z}/n\mathbb{Z}) : AB = BA \big\}
$$

is an abelian subgroup of $\mathrm{GL}_2(\mathbb{Z}/n\mathbb{Z})$. In other words, the matrices that commute with A also commute with each other.

Proof. It is easy to check that the indicated set is a subgroup of $\mathrm{GL}_2(\mathbb{Z}/n\mathbb{Z})$. We leave the verification to you. The hard part is to show that it is abelian. We are going to prove Lemma 6.21 one prime at a time.[4] In order to show that two numbers, or two matrices, are congruent modulo n, it suffices to show that they are congruent modulo ℓ^e for all prime powers ℓ^e dividing n. So it is enough to prove Lemma 6.21 in the case that $n = \ell^e$ is a prime power.

---

[4]This may remind you of our proof of the Nagell–Lutz theorem in Section 2.4. There we proved that a certain rational number a/d was an integer by checking, for each prime ℓ, that ℓ did not divide d. This idea of looking at one prime at a time, which in fancy language is called *localization*, is a powerful number theoretic tool. It is the algebraic equivalent of looking at a neighborhood of a point when you are studying real or complex analysis.

The idea of the proof is easy. Making a change-of-basis, we will put A into rational normal form

$$A = \begin{pmatrix} 0 & * \\ 1 & * \end{pmatrix}.$$

Then we explicitly describe all matrices that commute with such an A and check that they also commute with one another. The details are given in the following two sublemmas.

SubLemma 6.21′. *Let $A \in \mathrm{GL}_2(\mathbb{Z}/\ell^e\mathbb{Z})$ be a matrix that is not a scalar matrix modulo ℓ. Then there is a change-of-basis matrix $T \in \mathrm{GL}_2(\mathbb{Z}/\ell^e\mathbb{Z})$ that puts A into rational normal form,*

$$T^{-1}AT = \begin{pmatrix} 0 & * \\ 1 & * \end{pmatrix}.$$

SubLemma 6.21″. *Let $A = \left(\begin{smallmatrix} 0 & * \\ 1 & * \end{smallmatrix}\right) \in \mathrm{GL}_2(\mathbb{Z}/n\mathbb{Z})$. Then*

$$\left\{ B \in \mathrm{GL}_2(\mathbb{Z}/n\mathbb{Z}) : AB = BA \right\}$$

is an abelian subgroup of $\mathrm{GL}_2(\mathbb{Z}/n\mathbb{Z})$.

We start by proving Sublemma 6.21″ because the proof is just a calculation. We assume that A has the form $\left(\begin{smallmatrix} 0 & b \\ 1 & d \end{smallmatrix}\right)$ and ask which $B \in \mathrm{GL}_2(\mathbb{Z}/n\mathbb{Z})$ commute with A. Writing out the products AB and BA, we find that

$$AB = BA \quad \Longleftrightarrow \quad \begin{pmatrix} 0 & b \\ 1 & d \end{pmatrix}\begin{pmatrix} \alpha & \beta \\ \gamma & \delta \end{pmatrix} = \begin{pmatrix} \alpha & \beta \\ \gamma & \delta \end{pmatrix}\begin{pmatrix} 0 & b \\ 1 & d \end{pmatrix}$$

$$\Longleftrightarrow \quad \begin{pmatrix} b\gamma & b\delta \\ \alpha + d\gamma & \beta + d\delta \end{pmatrix} = \begin{pmatrix} \beta & b\alpha + d\beta \\ \delta & b\gamma + d\delta \end{pmatrix}.$$

Treating b and d as fixed quantities, we get four equations for the four variables $\alpha, \beta, \gamma, \delta$, but the equations are not independent. A little algebra shows that the general solution is

$$\beta = b\gamma \quad \text{and} \quad \delta = \alpha + d\gamma.$$

Hence for $A = \left(\begin{smallmatrix} 0 & b \\ 1 & d \end{smallmatrix}\right)$ with fixed b and d,

$$\left\{ B \in \mathrm{GL}_2(\mathbb{Z}/n\mathbb{Z}) : AB = BA \right\}$$
$$= \left\{ \begin{pmatrix} \alpha & b\gamma \\ \gamma & \alpha + d\gamma \end{pmatrix} \in \mathrm{GL}_2(\mathbb{Z}/n\mathbb{Z}) : \alpha, \gamma \in \mathbb{Z}/n\mathbb{Z} \right\}.$$

Now we check that the matrices in this set commute with one another. To do this, we just take two of them, multiply them together in both orders, and verify that the answers are the same:

$$\begin{pmatrix} \alpha & b\gamma \\ \gamma & \alpha + d\gamma \end{pmatrix} \begin{pmatrix} \alpha' & b\gamma' \\ \gamma' & \alpha' + d\gamma' \end{pmatrix} = \begin{pmatrix} \alpha' & b\gamma' \\ \gamma' & \alpha' + d\gamma' \end{pmatrix} \begin{pmatrix} \alpha & b\gamma \\ \gamma & \alpha + d\gamma \end{pmatrix}.$$

We leave it to you to do the multiplication. This completes the proof of Sublemma 6.21″.

Now we tackle Sublemma 6.21′. We write

$$A = \begin{pmatrix} a & b \\ c & d \end{pmatrix} \in \mathrm{GL}_2(\mathbb{Z}/n\mathbb{Z}).$$

Recall from linear algebra that to put a 2×2 matrix A into rational normal form, one takes a basis of the form $\{v, Av\}$. Then the columns of the change-of-basis matrix T are the two column vectors v and Av, after which one easily calculates that

$$AT = T \begin{pmatrix} 0 & * \\ 1 & * \end{pmatrix}.$$

This is what we will do, but there is a small difficulty in ensuring that the matrix T that we choose has an inverse. There are three cases that must be considered.

We have assumed that A is not a scalar matrix modulo ℓ, so at least one of the following three conditions is true:
  (i) $b \not\equiv 0 \pmod{\ell}$,
  (ii) $c \not\equiv 0 \pmod{\ell}$,
  (iii) $a \not\equiv d \pmod{\ell}$.
Corresponding to these three possibilities, we make the following choice for the matrix T:
  (i)   If $b \not\equiv 0 \pmod{\ell}$, then $T = \begin{pmatrix} 0 & b \\ 1 & d \end{pmatrix}$.
  (ii) If $b \equiv 0 \pmod{\ell}$ and $c \not\equiv 0 \pmod{\ell}$, then $T = \begin{pmatrix} 1 & a \\ 0 & c \end{pmatrix}$.
  (iii) If $b \equiv c \equiv 0 \pmod{\ell}$ and $a \not\equiv d \pmod{\ell}$, then $T = \begin{pmatrix} 1 & a+c \\ 1 & b+d \end{pmatrix}$.
Note that in all three cases we have $\det(T) \not\equiv 0 \pmod{\ell}$, so in all three cases $T \in \mathrm{GL}_2(\mathbb{Z}/\ell^e\mathbb{Z})$. For example, in case (iii),

$$\det(T) = (b+d) - (a+c) \equiv d - a \not\equiv 0 \pmod{\ell}.$$

Hence T is invertible, and since it is obvious in each case that $AT = T \left(\begin{smallmatrix} 0 & * \\ 1 & * \end{smallmatrix} \right)$, we conclude that

$$T^{-1}AT = \begin{pmatrix} 0 & * \\ 1 & * \end{pmatrix}.$$

So that completes the proof of Sublemma 6.21′.

Now we use the sublemmas to complete the proof of Lemma 6.21. Let $A \in \mathrm{GL}_2(\mathbb{Z}/\ell^e\mathbb{Z})$ be a matrix that is not a scalar matrix modulo ℓ. Using Sublemma 6.21′, we find a matrix $T \in \mathrm{GL}_2(\mathbb{Z}/\ell^e\mathbb{Z})$ so that $T^{-1}AT = \left(\begin{smallmatrix} 0 & * \\ 1 & * \end{smallmatrix} \right)$. Next, let $B, B' \in \mathrm{GL}_2(\mathbb{Z}/\ell^e\mathbb{Z})$ be matrices that commute with A,

$$AB = BA \quad \text{and} \quad AB' = B'A.$$

These formulas imply that

$$(T^{-1}AT)(T^{-1}BT) = (T^{-1}BT)(T^{-1}AT), \quad \text{and}$$
$$(T^{-1}AT)(T^{-1}B'T) = (T^{-1}B'T)(T^{-1}AT).$$

Sublemma 6.21″ tells us that $T^{-1}BT$ and $T^{-1}B'T$ commute,

$$(T^{-1}BT)(T^{-1}B'T) = (T^{-1}B'T)(T^{-1}BT).$$

Since T is invertible, this implies that $BB' = B'B$, which completes the proof of Lemma 6.21. $\qquad\Box$

Now we possess all of the tools needed to prove Theorem 6.19. We have the representation

$$\rho_n : \mathrm{Gal}\big(K_n/\mathbb{Q}(i)\big) \longrightarrow \mathrm{GL}_2(\mathbb{Z}/n\mathbb{Z})$$

and we have the matrix

$$A = \begin{pmatrix} a & b \\ c & d \end{pmatrix} \in \mathrm{GL}_2(\mathbb{Z}/n\mathbb{Z})$$

corresponding to the homomorphism $\phi : C[n] \to C[n]$. We showed in Lemma 6.20 that A is not equal to a scalar matrix modulo ℓ for all primes ℓ dividing n. Let $\sigma \in \mathrm{Gal}\big(K_n/\mathbb{Q}(i)\big)$ be any element of the Galois group. We verified that σ and ϕ commute in their action on $C[n]$, which implies that their matrices commute,

$$A\rho_n(\sigma) = \rho_n(\sigma)A.$$

Applying Lemma 6.21, we conclude that the matrices in the set

$$\big\{\rho_n(\sigma) : \sigma \in \mathrm{Gal}\big(K_n/\mathbb{Q}(i)\big)\big\}$$

commute with one another. Since the representation ρ_n is a homomorphism, it follows that

$$\rho_n(\sigma_1\sigma_2) = \rho_n(\sigma_1)\rho_n(\sigma_2) = \rho_n(\sigma_2)\rho_n(\sigma_1) = \rho_n(\sigma_2\sigma_1)$$
$$\text{for all } \sigma_1, \sigma_2 \in \mathrm{Gal}\big(K_n/\mathbb{Q}(i)\big).$$

Finally, we use Theorem 6.7 from Section 6.3, which says that the homomorphism ρ_n is one-to-one, to conclude that

$$\sigma_1\sigma_2 = \sigma_2\sigma_1 \quad \text{for all } \sigma_1, \sigma_2 \in \mathrm{Gal}\big(K_n/\mathbb{Q}(i)\big).$$

This prove that $\mathrm{Gal}\big(K_n/\mathbb{Q}(i)\big)$ is abelian, which completes the proof of Theorem 6.19. □

You may recall that in the case of abelian extensions of $\mathbb{Q}$, not only do all cyclotomic fields have abelian Galois groups, but it is also true that every extension with abelian Galois group is contained in a cyclotomic extension. A similar statement holds for abelian extensions of $\mathbb{Q}(i)$. The proof is too difficult for us to give, but we would be remiss if we failed to at least state this beautiful result.

Theorem 6.22. *Let* $C : y^2 = x^3 + x$ *be the elliptic curve that we have been studying in this section. Let* $F/\mathbb{Q}(i)$ *be a Galois extension of* $\mathbb{Q}(i)$ *of finite degree, and suppose that* $\mathrm{Gal}\big(F/\mathbb{Q}(i)\big)$ *is abelian. Then there is an integer* $n \geq 1$ *such that*

$$F \subset K_n = \mathbb{Q}(i)(C[n]).$$

Earlier we talked about Kronecker's dream of constructing extension fields with abelian Galois groups by using special values of complex analytic functions. We have now shown how to construct abelian extensions of $\mathbb{Q}(i)$ by taking the coordinates of points of finite order on the elliptic curve $y^2 = x^3 + x$. We conclude by briefly explaining how this construction is a realization of Kronecker's dream.

We begin by writing $C(\mathbb{C}) = \mathbb{C}/L$ and choosing generators for the lattice L, say $L = \mathbb{Z}\omega_1 + \mathbb{Z}\omega_2$, as described in Section 2.2. Then, as generators for $C[n]$, we may take

$$P_1 = \frac{\omega_1}{n} \quad \text{and} \quad P_2 = \frac{\omega_2}{n}.$$

Using P_1 and P_2, we get a representation

$$\rho_n : \mathrm{Gal}\big(K_n/\mathbb{Q}(i)\big) \longrightarrow \mathrm{GL}_2(\mathbb{Z}/n\mathbb{Z})$$

as usual.

The isomorphism $C(\mathbb{C}) \cong \mathbb{C}/L$ described in Section 2.2 uses the Weierstrass $\wp$ function,

$$\mathbb{C}/L \xrightarrow{\sim} C(\mathbb{C}), \quad z \longmapsto \big(\wp(z), \wp'(z)\big).$$

So the x and y-coordinates of points in $C(\mathbb{C})$ are the values of $\wp$ and $\wp'$. In particular, a point of order dividing n in $\mathbb{C}/L$ looks like

$$\frac{a_1\omega_1 + a_2\omega_2}{n} \quad \text{for some } a_1, a_2 \in \mathbb{Z}.$$

Hence K_n is generated by i and the numbers

$$\wp\left(\frac{a_1\omega_1 + a_2\omega_2}{n}\right) \quad \text{and} \quad \wp'\left(\frac{a_1\omega_1 + a_2\omega_2}{n}\right) \quad \text{for } 0 \le a_1, a_2 < n.$$

Since the K_n's are abelian extensions of $\mathbb{Q}(i)$, we have realized one part of Kronecker's Jugendtraum; we have generated abelian extensions of $\mathbb{Q}(i)$ using special values of meromorphic functions.

But more is true. We can use the representation ρ_n to describe how elements of $\mathrm{Gal}\big(K_n/\mathbb{Q}(i)\big)$ act on these special values. Thus

$$\sigma\left(\wp\left(\frac{a_1\omega_1 + a_2\omega_2}{n}\right)\right) = \sigma\big(x(a_1P_1 + a_2P_2)\big)$$

$$= x\big(a_1\sigma(P_1) + a_2\sigma(P_2)\big)$$

$$= \wp\left(\frac{(a_1\alpha_\sigma + a_2\beta_\sigma)\omega_1}{n} + \frac{(a_1\gamma_\sigma + a_8\delta_\sigma)\omega_2}{n}\right),$$

and similarly for $\wp'$. Alternatively, letting $t_1 = \omega_1/n$ and $t_2 = \omega_2/n$ be generators for the points of order dividing n, we can rewrite this last formula using matrix notation

$$\sigma\left(\wp\left((t_1\ t_2)\begin{pmatrix} a_1 \\ a_2 \end{pmatrix}\right)\right) = \wp\left((t_1\ t_2)\,\rho_n(\sigma)\begin{pmatrix} a_1 \\ a_2 \end{pmatrix}\right).$$

This formula, and the analogous formula for $\wp'$, convert the complicated algebraic action of $\mathrm{Gal}\big(K_n/\mathbb{Q}(i)\big)$ on K_n into a simple linear algebra matrix multiplication. They provide a concrete realization of Kronecker's Jugendtraum for the field $\mathbb{Q}(i)$.

6.6 Elliptic Curves and Fermat's Last Theorem

Fermat's Last Theorem is the assertion that for every integer $n \geq 3$ the equation

$$A^n + B^n = C^n$$

has no solutions in non-zero integers A, B, and C. The study of Fermat's equation has a long and storied history, starting from Fermat's marginal note in his copy of Diophantus' *Arithmetica*, where he asserts that he has "a truly marvelous proof of this [fact], which this margin is too narrow to contain." It seems unlikely that Fermat had a valid proof, but he did give a proof for the case $n = 4$. Over the succeeding centuries, proofs were given for some other small values of n, and there is also a vast literature that deals with special cases and weaker statements. For example, it was proven that there are no solutions (A, B, C) with $n = p \geq 3$ prime and satisfying $p \nmid ABC$ (the so-called first case) if either of the following statements is true:

- $2p + 1$ is also prime,   Sophie Germain, $\sim$ 1820.

- $2^p \not\equiv 1 \pmod{p^2}$,   Weiferich, $\sim$ 1910.[5]

The Fermat equation defines a smooth projective curve in $\mathbb{P}^2$, but as soon as n is at least 4, it is not an elliptic curve. More precisely, it is a curve of genus $g = \frac{1}{2}(n-1)(n-2)$, which means that the complex solutions to the Fermat equation form a g-holed torus, while we know that an elliptic curve is a 1-holed torus. So it is not clear that there are any connections between Fermat's equation and elliptic curves.

Yves Hellegouarch and Gerhard Frey independently noted that solutions to Fermat's equation could be used to construct elliptic curves with interesting properties. To do this, they took a putative solution (A, B, C) to Fermat's equation and used it to create the elliptic curve

$$E_{A,B,C} : y^2 = x(x - A^n)(x + B^n).$$

Initially the focus was on points of finite order. Then, in the mid-1980s, Frey noted that $E_{A,B,C}$ has such unusual properties that he thought it unlikely that $E_{A,B,C}$ could be "modular," a term whose definition we defer until later in this section. Since there was at the time a Modularity Conjecture of

---

[5] A heuristic argument suggests that the number of primes $p \leq T$ that satisfy the congruence $2^p \equiv 1 \pmod{p^2}$ should be roughly $\log \log T$. As of 2015, the only primes known to have this property are $p = 1093$ and $p = 3511$.

Taniyama and Shimura, later extended by Weil, asserting that every rational elliptic curve is modular, this suggested a completely new, two-step approach to proving Fermat's Last Theorem.

(I) Prove that $E_{A,B,C}$ is not modular.

(II) Prove that all (or at least, sufficiently many) rational elliptic curves are modular.

Building on ideas of Serre, in 1986 Ken Ribet proved that $E_{A,B,C}$ is not modular, thereby completing Step (I). Ribet's proof of Step (I) is difficult and uses many deep tools, but it was widely acknowledged at the time that Step (II) was likely to be at least an order of magnitude more difficult to prove, and few people thought that it would be done in the foreseeable future. But Andrew Wiles, inspired by Ribet's result and having a life-long fascination with Fermat's last theorem, spent the next 6 years working on the modularity conjecture without telling anyone in the mathematical community of his work. Then, in 1993, Wiles gave a series of lectures in which he announced a proof of a sufficient part of the modularity conjecture to imply Fermat's last theorem. Unfortunately, on further scrutiny a significant gap was discovered in the argument. Wiles spent the next year devising an alternative argument to fill the gap, and with the some assistance from Richard Taylor, the proof of semistable modularity and Fermat's last theorem was submitted for publication in October 1994 and appeared in print as a pair of articles in 1995 [53, 60].

The use of the so-called *Frey curve* $E_{A,B,C}$ shows how Fermat's last theorem may be reduced to a question, or rather two questions, about elliptic curves. The proofs of (I) and (II) are far beyond the scope of this book, but in the rest of this section we try to give some flavor of what it means for an elliptic curve to be modular and why the Frey curves are so strange as to be non-modular.

We start with the Frey curves. The roots of the cubic polynomial

$$x(x - A^n)(x + B^n)$$

are 0, A^n, and $-B^n$, so the discriminant of the Frey curve $E_{A,B,C}$ is

$$D = (A^n)^2(-B^n)^2(A^n + B^n)^2 = (ABC)^{2n}.$$

This already marks the Frey curve as being special, since its discriminant is a large perfect power. We also see that $E_{A,B,C}$ modulo p is singular precisely for the primes p dividing ABC.[6] More precisely, factoring out any common factors of A, B, C, we may assume that $\gcd(A, B, C) = 1$, and then we find that when the Frey curve is reduced modulo p for $p \mid ABC$, it acquires a node whose slopes are defined over $\mathbb{F}_p$, i.e., it looks like Figure 1.13, rather than Figure 1.14 or Figure 1.15. Of course, it requires some mental agility to say that an elliptic curve mod p "looks like" the graph of an elliptic curve in $\mathbb{R}^2$! The technical terminology for the condition that the singular reductions have nodes is to say that $E_{A,B,C}$ has *semi-stable reduction*, and if the tangent directions are rational, the semi-stable reduction is said to be *split*. Wiles proved that every rational semi-stable elliptic curve is modular. This, combined with Ribet's theorem, is enough to prove that the Frey curves $E_{A,B,C}$ cannot exist, since the Frey curves have semi-stable reduction. Later, a number of mathematicians extended Wiles argument, and the modularity conjecture for all rational elliptic curves was established by Breuil, Conrad, Diamond, and Taylor [7] in 2001.

We next take up the topics of L-series and modular forms. Let E be an elliptic curve

$$E : y^2 = x^3 + ax^2 + bx + c$$

with integer coefficients. As in Section 4.3, for each odd prime p we can reduce the coefficients of E to obtain a curve

$$\tilde{E}_p : y^2 = x^3 + \tilde{a}x^2 + \tilde{b}x + \tilde{c} \quad \text{over the finite field } \mathbb{F}_p.$$

The curve $\tilde{E}_p$ is non-singular, and hence an elliptic curve, if and only if p does not divide the discriminant D. In any case, we may consider the set of points on $\tilde{E}_p$ with coordinates in $\mathbb{F}_p$,

$$\tilde{E}_p(\mathbb{F}_p) = \left\{ (\tilde{x}, \tilde{y}) \in \mathbb{F}_p^2 : \tilde{y}^2 = \tilde{x}^3 + \tilde{a}\tilde{x}^2 + \tilde{b}\tilde{x} + \tilde{c} \right\} \cup \{\tilde{\mathcal{O}}\}.$$

If $\tilde{E}_p$ is non-singular, then Hasse's theorem (Theorem 4.1) says that

$$\tilde{E}_p(\mathbb{F}_p) = p + 1 - \epsilon_p \quad \text{with} \quad |\epsilon_p| \le 2\sqrt{p}.$$

We may view the set of numbers $\{\epsilon_p\}$ as a record that describes the reduction of E modulo (good) primes. Whenever mathematicians have a list of integers that describe some phenomenon, they like to encapsulate all of the

---

[6]In its present form, the Frey equation may be quite badly singular when reduced modulo 2, but a change of variables takes care of the problem. For ease of exposition, we will mostly ignore the prime 2 in our discussion.

data by encoding it into a series. If the data $a_1, a_2, a_3, \ldots$ is indexed by the natural numbers, it is standard practice to use the associated Dirichlet series $\sum a_n/n^s$, where s is a complex variable. But when the series is naturally indexed by primes, as is our elliptic curve series, then it is better to use a slightly more complicated definition based on the multiplicative nature of prime numbers.

Before presenting the elliptic curve series, we briefly digress to recall the Riemann zeta function, which is simply the Dirichlet series for the sequence $1, 1, 1, \ldots$, i.e.,

$$\zeta(s) = \sum_{n=1}^{\infty} \frac{1}{n^s}. \tag{ζ_1}$$

This formula expresses $\zeta(s)$ as a sum, but there is another formula for $\zeta(s)$ that expresses it as an Euler product over primes,

$$\zeta(s) = \prod_{p \text{ prime}} \left(1 - \frac{1}{p^s}\right)^{-1}. \tag{ζ_2}$$

The series (ζ_1) and product (ζ_2) converge for all complex numbers satisfying $\mathrm{Re}(s) > 1$, and the fact that they are equal is equivalent to the statement that every positive integer is uniquely expressible as a product of primes (up to rearranging the factors). It is also known that the function $\zeta(s)$ can be extended to a meromorphic function on all of $\mathbb{C}$. Many of the deepest theorems concerning the distribution of primes come from comparing these two formulas for $\zeta(s)$.

Returning now to our elliptic curve E, we use its list of ϵ_p values to define the *L-function of E* as the product[7]

$$L(E, s) = \prod_{p \text{ prime}} \left(1 - \frac{\epsilon_p}{p^s} + \frac{1}{p^{2s-1}}\right)^{-1}. \tag{$*$}$$

---

[7]For ease of exposition, we have, and will continue to, ignore a number of technical issues. First, we should take a "minimal" equation for E, which roughly means that there is no change of variables that makes the discriminant smaller while keeping integer coefficients. Second, we completely ignore the prime 2. Third, for primes p that divide D, we should take $\epsilon_p \in \{-1, 0, 1\}$, with the exact value depending on whether $\tilde{E}$ has a node or cusp, and if a node, whether the tangent slopes are in $\mathbb{F}_p$. Then the corresponding factor in the L-series is $(1 - \epsilon_p p^{-s})^{-1}$.

This looks complicated, but it's not that bad. We can use the geometric series and the binomial theorem to expand

$$\left(1 - \frac{\epsilon_p}{p^s} + \frac{1}{p^{2s-1}}\right)^{-1} = \sum_{k=0}^{\infty} \left(\frac{\epsilon_p}{p^s} - \frac{1}{p^{2s-1}}\right)^k$$

$$= \sum_{k=0}^{\infty} \sum_{i=0}^{k} \binom{k}{i} \left(\frac{\epsilon_p}{p^s}\right)^i \left(\frac{-1}{p^{2s-1}}\right)^{k-i}.$$

If we then take the product over primes and combine terms that end up with the same power of n^s in the denominator, then we get a Dirichlet series

$$L(E,s) = \sum_{n=1}^{\infty} \frac{\epsilon_n}{n^s}. \tag{$**$}$$

We leave you to check that for primes p, the values of ϵ_p in $(*)$ and $(**)$ are consistant.

And if you don't like all of this algebra, Exercise 6.24 describes another way to get the coefficients of the Dirichlet series $(**)$. For primes p, use the value of ϵ_p coming from counting points in $\tilde{E}(\mathbb{F}_p)$. For higher powers of p, use the recursion formula

$$\epsilon_{p^{k+1}} = \epsilon_{p^k} \epsilon_p - p \epsilon_{p^{k-1}} \quad \text{for } k \geq 1.$$

And for the other coefficients, use the fact that the map $n \to \epsilon_n$ is a multiplicative function, so

$$\epsilon_n = \epsilon_{p_1^{k_1}} \cdots \epsilon_{p_t^{k_t}} \quad \text{for } n = p_1^{k_1} \cdots p_t^{k_t} \text{ with } p_1, \ldots, p_t \text{ distinct primes.}$$

It is not hard, using the Hasse–Weil estimate

$$|\epsilon_p| \leq 2\sqrt{p},$$

to prove that the Dirichlet series for $L(E,s)$ converges for all complex numbers s in the half-plane $\text{Re}(s) > \frac{3}{2}$. The Modularity Theorem (which we have not stated) has the following incredible consequences for $L(E,s)$, where the functional equation uses the classical Γ-function $\Gamma(s) = \int_0^{\infty} t^{s-1} e^{-t} \, dt$.

Theorem 6.23. [7, 53, 60]
(a) *The Dirichlet series $L(E,s)$ extends to a holomorphic function on all of $\mathbb{C}$.*

(b) *There is an integer N_E, called the* conductor of E, *so that the function*

$$\xi(E, s) = N_E^{s/2}(2\pi)^{-s}\Gamma(s)L(E, s)$$

satisfies the functional equation[8]

$$\xi(E, s) = \pm\xi(E, 2 - s) \quad \textit{for all } s \in \mathbb{C}.$$

The conductor N_E is similar to the discriminant D in that it is a product of the primes where $\tilde{E}_p$ is singular. In particular, if E is semi-stable, then N_E is simply the product of the primes dividing D.

Before describing the modularity conjecture, we cannot resist discussing one more aspect of the L-function $L(E, s)$. We emphasize that $L(E, s)$ is built up solely using information about the reduced curves $\tilde{E}_p$. A fundamental conjecture says that this mod p information suffices to determine the rank of the group $E(\mathbb{Q})$ of rational points.

Conjecture 6.24. (Birch and Swinnerton-Dyer) *Let E be a rational elliptic curve, and let $L(E, s)$ be its L-function. The order of vanishing of $L(E, s)$ at $s = 1$ is equal to the rank of the group of rational points $E(\mathbb{Q})$,*

$$\operatorname*{ord}_{s=1} L(E, s) = \operatorname{rank} E(\mathbb{Q}).$$

There is a further part of the conjecture which says that if $L(E, s)$ is expanded as a Taylor series around $s = 1$, say

$$L(E, s) = c_E(s - 1)^{\operatorname{rank} E(\mathbb{Q})} + \cdots,$$

then the value of the leading coefficient c_E encapsulates a great deal of information about E, including the size of a set of generators for $E(\mathbb{Q})$. But in order to claim the \$1,000,000 Millenium Prize, it is "enough" to prove that the order of vanishing of $L(E, s)$ is as described in Conjecture 6.24.

Suppose that $E(\mathbb{Q})$ has positive rank. The original motivation for Conjecture 6.24 was that when the infinitely many points in $E(\mathbb{Q})$ are reduced

---

[8]You may have seen that the Riemann ζ-function similarly has a meromorphic continuation to $\mathbb{C}$ and satisfies a functional equation. Setting $\xi(s) = \frac{1}{2}\pi^{-s/2}s(s - 1)\Gamma(s/2)\zeta(s)$, the functional equation for $\zeta(s)$ says that $\xi(s) = \xi(1 - s)$.

modulo p, they tend to make $\#\tilde{E}_p(\mathbb{F}_p)$ somewhat larger than one would expect on average. This suggests looking at the product

$$\prod_{p \leq T} \frac{p}{\#\tilde{E}_p(\mathbb{F}_p)}$$

and seeing what happens as T grows. If $\#\tilde{E}_p(\mathbb{F}_p)$ is equally likely to be greater than and less than p, then one might expect the limit to be positive, but if $\#\tilde{E}_p(\mathbb{F}_p)$ is biased to be larger than p, then one might expect the limit to equal 0. This heuristic and numerical evidence led Birch and Swinnerton-Dyer to conjecture that

$$\operatorname{rank} E(\mathbb{Q}) \geq 1 \quad \Longleftrightarrow \quad \lim_{T \to \infty} \prod_{p \leq T} \frac{p}{\#\tilde{E}_p(\mathbb{F}_p)} = 0. \tag{$\dagger$}$$

In order to relate this conjecture to the L-series of E, we note that if we blindly substitute $s = 1$ into the infinite product ($*$) defining $L(E, s)$ (which is completely unjustified!), we obtain the product

$$L(E, 1) \text{ "=" } \prod_{p \text{ prime}} \left(1 - \frac{\epsilon_p}{p} + \frac{1}{p}\right)^{-1} = \prod_{p \text{ prime}} \frac{p}{p - \epsilon_p + 1} = \prod_{p \text{ prime}} \frac{p}{\#\tilde{E}_p(\mathbb{F}_p)}.$$

So based on ($\dagger$), it is not unreasonable to guess that $E(\mathbb{Q})$ has positive rank if and only if $L(E, 1) = 0$.

Now we're ready to define modularity and to state the modularity theorem. We fix an integer $N \geq 1$, called the *level*, and we let $\Gamma_0(N)$ be the modular subgroup of $\operatorname{SL}_2(\mathbb{Z})$ defined by

$$\Gamma_0(N) = \left\{ \begin{pmatrix} a & b \\ c & d \end{pmatrix} : c \equiv 0 \pmod{N} \right\}.$$

Let $f(z)$ be a function given by a series of the form

$$f(z) = \sum_{n=1}^{\infty} c_n e^{2\pi i n z}.$$

We assume that the c_n do not grow too quickly, more precisely, we assume that there are constants κ and ν so that $|c_n| \leq \kappa n^\nu$ for all $n \geq 1$. Then the series for $f(z)$ converges and defines a holomorphic function for all z in the upper half-plane

$$\mathfrak{H} = \left\{ z \in \mathbb{C} : \operatorname{Im}(z) > 0 \right\}.$$

Such a function $f(z)$ is called a *modular cusp form of level N* if it satisfies[9]

$$f\left(\frac{az+b}{cz+d}\right) = (cz+d)^2 f(z) \text{ for all } \begin{pmatrix} a & b \\ c & d \end{pmatrix} \in \Gamma_0(N) \text{ and all } z \in \mathfrak{H}.$$

Theorem 6.25. *[7, 53, 60] Let E be a rational elliptic curve and let $L(E,s) = \sum \epsilon_n/n^s$ be the L-series associated to E. We use the coefficients of $L(E,s)$ to define a function*

$$f_E(z) = \sum_{n=1}^{\infty} \epsilon_n e^{2\pi i n z}.$$

Then $f_E(z)$ is a modular cusp form of level N_E, where N_E is the conductor of E as described in Theorem 6.23(b).

A rational elliptic curve is said to be *modular* if its L-function has the property described in Theorem 6.25, and the modularity conjecture of Shimura, Taniyama, and Weil had been that every rational elliptic curve is modular. Wiles, with one argument joint with Taylor, proved that this is true for every rational elliptic curve having semi-stable reduction, i.e., for curves such that $\tilde{E}_p$ has a node for all primes p dividing the discriminant D. Breuil, Conrad, Diamond, and Taylor subsequently completed the proof for all rational elliptic curves.

So how does the Modularity Theorem help to prove Fermat's Last Theorem. It tells us that the Frey curve $E_{A,B,C}$ associated to a putative solution (A, B, C) to the Fermat equation is modular. We note that given any Fermat solution for exponent n, if $m \mid n$, then we get a solution for exponent m via

$$(A^{n/m})^m + (B^{n/m})^m = (C^{n/m})^m.$$

It thus suffices to prove Fermat's Last Theorem for exponent 4, which Fermat himself did, and for prime exponents ℓ. Further, since the first few values of ℓ

---

[9]We have omitted a technical condition that $f(z)$ vanish at every cusp. A more intrinsic way to describe the transformation formula, which may be less mysterious, is to say that the differential form $f(z)\,dz$ is invariant under the transformation sending z to $(az+b)/(cz+d)$. More precisely, we want $f(z)\,dz$ to be a well-defined differential form on (a smooth completion of) the quotient $\mathfrak{H}/\Gamma_0(N)$. We also mention that our modular forms have weight 2, and that there also exist modular forms of other weights, where one replaces the $(cz + d)^2$ with $(cz + d)^k$ for some other value of k.

were done in the nineteenth century, we may assume that $\ell \geq 5$. So our goal is to derive a contradiction assuming that there is a solution to

$$A^\ell + B^\ell = C^\ell \quad \text{for some prime } \ell \geq 5.$$

The Modularity Theorem tells us that the Frey curve $E_{A,B,C}$ has an associated modular cusp form $f_{E_{A,B,C}}(z)$ of level $N_{E_{A,B,C}}$. To complete the proof of Fermat's Last Theorem, we use Ribet's level-lowering theorem.[10]

Theorem 6.26. (Ribet [38]) *Let E be a rational elliptic curve of conductor N and discriminant D, so $f_E(z)$ is a modular cusp form of level N. Let $\ell \geq 5$ be prime, and let $p_1, \ldots, p_r \geq 3$ be primes at which E has semi-stable reduction. (Thus $p_i \mid N$ and $p_i^2 \nmid N$.) Suppose further that*

$$\mathrm{ord}_{p_i}(D) \equiv 0 \pmod{\ell} \quad \text{for all } 1 \leq i \leq r.$$

Then there exists a non-zero modular cusp form $g(z)$ of level $N/p_1 \cdots p_r$, i.e., a modular cusp form $g(z)$ satisfying

$$g\left(\frac{az+b}{cz+d}\right) = (cz+d)^2 g(z) \quad \text{for all} \quad \begin{pmatrix} a & b \\ c & d \end{pmatrix} \in \Gamma_0(N/p_1 \cdots p_r).$$

The key here is that the group $\Gamma_0(N/p_1 \cdots p_r)$ is larger than the group $\Gamma_0(N)$, so the modular cusp form g has the modular transformation property for far more matrices than the modular cusp form f.

As we noted earlier, the Frey curve is semi-stable at every odd prime of bad reduction, and its discriminant is $D = (ABC)^{2\ell}$, so

$$\mathrm{ord}_p(D) = 2\ell\, \mathrm{ord}_p(ABC) \equiv 0 \pmod{\ell}$$

for all (odd) primes. This means that we can apply Ribet's theorem to $E_{A,B,C}$ with $\{p_1, \ldots, p_r\}$ equal to the set of all odd primes dividing N. It turns out that N is always even, since one of A, B, C is necessarily even, and $4 \nmid N$ since $E_{A,B,C}$ is semi-stable. So Ribet's theorem says that there is a non-zero modular cusp form $g(z)$ of level 2, i.e., $g(z)$ satisfies the modular transformation formula for all matrices $\begin{pmatrix} a & b \\ c & d \end{pmatrix} \in \Gamma_0(2)$. But it turns out that there are no non-zero modular cusp forms of level 2, or indeed of any level smaller than 11. Why not? The reason comes from geometry. Let $f(z)$ be a modular cusp form of level N. Then as indicated earlier, the differential form

---

[10]What we have stated is a consequence of Ribet's theorem, whose full statement requires concepts and terminology that would take too long to develop here.

$f(z)\,dz$ extends to a holomorphic differential form on a smooth completion of the quotient space $\mathfrak{H}/\Gamma_0(N)$. This Riemann surface is denoted $X_0(N)$. It is not hard to compute the genus of $X_0(N)$, and it turns out that if $N \leq 10$, then $X_0(N)$ has genus 0, which means that it looks like a sphere. And spheres have no holomorphic differential forms. This can be checked either by a direct calculation, or via the Riemann–Roch theorem, which says that the space of holomorphic differential forms on a smooth Riemann surface of genus g is a vector space of dimension g. And this contradiction completes the proof of Fermat's last theorem.

The series $L(E, s)$ and the modular cusp form $f_E(z)$ associated to an elliptic curve E are intimately related to the representations that we've studied in this chapter. We recall that for each integer n there is a homomorphism

$$\rho_n : \mathrm{Gal}\big(\mathbb{Q}\big(E[n]\big)/\mathbb{Q}\big) \longrightarrow \mathrm{GL}_2(\mathbb{Z}/n\mathbb{Z}), \qquad \rho_n(\sigma) = \begin{pmatrix} \alpha_\sigma & \beta_\sigma \\ \gamma_\sigma & \delta_\sigma \end{pmatrix}.$$

The particular matrix that we get depends on the choice of a basis for $E[n]$, but the trace and the determinant of $\rho_n(\sigma)$ do not.

We now need to use a few concepts from algebraic number theory. Let p be a prime not dividing n, let $\mathfrak{O}$ be the ring of integers of $\mathbb{Q}\big(E[n]\big)$, let $\mathfrak{P}$ by a prime of $\mathfrak{O}$ lying over p, and let $\mathcal{D}_{\mathfrak{P}} \subset \mathrm{Gal}\big(\mathbb{Q}\big(E[n]\big)/\mathbb{Q}\big)$ be the decomposition group of $\mathfrak{P}$.[11] There is an element $\sigma_{\mathfrak{P}} \in \mathcal{D}_{\mathfrak{P}}$, called the p-power Frobenius element, that is characterized by the property that

$$\sigma_{\mathfrak{P}}(\alpha) \equiv \alpha^p \pmod{p} \quad \text{for all } \alpha \in \mathfrak{O}.$$

If we suppose further that $\tilde{E} \bmod p$ is non-singular, then it turns out that the representation ρ_n is closely connected to the collection of values $\epsilon_p = p + 1 - \#\tilde{E}_p$ used in building the L-function and the modular cusp form of E. This connection is via the congruence (cf. [49, V.2.6])

$$\mathrm{Trace}\big(\rho_n(\sigma_{\mathfrak{P}})\big) \equiv \epsilon_p \pmod{n}.$$

In particular, if $n > 4\sqrt{p}$, then the value of $\rho_n(\sigma_{\mathfrak{P}})$ completely determines ϵ_p, since we know from Hasse's theorem that $|\epsilon_p| < 2\sqrt{p}$.

Suppose now that p is a prime such that $\tilde{E}_p$ has a node with slopes defined over $\mathbb{F}_p$, and let ℓ be a prime different from p. Then a construction due to the

---

[11]We recall that the decomposition group of $\mathfrak{P}$ is the set of $\sigma \in \mathrm{Gal}\big(\mathbb{Q}(E[n])/\mathbb{Q}\big)$ such that $\sigma(\mathfrak{P}) = \mathfrak{P}$.

second author implies that there is a basis for $E[\ell]$ so that the representation on the ℓ-torsion has the special form

$$\rho_\ell(\sigma) = \begin{pmatrix} \alpha_\sigma & \beta_\sigma \\ 0 & 1 \end{pmatrix} \quad \text{for all } \sigma \in \mathcal{D}_{\mathfrak{P}}.$$

(See [48, Chapter V] or [51].) More precisely, there is an ℓ'th root of unity ζ and a number q so that α_σ and β_σ are determined by the formulas

$$\sigma(\zeta) = \zeta^{\alpha_\sigma} \quad \text{and} \quad \sigma(q^{1/\ell}) = \zeta^{\beta_\sigma} q^{1/\ell} \quad \text{for all } \sigma \in \mathcal{D}_{\mathfrak{P}}.$$

Further, the theory shows that

$$\operatorname{ord}_p(q) = \operatorname{ord}_p(D),$$

where as usual D is the discriminant of E.

Now consider a Frey curve $E = E_{A,B,C}$ coming from a solution to the Fermat equation. If $p \mid D$, then we saw earlier that

$$\operatorname{ord}_p(q) = \operatorname{ord}_p(D) = 2\ell \operatorname{ord}_p(ABC),$$

so the power of p dividing q is an ℓ'th-power. This is true for every $p \mid D$, so we can write

$$q = q_1 q_2^\ell \quad \text{with } \gcd(q_1, D) = 1.$$

Hence when we compute $q^{1/\ell}$, we only need to take the ℓ'th root of something that is relatively prime to D. This is reasonably benign,[12] and indeed, it comes close to implying that $E_{A,B,C}$ has non-singular reduction modulo every prime p dividing D.[13] And although it does not, in fact, imply that $\tilde{E}_{A,B,C} \bmod p$ is non-singular, it does provide just enough ammunition for the proof of Ribet's level-lowering theorem via the connection between the modular cusp form f_E and the representations ρ_n.

---

[12]In mathematical terminology, the extension generated by $q^{1/\ell}$ is unramified at every prime p dividing D.

[13]The criterion of Néron–Ogg–Shafarevich [49, VII.7.1] says that if for every $t \geq 1$, the representation ρ_{ℓ^t} on $\mathcal{D}_p$ is unramified at p, then $\tilde{E}_p$ is non-singular. For the Frey curve, we know this property for $t = 1$, which is a start.

Exercises

6.1. The *discriminant* of a monic polynomial

$$f(X) = (X - \alpha_1)(X - \alpha_2) \cdots (X - \alpha_n)$$

is defined to be
$$\mathrm{Disc}(f) = \prod_{1 \le j < i \le n} (\alpha_i - \alpha_j)^2.$$

(a) Prove that
$$\mathrm{Disc}(f) = (-1)^{(n^2-n)/2} \prod_{i=1}^{n} f'(\alpha_i).$$

(b) Let $f(X) = X^n - 1$. Prove that
$$\mathrm{Disc}(f) = (-1)^{(n-1)(n-2)/2} n^n.$$

(c) Let ζ be a primitive n'th root of unity. Prove that the cyclotomic field $\mathbb{Q}(\zeta)$ contains $\sqrt{\mathrm{Disc}(f)}$.

(d) Let ζ' be a primitive $4n$'th root of unity. Use (c) to prove that $\mathbb{Q}(\sqrt{n})$ is contained in the cyclotomic field $\mathbb{Q}(\zeta')$.

(e) In Section 6.1 we used Gauss sums to prove (d) when $n = p$ is prime. Give an alternative proof of (d) by using the fact that it is true for primes and that $\mathbb{Q}(\sqrt{n})$ is the compositum of the fields $\mathbb{Q}(\sqrt{p})$ for all primes dividing n.

This exercise shows that every quadratic extension of $\mathbb{Q}$ is contained in a cyclotomic extension, thereby proving the Kronecker–Weber theorem for extensions $K/\mathbb{Q}$ satisfying $\mathrm{Gal}(K/\mathbb{Q}) = \mathbb{Z}/2\mathbb{Z}$.

6.2. (a) Suppose that $\lambda(z)$ is a polynomial

$$\lambda(z) = a_0 z^n + a_1 z^{n-1} + \cdots + a_{n-1} z + a_n$$

of degree n such that $\lambda : \mathbb{C}^* \to \mathbb{C}^*$ is a homomorphism. Prove that $\lambda(z) = z^n$.

(b) Suppose that $\lambda(z)$ is a meromorphic function such that $\lambda : \mathbb{C}^* \to \mathbb{C}^*$ is a homomorphism. Prove that $\lambda(z) = z^n$ for some $n \in \mathbb{Z}$.

6.3. Let C be a rational elliptic curve, and let K be a Galois extension of $\mathbb{Q}$.

(a) Prove that for all $P \in C(K)$ and all $\sigma, \tau \in \mathrm{Gal}(K/\mathbb{Q})$,

$$\tau\bigl(\sigma(P)\bigr) = (\tau\sigma)(P).$$

This is a sort of associative law. The mathematical terminology is that the group $\mathrm{Gal}(K/\mathbb{Q})$ *acts on* the abelian group $C(K)$. Group actions are very important in many areas of mathematics.

(b) Prove that for all $P \in C(K)$ and all $\sigma \in \text{Gal}(K/\mathbb{Q})$,

$$\sigma(2P) = 2\sigma(P).$$

N.B. Do not just quote Proposition 6.3(d) from Section 6.2. When we proved that proposition, this is one of the cases that we left for you to prove.

6.4. The sequence of *division polynomials* $\psi_n \in \mathbb{Z}[x, y]$ for the elliptic curve

$$C : y^2 = x^3 + x$$

are defined recursively by the following rules:

$$\psi_1 = 1,$$
$$\psi_2 = 2y,$$
$$\psi_3 = 3x^4 + 6x^2 - 1,$$
$$\psi_4 = 4y(x^6 + 5x^4 - 5x^2 - 1),$$
$$\psi_{2n+1} = \psi_{n+2}\psi_n^3 - \psi_{n-1}\psi_{n+1}^3 \quad \text{for } n \geq 2,$$
$$2y\psi_{2n} = \psi_n(\psi_{n+2}\psi_{n-1}^2 - \psi_{n-2}\psi_{n+1}^2) \quad \text{for } n \geq 3.$$

Further define ϕ_n and ω_n by

$$\phi_n = x\psi_n^2 - \psi_{n+1}\psi_{n-1},$$
$$4y\omega_n = \psi_{n+2}\psi_{n-1}^2 - \psi_{n-2}\psi_{n+1}^2.$$

(a) Prove that all of the ψ_n, ϕ_n, and ω_n are in $\mathbb{Z}[x, y]$. (Note that the only potential problem is that the recursive definition of ψ_{2n} appears to require dividing by $2y$.)

(b) If n is odd, prove that ψ_n, ϕ_n, and $y^{-1}\omega_n$ are in $\mathbb{Z}[x, y^2]$, so replacing y^2 with $x^3 + x$, we may view them as being in $\mathbb{Z}[x]$. Similarly, if n is even, prove that ψ_n, ϕ_n, and ω_n are in $\mathbb{Z}[x, y^2]$, hence may be viewed as being in $\mathbb{Z}[x]$.

(c) Show that, as polynomials in x, we have

$$\phi_n(x) = x^{n^2} + \text{lower order terms},$$
$$\psi_n(x)^2 = n^2 x^{n^2-1} + \text{lower order terms}.$$

(d) Let $P = (x, y) \in C$. Prove that

$$nP = \left(\frac{\phi_n(P)}{\psi_n(P)^2}, \frac{\omega_n(P)}{\psi_n(P)^3} \right).$$

(e) Prove that $\psi_n(x)^2$ has no double roots in $\mathbb{C}$. Prove that $\psi_n(x)^2$ and $\phi_n(x)$ have no common roots in $\mathbb{C}$.

(f) Let $P = (x, y) \in C(\mathbb{C})$. Prove that $nP = \mathcal{O}$ if and only if $\psi_n(x)^2 = 0$.

(g) Prove that for every n, the group $C[n]$ contains n^2 points. Deduce that

$$C[n] \cong \mathbb{Z}/n\mathbb{Z} \oplus \mathbb{Z}/n\mathbb{Z}.$$

6.5. * Redo the previous exercise for the elliptic curve

$$C : y^2 = x^3 + bx + c.$$

Everything will be the same except that ψ_3 and ψ_4 are given by the formulas

$$\psi_3 = 3x^4 + 6bx^2 + 12cx - b^2,$$
$$\psi_4 = 4y(x^6 + 5bx^4 + 20cx^3 - 5b^2x^2 - 4bcx - 8c^2 - b^3).$$

6.6. Let C be a rational elliptic curve and let $m, n \geq 1$ be integers.
(a) If $\gcd(m, n) = 1$, prove that $\mathbb{Q}(C[mn])$ is equal to the compositum of the fields $\mathbb{Q}(C[m])$ and $\mathbb{Q}(C[n])$, where we recall that the compositum of two fields K_1 and K_2 is the smallest field containing both K_1 and K_2.
(b) More generally, let $\ell = \operatorname{LCM}(m, n)$. Prove that $\mathbb{Q}(C[\ell])$ is the compositum of the fields $\mathbb{Q}(C[m])$ and $\mathbb{Q}(C[n])$.

6.7. Let R be a commutative ring with multiplicative identity. Let A be an $r \times r$ matrix with coefficients in R.
(a) If $\det(A)$ is a unit in R, prove that there is a matrix B with coefficients in R such that $AB = I$.
(b) Conversely, if there is a matrix B with coefficients in R such that $AB = I$, prove that $\det(A)$ is a unit in R.

6.8. Let A be an abelian group, and define $\operatorname{End}(A)$ to be the set of homomorphisms from A to itself,

$$\operatorname{End}(A) = \{\text{homomorphisms } A \to A\}.$$

Define an addition and multiplication on $\operatorname{End}(A)$ by the rules

$$(g + h)(\alpha) = g(\alpha) + h(\alpha) \quad \text{and} \quad (gh)(\alpha) = g(h(\alpha)).$$

N.B. $(gh)(\alpha)$ is not equal to the product $g(\alpha)h(\alpha)$. In this exercise you will verify that $\operatorname{End}(A)$ is a ring, called the *endomorphism ring of* A.
(a) Prove that $g + h \in \operatorname{End}(A)$ and that $gh \in \operatorname{End}(A)$.
(b) Prove that these addition and multiplication rules make $\operatorname{End}(A)$ into a (not necessarily commutative) ring. What is the multiplicative identity of this ring?
(c) The *automorphism group of* A is the unit group of the ring $\operatorname{End}(A)$. Prove that the elements of $\operatorname{Aut}(A)$ are isomorphisms from A to itself.
(d) Give an example to show that if A is non-abelian, then $\operatorname{End}(A)$ is not a ring. (*Hint.* The distributive law may fail.)

6.9. (a) Let A be a cyclic group of order n. Prove that

$$\operatorname{End}(A) \cong \mathbb{Z}/n\mathbb{Z} \quad \text{and} \quad \operatorname{Aut}(A) \cong (\mathbb{Z}/n\mathbb{Z})^*.$$

(b) Let A be a direct sum of r cyclic groups of order n. Prove that $\mathrm{End}(A)$ may be naturally identified with the ring of r-by-r matrices with coefficients in $\mathbb{Z}/n\mathbb{Z}$.

(c) With the identification in (b), prove that

$$\mathrm{Aut}(A) \cong \mathrm{GL}_r(\mathbb{Z}/n\mathbb{Z}).$$

6.10. Let C be a rational elliptic curve. Prove that there is a one-to-one homomorphism

$$\mathrm{Gal}\big(\mathbb{Q}(C[n])/\mathbb{Q}\big) \longrightarrow \mathrm{Aut}\big(C[n]\big)$$

defined by the rule that $\sigma \in \mathrm{Gal}\big(\mathbb{Q}(C[n])/\mathbb{Q}\big)$ goes to the map

$$C[n] \longrightarrow C[n], \quad P \longmapsto \sigma(P).$$

Further, show that $\mathrm{Aut}\big(C[n]\big) \cong \mathrm{GL}_2(\mathbb{Z}/n\mathbb{Z})$, thereby recovering the representation $\rho_n : \mathrm{Gal}\big(\mathbb{Q}(C[n])/\mathbb{Q}\big) \to \mathrm{GL}_2(\mathbb{Z}/n\mathbb{Z})$ from Section 6.3.

6.11. Let F be a field, and let V be an F-vector space of dimension r. Let

$$\mathrm{Aut}_F(V) = \left\{ \begin{array}{c} \text{one-to-one and onto } F\text{-linear} \\ \text{transformations } V \to V \end{array} \right\}.$$

Prove that $\mathrm{Aut}_F(V)$ is isomorphic to $\mathrm{GL}_r(F)$, the group of invertible $r \times r$ matrices with coefficients in F.

6.12. Let C be the elliptic curve $y^2 = x^3 + x$. The points $P_1 = (i,0)$ and $P_2 = (-i,0)$ are generators of $C[2]$, cf. Example 6.9 in Section 6.3. The Galois group $\mathrm{Gal}\big(\mathbb{Q}(C[2])/\mathbb{Q}\big)$ consists of two elements, the identity σ_0 and complex conjugation σ_1. What is the matrix $\rho_2(\sigma_1) \in \mathrm{GL}_2(\mathbb{Z}/2\mathbb{Z})$ if the representation ρ_2 is defined using P_1 and P_2 as generators for $C[2]$?

6.13. For each of the following curves, determine $\mathrm{Gal}\big(\mathbb{Q}(C[2])/\mathbb{Q}\big)$, the Galois group of the extension of $\mathbb{Q}$ generated by the points of order two.

(a) $y^2 = x^3 - x$.

(b) $y^2 = x^3 - x - 2$.

(c) $y^2 = x^3 + x - 2$.

(d) $y^2 = x^3 - 3x + 1$.

6.14. For each of the elliptic curves in the previous exercise, choose a basis for $C[2]$ and write down matrices $\rho_2(\sigma)$ for each element σ in $\mathrm{Gal}\big(\mathbb{Q}(C[2])/\mathbb{Q}\big)$, as we did in Examples 6.8, 6.9, and 6.10 in Section 6.3.

6.15. Let $\mathbb{C}/L$ be an elliptic curve with a complex multiplication,

$$f : \mathbb{C}/L \longrightarrow \mathbb{C}/L, \quad f(z) = cz.$$

(a) Prove that there are integers A and B such that

$$c^2 + Ac + B = 0.$$

(*Hint.* Write $L = \mathbb{Z}\omega_1 + \mathbb{Z}\omega_2$ and use the fact that both $c\omega_1$ and $c\omega_2$ are in L.)

(b) Prove that the integers A and B satisfy $A^2 < 4B$.

(c) Prove that the field $\mathbb{Q}(c)$ is a degree 2 extension of $\mathbb{Q}$ and that $\mathbb{Q}(c)$ is not contained in $\mathbb{R}$, i.e., prove that $\mathbb{Q}(c)$ is an imaginary quadratic field.

6.16. (a) Let C be an elliptic curve. Define the *endomorphism ring of C* to be

$$\text{End}(C) = \{\text{endomorphisms } C \to C\}.$$

Note that this is a little different from the endomorphism ring of C considered as an abelian group, because we are not taking all group homomorphisms from C to itself, but only those defined by rational functions. In other words, $\text{End}(C)$ is the set of *algebraic* endomorphisms of C. Prove that the addition and multiplication rules

$$(\phi_1 + \phi_2)(P) = \phi_1(P) + \phi_2(P) \quad \text{and} \quad (\phi_1\phi_2)(P) = \phi_1\big(\phi_2(P)\big)$$

make $\text{End}(C)$ into a ring.

(b) Let $L \subset \mathbb{C}$ be a lattice. Define a set of complex numbers R_L by

$$R_L = \{c \in \mathbb{C} : cL \subset L\}.$$

Prove that R_L is a ring.

(c) Let $C(\mathbb{C}) = \mathbb{C}/L$ be an elliptic curve. For each $\phi \in \text{End}(C)$, we showed in Section 6.4 that ϕ corresponds to a map

$$f : \mathbb{C}/L \longrightarrow \mathbb{C}/L, \quad f(z) = c_\phi z,$$

where $c_\phi \in \mathbb{C}$ is uniquely determined by ϕ and satisfies $c_\phi L \subset L$. Prove that the map

$$\text{End}(C) \longrightarrow R_L, \quad \phi \longmapsto c_\phi,$$

is a one-to-one homomorphism of rings.

(d) * Prove that the homomorphism in (c) is an isomorphism.

6.17. Let C be the elliptic curve $y^2 = x^3 + x$, and let $K_n = \mathbb{Q}(i)\big(C[n]\big)$ be the field considered in Section 6.5. We proved that K_n is a Galois extension of $\mathbb{Q}$.

(a) Let $\tau : \mathbb{C} \to \mathbb{C}$ be complex conjugation, which we may consider to be an element of $\text{Gal}(K_n/\mathbb{Q})$ by fixing an inclusion $K_n \subset \mathbb{C}$. Prove that every element of $\text{Gal}(K_n/\mathbb{Q})$ can be written uniquely in the form $\sigma = st$ with $s \in \text{Gal}\big(K_n/\mathbb{Q}(i)\big)$ and $t \in \{e, \tau\}$.

(b) Prove that for all $s \in \text{Gal}\big(K_n/\mathbb{Q}(i)\big)$ there is an integer $m \in (\mathbb{Z}/n\mathbb{Z})^*$ such that

$$(s\tau s\tau^{-1})(P) = mP \quad \text{for all } P \in C[n].$$

In other words, the matrix describing the action of $s\tau s\tau^{-1}$ on $C[n]$ is the diagonal matrix $\left(\begin{smallmatrix} m & 0 \\ 0 & m \end{smallmatrix}\right)$.

(c) Use (b) to prove that $\text{Gal}(K_n/\mathbb{Q})$ is abelian if and only if for every element $s \in \text{Gal}(K_n/\mathbb{Q}(i))$ there is an integer m such that

$$s^2(P) = mP \quad \text{for all } P \in C[n].$$

6.18. Let C be the elliptic curve $C : y^2 = x^3 + x$, and let

$$\beta = \sqrt[4]{\frac{8\sqrt{3} - 12}{9}};$$

see Example 6.6 in Section 6.2.

(a) Prove that the minimal polynomial of β over $\mathbb{Q}$ is

$$27x^8 + 72x^4 - 16 = 0.$$

(b) Prove that $\mathbb{Q}(C[3]) = \mathbb{Q}(\beta, i)$.
(c) Compute the Galois group of $\mathbb{Q}(\beta, i)$ over $\mathbb{Q}(i)$. In particular, verify that it is abelian.

6.19. Let C be the elliptic curve

$$C : y^2 = x^3 + 1.$$

For each integer $n \geq 1$, let

$$K_n = \mathbb{Q}(\sqrt{-3})(C[n])$$

be the extension field of $\mathbb{Q}(\sqrt{-3})$ generated by the coordinates of the points of order n. Note that C has complex multiplication; cf. Example 6.14 in Section 6.4.

(a) Prove that K_n is a Galois extension of $\mathbb{Q}$.
(b) Prove that

$$\text{Gal}(K_n/\mathbb{Q}(\sqrt{-3}))$$

is abelian.

6.20. Let C be the elliptic curve

$$C : y^2 = x^3 + 4x^2 + 2x.$$

(a) Prove that the formula

$$\phi(P) = \begin{cases} \left(\dfrac{-y^2}{2x^2}, \dfrac{-y(x^2 - 2)}{2\sqrt{-2}x^2} \right), & \text{if } P = (x, y) \neq (0, 0), \\ \mathcal{O} & \text{if } P = (0, 0) \text{ or } P = \mathcal{O}. \end{cases}$$

is an endomorphism $\phi : C \to C$.

(b) Prove that C has complex multiplication. (*Hint.* What is the kernel of ϕ?)

(c) Let

$$K_n = \mathbb{Q}(\sqrt{-2})(C[n])$$

be the extension field of $\mathbb{Q}(\sqrt{-2})$ generated by the coordinates of the points of order n. Prove that K_n is a Galois extension of $\mathbb{Q}$.

(d) Prove that

$$\mathrm{Gal}\big(K_n/\mathbb{Q}(\sqrt{-2})\big)$$

is abelian.

6.21. Let C be the elliptic curve $C : y^2 = x^3 + x$, let L be the lattice $\mathbb{Z} + \mathbb{Z}i$, and let

$$g_3 = 140 \sum_{\omega \in L,\, \omega \neq 0} \frac{1}{\omega^6}$$

be the quantity that we defined in Section 2.2.

(a) Prove that $g_3 = 0$. (*Hint.* If $\omega \in L$, then $i\omega$ is also in L.)

(b) Prove that there is a complex number γ so that the map

$$\mathbb{C}/L \longrightarrow C(\mathbb{C}), \quad z \longmapsto \big(4\gamma^2 \wp(z), 4\gamma^3 \wp'(z)\big)$$

is an isomorphism, where $\wp$ is the Weierstrass $\wp$ function described in Section 2.2.

(c) Show that the complex multiplication map

$$C(\mathbb{C}) \longrightarrow C(\mathbb{C}), \quad (x, y) \longmapsto (-x, -iy),$$

corresponds to the map

$$\mathbb{C}/L \longrightarrow \mathbb{C}/L, \quad z \longmapsto iz.$$

In other words, verify the formulas

$$\wp(iz) = -\wp(z) \quad \text{and} \quad \wp'(iz) = -i\wp'(iz).$$

6.22. Let C be a rational elliptic curve, let $\{P_1, P_2\}$ be a basis for $C[n]$, and let

$$\rho_n : \mathrm{Gal}\big(\mathbb{Q}(C[n])/\mathbb{Q}\big) \longrightarrow \mathrm{GL}_2(\mathbb{Z}/n\mathbb{Z})$$

be the associated representation. Now let $\{P_1', P_2'\}$ be another basis for $C[n]$, and let

$$\rho_n' : \mathrm{Gal}\big(\mathbb{Q}(C[n])/\mathbb{Q}\big) \longrightarrow \mathrm{GL}_2(\mathbb{Z}/n\mathbb{Z})$$

be the representation defined using this new basis. Prove that there is a matrix $U \in \mathrm{GL}_2(\mathbb{Z}/n\mathbb{Z})$ so that

$$\rho_n'(\sigma) = U^{-1}\rho_n(\sigma)U \quad \text{for all } \sigma \in \mathrm{Gal}\big(\mathbb{Q}(C[n])/\mathbb{Q}\big).$$

6.23. Let ϵ and $p > 0$ be arbitrary real numbers, and factor

$$1 - \epsilon T + pT^2 = (1 - \alpha T)(1 - \beta T)$$

using complex numbers α and β.

(a) Prove that

$$\frac{1}{1 - \epsilon T + pT^2} = \sum_{k=0}^{\infty} \frac{\alpha^{k+1} - \beta^{k+1}}{\alpha - \beta} T^k.$$

(*Hint.* Partial fractions and the geometric series.)

(b) Prove that the Taylor coefficients

$$\epsilon_k := \frac{\alpha^{k+1} - \beta^{k+1}}{\alpha - \beta}$$

from (a) satisfy the recursion

$$\epsilon_1 = \epsilon \quad \text{and} \quad \epsilon_{k+1} = \epsilon_k \epsilon - p\epsilon_{k-1} \quad \text{for } k \geq 1.$$

6.24. Let E be a rational elliptic curve, let $\#\tilde{E}(\mathbb{F}_p) = p + 1 - \epsilon_p$, and let $L(E, s)$ be the L-function of E defined by the product

$$L(E, s) = \prod_{p \text{ prime}} \left(1 - \frac{\epsilon_p}{p^s} + \frac{1}{p^{2s-1}}\right)^{-1}. \tag{†}$$

Expanding the product, write

$$L(E, s) = \sum_{n=1}^{\infty} \frac{\epsilon_n}{n^s} \tag{††}$$

as a Dirichlet series.

(a) Prove that:

   (i) The definitions of ϵ_p in (†) and (††) are consistent.

   (ii) $\epsilon_{p^{k+1}} = \epsilon_{p^k} \epsilon_p - p\epsilon_{p^{k-1}}$ for all $k \geq 1$.

   (iii) $\epsilon_{mn} = \epsilon_m \epsilon_n$ for all indices satisfying $\gcd(m, n) = 1$.

   (*Hint.* Use Exercise 6.23.)

(b) Factor

$$1 - \epsilon_p T + pT^2 = (1 - \alpha_p T)(1 - \beta_p T) \quad \text{with } \alpha_p, \beta_p \in \mathbb{C}.$$

Show that the Hasse–Weil estimate $|\epsilon_p| \leq 2\sqrt{p}$ (Theorem 4.1) implies that

$$|\alpha_p| = |\beta_p| = \sqrt{p}.$$

(c) Prove that

$$|\epsilon_{p^k}| \le (k+1)p^{k/2},$$

and use this to prove that

$$|\epsilon_n| \le d(n)\sqrt{n},$$

where $d(n) = \sum_{d|n} 1$ is the number of distinct divisors of n.

(d) Use (c) to prove that the Dirichlet series $L(E, s) = \sum \epsilon_n/n^s$ converges for all s in the half-plane $\mathrm{Re}(s) > \frac{3}{2}$.

6.25. The first few ϵ_p values for the elliptic curve $E : y^2 = x^3 + x^2 + x + 1$ are

p	2	3	5	7	11	13	17	19	23	29
ϵ_p	0	-2	-2	-4	2	-2	-2	-2	4	6

Use these values and the formulas in Exercise 6.24(a) to compute ϵ_n for all $n \le 30$.

Appendix A

Projective Geometry

In this appendix we summarize the basic properties of the projective plane and projective curves that are used elsewhere in this book. For further reading about projective algebraic geometry, the reader might profitably consult Brieskorn–Knörrer [8], Fulton [16], Harris [18], or Reid [37]. More high-powered accounts of modern algebraic geometry are given in Hartshorne [20] and Griffiths–Harris [17].

A.1 Homogeneous Coordinates and the Projective Plane

There are many ways to construct the projective plane. We describe two constructions, one algebraic and one geometric, since each in its own way provides enlightenment.

We begin with a famous problem from number theory, namely the solution of the equation

$$x^N + y^N = 1 \qquad \text{(Fermat Equation \#1)}$$

in rational numbers x and y. Suppose that we have found a solution, say $x = a/c$ and $y = b/d$, where we write fractions in lowest terms and with positive denominators. Substituting and clearing denominators gives the equation

$$a^N d^N + b^N c^N = c^N d^N.$$

© Springer International Publishing Switzerland 2015
J.H. Silverman, J.T. Tate, *Rational Points on Elliptic Curves*,
Undergraduate Texts in Mathematics, DOI 10.1007/978-3-319-18588-0

It follows that $c^N \mid a^N d^N$, but $\gcd(a, c) = 1$ by assumption, so we conclude that $c^N \mid d^N$, and hence $c \mid d$. Similarly $d^N \mid b^N c^N$ and $\gcd(b, d) = 1$, which implies that $d \mid c$. Therefore $c = \pm d$, and since we've assumed that c and d are positive, we find that $c = d$. Thus any solution to Fermat Equation #1 in rational numbers has the form $(a/c, b/c)$, and thus gives a solution in integers (a, b, c) to the homogeneous equation

$$X^N + Y^N = Z^N \qquad \text{(Fermat Equation #2)}$$

Conversely, any integer solution (a, b, c) to the second Fermat equation with $c \neq 0$ gives a rational solution $(a/c, b/c)$ to the first. However, different integer solutions (a, b, c) may lead to the same rational solution. For example, if (a, b, c) is an integer solution to Fermat Equation #2, then for any integer t, the triple (ta, tb, tc) is also a solution, and clearly (a, b, c) and (ta, tb, tc) give the same rational solutions to Fermat Equation #1. The moral is that in solving Fermat Equation #2, we should really treat triples (a, b, c) and (ta, tb, tc) as being the same solution, at least for non-zero t. This leads to the notion of *homogeneous coordinates*, which we describe in more detail later.

There is one more observation that we wish to make before leaving this example, namely the "problem" that Fermat Equation #2 may have some integer solutions that do not correspond to rational solutions of Fermat Equation #1. First, the point $(0, 0, 0)$ is always a solution of the second equation, but this solution is so trivial that we will just discard it. Second, and potentially more serious, is the fact that if N is odd, then Fermat Equation #2 has the solutions $(1, -1, 0)$ and $(-1, 1, 0)$ that do not give solutions to Fermat Equation #1. To see what is happening, suppose that we take a sequence of solutions

$$(a_1, b_1, c_1), (a_2, b_2, c_2), (a_3, b_3, c_3), \ldots$$

such that

$$(a_i, b_i, c_i) \longrightarrow (1, -1, 0) \quad \text{as} \quad i \longrightarrow \infty.$$

Of course, we cannot do this with integer solutions, so now we let the a_i, b_i, c_i's be real numbers. The corresponding solutions to Fermat Equation #1 are $(a_i/c_i, b_i/c_i)$, and we see that these solutions approach (∞, ∞) as $(a_i, b_i, c_i) \to (1, -1, 0)$. In other words, the extra solutions $(1, -1, 0)$ and $(-1, 1, 0)$ to Fermat Equation #2 somehow correspond to solutions to Fermat Equation #1 that lie "at infinity." As we will see, the theory of solutions to polynomial equations becomes neater and clearer if we treat these extra points "at infinity" just as we treat all other points.

We are now ready for our first definition of the projective plane, which is essentially an algebraic definition. We define the *projective plane* to be the set of triples $[a, b, c]$ with a, b, c not all zero, but we consider two triples $[a, b, c]$ and $[a', b', c']$ to be the same point if there is a non-zero t such that

$$a = ta', \quad b = tb', \quad c = tc'.$$

We denote the projective plane by $\mathbb{P}^2$. In other words, we define an equivalence relation $\sim$ on the set of triples $[a, b, c]$ by the rule

$$[a, b, c] \sim [a', b', c'] \quad \text{if } a = ta', b = tb', c = tc' \text{ for some non-zero } t.$$

Then $\mathbb{P}^2$ is the set of equivalence classes of triples $[a, b, c]$, except that we exclude the triple $[0, 0, 0]$. Thus

$$\mathbb{P}^2 = \frac{\{[a, b, c] : a, b, c \text{ are not all zero}\}}{\sim}.$$

The numbers a, b, c are called *homogeneous coordinates* for the point $[a, b, c]$ in $\mathbb{P}^2$. More generally, for any integer $n \geq 1$, we define *projective n-space* to be the set of equivalence classes of homogeneous $n + 1$-tuples,

$$\mathbb{P}^n = \frac{\{[a_0, a_1, \ldots, a_n] : a_0, \ldots, a_n \text{ not all zero}\}}{\sim},$$

where

$$[a_0, \ldots, a_n] \sim [a'_0, \ldots, a'_n] \quad \text{if } a_0 = ta'_0, \ldots, a_n = ta'_n \text{ for some non-zero } t.$$

We eventually want to do geometry in projective space, so we need to define some geometric objects. In the next section we study quite general curves, but for the moment we are content to describe lines in $\mathbb{P}^2$. We define a *line in* $\mathbb{P}^2$ to be the set of points $[a, b, c] \in \mathbb{P}^2$ whose coordinates satisfy an equation of the form

$$\alpha X + \beta Y + \gamma Z = 0$$

for some constants α, β, γ not all zero. Note that if $[a, b, c]$ satisfies such an equation, then so does $[ta, tb, tc]$ for any t, so to check if a point of $\mathbb{P}^2$ is on a given line, one can use any homogeneous coordinates for the point.

In order to motivate our second description of the projective plane, we consider a geometric question. It is well-known that two points in the usual (x, y)-plane determine a unique line, namely the line that goes through them. Similarly, two lines in the plane determine a unique point, namely the point where they intersect, unless the two lines happen to be parallel. From both an aesthetic and a practical viewpoint, it would be nice to provide these poor parallel lines with an intersection point of their own. Since the plane itself doesn't contain the requisite points, we add on extra points by fiat. How many extra points do we need? For example, would it suffice to use one extra point P and decree that any two parallel lines intersect at P? The answer is no, and here's why.

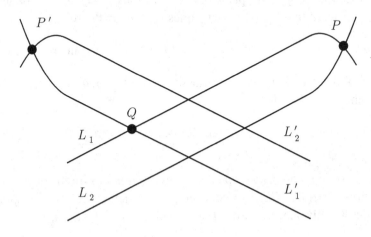

Figure A.1: Parallel lines with intersection points "at infinity"

Let L_1 and L_2 be parallel lines, and let P be the extra point where they intersect. Similarly, let L_1' and L_2' be parallel lines that intersect at the extra point P', as illustrated in Figure A.1. Suppose that L_1 and L_1' are not parallel. Then L_1 and L_1' already intersect at some ordinary point, say $L_1 \cap L_1' = \{Q\}$. But two lines are allowed to have only one point in common, so it follows that the points $P \in L_1$ and $P' \in L_1'$ must be distinct. So we really need to add an extra point for each distinct direction in the ordinary plane, and then we decree that a line L consists of its usual points together with the extra point determined by its direction.

This leads to our second definition of the projective plane, this time in purely geometric terms. For simplicity, we denote the usual *Euclidean plane* (also called the *affine plane*) by

$$\mathbb{A}^2 = \big\{(x, y): x \text{ and } y \text{ are numbers}\big\}.$$

Then we define the projective plane to be

$$\mathbb{P}^2 = \mathbb{A}^2 \cup \{\text{the set of directions in } \mathbb{A}^2\},$$

where *direction* is a non-oriented notion. Two lines have the same direction if and only if they are parallel. Logically we could define a direction in this sense to be an equivalence class of parallel lines, that is, a direction is a collection of all lines parallel to a given line. The extra points in $\mathbb{P}^2$ associated to directions, that is the points in $\mathbb{P}^2$ that are not in $\mathbb{A}^2$, are often called *points at infinity*.

As indicated earlier, a line in $\mathbb{P}^2$ then consists of a line in $\mathbb{A}^2$ together with the point at infinity specified by its direction. The intersection of two parallel lines is the point at infinity corresponding to their common direction. Finally, the set of all points at infinity is itself considered to be a line, which we denote by L_∞, and the intersection of any other line L with L_∞ is the point at infinity corresponding to the direction of L. With these conventions, it is easy to see that there is a unique line going through any two distinct points of $\mathbb{P}^2$, and further that any two distinct lines in $\mathbb{P}^2$ intersect in exactly one point. So the projective plane in this geometric incarnation eliminates the need to make a distinction between parallel and non-parallel lines. In fact, $\mathbb{P}^2$ has no parallel lines at all.

We now have two definitions of the projective plane, so it behooves us to show that they are equivalent. First we need a more analytic description of the set of directions in $\mathbb{A}^2$. One way to describe these directions is to use the set of lines in $\mathbb{A}^2$ that go through the origin, since every line in $\mathbb{A}^2$ is parallel to a unique line through the origin. Now the lines through the origin are given by equations

$$Ay = Bx$$

with A and B not both zero. However, it is possible for two pairs to give the same line. More precisely, the pairs (A, B) and (A', B') give the same line if and only if there is a non-zero t such that $A = tA'$ and $B = tB'$. Thus the set of directions in $\mathbb{A}^2$ is naturally described by the points $[A, B]$ of the projective line $\mathbb{P}^1$. This allows us to write our second description of $\mathbb{P}^2$ in the form

$$\mathbb{P}^2 = \mathbb{A}^2 \cup \mathbb{P}^1.$$

A point $[A, B] \in \mathbb{P}^1 \subset \mathbb{P}^2$ corresponds to the direction of the line $Ay = Bx$.

How is this related to the definition of $\mathbb{P}^2$ in terms of homogeneous coordinates? Recall that in our original example we associated a point $(x, y) \in \mathbb{A}^2$ with the point $[x, y, 1] \in \mathbb{P}^2$, and similarly a point $[a, b, c] \in \mathbb{P}^2$ with $c \neq 0$ was associated to the point $(a/c, b/c) \in \mathbb{A}^2$. And the remaining points in $\mathbb{P}^2$, namely those with $c = 0$, just give a copy of $\mathbb{P}^1$. In other words, the maps given in Table A.1 show how to identify our two definitions of the projective plane. It is easy to check that these two maps are inverses. For example, if $c \neq 0$, then

$$[a, b, c] \longmapsto (a/c, b/c) \longmapsto [a/c, b/c, 1] = [a, b, c].$$

We leave the remaining verifications to you.

Each of our definitions of the projective plane came with a description of what constitutes a line, so we should also check that the lines match up

Algebraic definition of $\mathbb{P}^2$	Geometric definition of $\mathbb{P}^2$
$\dfrac{\left\{[a, b, c] : a, b, c \text{ not all zero}\right\}}{\sim}$ $\longleftrightarrow$	$\mathbb{A}^2 \cup \mathbb{P}^1$
$[a, b, c]$ $\longrightarrow$	$\begin{cases}(a/c, b/c) \in \mathbb{A}^2 & \text{if } c \neq 0 \\ [a, b] \in \mathbb{P}^1 & \text{if } c = 0\end{cases}$
$[x, y, 1]$ $\longleftarrow$ $[A, B, 0]$ $\longleftarrow$	$(x, y) \in \mathbb{A}^2$ $[A, B] \in \mathbb{P}^1$

Table A.1: Maps identifying two descriptions of $\mathbb{P}^2$

properly. For example, a line L in $\mathbb{P}^2$ using homogeneous coordinates is the set of solutions $[a, b, c]$ to an equation

$$\alpha X + \beta Y + \gamma Z = 0.$$

Suppose first that α and β are not both zero. Then any point $[a, b, c] \in L$ with $c \neq 0$ is sent to the point

$$(a/c, b/c) \quad \text{on the line} \quad \alpha x + \beta y + \gamma = 0 \quad \text{in } \mathbb{A}^2.$$

And the point $[-\beta, \alpha, 0] \in L$ is sent to the point $[-\beta, \alpha] \in \mathbb{P}^1$, which corresponds to the direction of the line $-\beta y = \alpha x$. This is exactly right, since the line $-\beta y = \alpha x$ is precisely the line going through the origin that is parallel to the line $\alpha x + \beta y + \gamma = 0$. This takes care of all lines except for the line $Z = 0$ in $\mathbb{P}^2$. But the line $Z = 0$ is sent to the line in $\mathbb{A}^2 \cup \mathbb{P}^1$ consisting of all of the points at infinity. So the lines in our two descriptions of $\mathbb{P}^2$ are consistent.

A.2 Curves in the Projective Plane

An *algebraic curve* in the affine plane $\mathbb{A}^2$ is defined to be the set of solutions to a polynomial equation in two variables

$$f(x, y) = 0.$$

For example, the equation $x^2 + y^2 - 1 = 0$ is a circle in $\mathbb{A}^2$, and $2x - 3y^2 + 1 = 0$ is a parabola.

In order to define curves in the projective plane $\mathbb{P}^2$, we need to use polynomials in three variables, since points in $\mathbb{P}^2$ are represented by homogeneous triples. But there is the further difficulty that each point in $\mathbb{P}^2$ can be represented by many different homogeneous triples. It thus makes sense to look only at polynomials $F(X, Y, Z)$ with the property that if $F(a, b, c) = 0$, then $F(ta, tb, tc) = 0$ for all t. These turn out to be the homogeneous polynomials, and we use them to define curves in $\mathbb{P}^2$.

More formally, a polynomial $F(X, Y, Z)$ is called a *homogeneous polynomial of degree d* if it satisfies the identity

$$F(tX, tY, tZ) = t^d F(X, Y, Z).$$

This identity is equivalent to the statement that F is a linear combination of monomials $X^i Y^j Z^k$ with $i + j + k = d$. We define a *projective curve C* in the projective plane $\mathbb{P}^2$ to be the set of solutions to a polynomial equation

$$C : F(X, Y, Z) = 0,$$

where F is a non-constant homogeneous polynomial. We also call C an *algebraic curve*, or sometimes just a *curve* if it is clear that we are working in $\mathbb{P}^2$. The *degree of the curve C* is the degree of the polynomial F. For example,

$$C_1 : X^2 + Y^2 - Z^2 = 0 \quad \text{and} \quad C_2 : Y^2 Z - X^3 - XZ^2 = 0$$

are projective curves, where C_1 has degree 2 and C_2 has degree 3.

In order to check whether a point $P \in \mathbb{P}^2$ is on the curve C, we can take any homogeneous coordinates $[a, b, c]$ for P and check whether $F(a, b, c) = 0$. This is true because any other homogeneous coordinates for P look like $[ta, tb, tc]$ for some non-zero t. Then $F(a, b, c)$ and $F(ta, tb, tc) = t^d F(a, b, c)$ are either both zero or both non-zero.

This tells us what a projective curve is when we use the definition of $\mathbb{P}^2$ by homogeneous coordinates. It is very illuminating to relate this to the description of $\mathbb{P}^2$ as $\mathbb{A}^2 \cup \mathbb{P}^1$ where $\mathbb{A}^2$ is the usual affine plane, and the points at infinity, i.e., the points in $\mathbb{P}^1$, correspond to the directions in $\mathbb{A}^2$. Let $C \subset \mathbb{P}^2$ be a curve given by a homogeneous polynomial of degree d,

$$C : F(X, Y, Z) = 0.$$

If $P = [a, b, c] \in C$ is a point of C with $c \neq 0$, then according to the identification $\mathbb{P}^2 \leftrightarrow \mathbb{A}^2 \cup \mathbb{P}^1$ described in Table A.1 in Section A.1, the point $P \in C \subset \mathbb{P}^2$ corresponds to the point

$$\left(\frac{a}{c}, \frac{b}{c}\right) \in \mathbb{A}^2 \subset \mathbb{A}^2 \cup \mathbb{P}^1.$$

On the other hand, combining $F(a, b, c) = 0$ with the fact that F is homogeneous of degree d shows that

$$0 = \frac{1}{c^d} F(a, b, c) = F\left(\frac{a}{c}, \frac{b}{c}, 1\right).$$

In other words, if we define a new, non-homogeneous, polynomial $f(x, y)$ by the formula

$$f(x, y) = F(x, y, 1),$$

then we get a map

$$\{[a, b, c] \in C : c \neq 0\} \longrightarrow \{(x, y) \in \mathbb{A}^2 : f(x, y) = 0\},$$
$$[a, b, c] \longmapsto (a/c, b/c).$$

And it is easy to see that this map is one-to-one and onto, since if $(r, s) \in \mathbb{A}^2$ satisfies the equation $f(x, y) = 0$, then clearly $[r, s, 1] \in C$. We call the curve $f(x, y) = 0$ the *affine part* of the projective curve C.

It remains to look at the points $[a, b, c] \in C$ with $c = 0$ and describe them geometrically in terms of the affine part of C. The points $[a, b, 0]$ on C satisfy the equation $F(X, Y, 0) = 0$, and they are sent to points at infinity $[a, b] \in \mathbb{P}^1$ in $\mathbb{A}^2 \cup \mathbb{P}^1$. We claim that these points, which recall are really directions in $\mathbb{A}^2$, correspond to the limiting tangent directions of the affine curve $f(x, y) = 0$ as we move along the affine curve out to infinity. In other words, and this is really the intuition to keep in mind, an affine curve $f(x, y)$ is somehow "missing" some points that lie out at infinity, and the points that are missing are the limiting directions as one moves along the curve out toward infinity.

Rather than giving a general proof we illustrate the idea with two examples. First we consider the line

$$L : \alpha X + \beta Y + \gamma Z = 0,$$

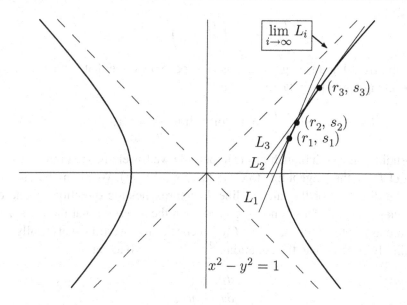

Figure A.2: Points at infinity are limits of tangent directions

say with $\alpha \neq 0$. The affine part of L is the line $L_0 : \alpha x + \beta y + 1 = 0$ in $\mathbb{A}^2$. The points at infinity on L correspond to the points with $Z = 0$. There is only one such point, namely $[-\beta, \alpha, 0]$, which corresponds to the point at infinity $[-\beta, \alpha] \in \mathbb{P}^1$, which in turn corresponds to the direction $-\beta y = \alpha x$ in $\mathbb{A}^2$. This direction is exactly the direction of the line L_0. Thus L consists of the affine line L_0, together with the single point at infinity corresponding to the direction of L_0.

Next we look at the projective curve

$$C : X^2 - Y^2 - Z^2 = 0.$$

There are two points on C with $Z = 0$, namely $[1, 1, 0]$ and $[1, -1, 0]$. These two points correspond, respectively, to the points at infinity $[1, 1], [1, -1] \in \mathbb{P}^1$, or equivalently to the directions $y = x$ and $y = -x$ in $\mathbb{A}^2$. The affine part of C is the hyperbola

$$C_0 : x^2 - y^2 - 1 = 0.$$

Suppose that we take a sequence of points $(r_1, s_1), (r_2, s_2), \ldots$ on C_0 such that these points tend toward infinity along one of the branches of the hyperbola. (Note that there are four choices of direction, since we can let $r_i \to \infty$ or $r_i \to -\infty$, and similarly $s_i \to \infty$ or $s_i \to -\infty$.) If we rewrite $r_i^2 - s_i^2 - 1 = 0$ as

$$\left(\frac{r_i}{s_i} - 1\right)\left(\frac{r_i}{s_i} + 1\right) = \frac{1}{s_i^2},$$

then the right-hand side goes to 0 as $i \to \infty$. So we see that if we travel out to ∞ along the hyperbola, then either

$$\lim_{i \to \infty} \frac{r_i}{s_i} = 1 \quad \text{or} \quad \lim_{i \to \infty} \frac{r_i}{s_i} = -1,$$

depending on which branch of the hyperbola we travel on; see Figure A.2.

Let L_i be the tangent line to C_0 at the point (r_i, s_i). We claim that as $i \to \infty$, the direction of the tangent line L_i approaches the direction of one of the lines $y = \pm x$. This is nothing more than the assertion that the lines $y = \pm x$ are asymptotes for the curve C_0. To check this assertion analytically, we implicitly differentiate the equation $x^2 - y^2 - 1 = 0$ to get

$$\frac{dy}{dx} = \frac{x}{y},$$

and hence

$$(\text{slope of } L_i) = \left(\text{slope of } C_0 \text{ at } (r_i, s_i)\right) = \frac{r_i}{s_i} \xrightarrow[i \to \infty]{} \pm 1.$$

The preceding discussion shows that if we start with a projective curve $C : F(X, Y, Z) = 0$, then we can write C as the union of its affine part C_0 and its points at infinity. Here C_0 is the affine curve given by the equation

$$C_0 : f(x, y) = F(x, y, 1) = 0,$$

and the points at infinity are the points with $Z = 0$, which correspond to the limiting directions of the tangent lines to C_0. The process of replacing the homogeneous polynomial $F(X, Y, Z)$ by the inhomogeneous polynomial $f(x, y) = F(x, y, 1)$ is called *dehomogenization* (*with respect to the variable Z*). We would now like to reverse this process.

Thus suppose that we begin with an affine curve C_0 given by an equation $f(x, y) = 0$. We want to find a projective curve C whose affine part is C_0, or equivalently, we want to find a homogeneous polynomial $F(X, Y, Z)$ so that $F(x, y, 1) = f(x, y)$. This is easy to do, although we want to be careful not to also include the line at infinity in our curve. If we write the polynomial $f(x, y)$ as $\sum a_{ij} x^i y^j$, then the *degree of f* is defined to be the largest value of $i + j$ for which the coefficient a_{ij} is not zero. For example,

$$\deg(x^2 + xy + x^2 y^2 + y^3) = 4 \quad \text{and} \quad \deg(y^2 - x^3 - ax^2 - bx - c) = 3.$$

Then the *homogenization* of a polynomial $f(x, y) = \sum a_{ij} x^i y^j$ of degree d is defined to be

$$F(X, Y, Z) = \sum_{i,j} a_{ij} X^i Y^j Z^{d-i-j}.$$

It is clear from this definition that F is homogeneous of degree d and that $F(x, y, 1) = f(x, y)$. Further, our choice of d ensures that $F(X, Y, 0)$ is not identically zero, so the curve defined by $F(X, Y, Z) = 0$ does not contain the entire line at infinity. Thus using homogenization and dehomogenization, we obtain a one-to-one correspondence between affine curves and projective curves that do not contain the line at infinity.

We should also mention that there is nothing sacred about the variable Z. We could just as well dehomogenize a curve $F(X, Y, Z)$ with respect to one of the other variables, say Y, to get an affine curve $F(x, 1, z) = 0$ in the affine xz-plane. It is sometimes convenient to do this if we are especially interested in one of the points at infinity on the projective curve C. In essence, what we are doing is taking a different line, in this case the line $Y = 0$, and making it into the "line at infinity." An example should make this clearer. Suppose that we want to study the curve

$$C : Y^2 Z - X^3 - Z^3 = 0 \quad \text{and the point} \quad P = [0, 1, 0] \in C.$$

If we dehomogenize with respect to Z, then the point P becomes a point at infinity on the affine curve $y^2 - x^3 - 1 = 0$. So instead we dehomogenize with respect to Y, which means setting $Y = 1$. We then get the affine curve

$$z - x^3 - z^3 = 0,$$

and the point P becomes the point $(x, z) = (0, 0)$. In general, by taking different lines to be the line at infinity, we can break a projective curve C up into a lot of overlapping affine parts, and then these affine parts can be "glued" together to form the entire projective curve.

Up to now we have been working with polynomials without worrying overmuch about what the coefficients of our polynomials look like, and similarly we've talked about solutions of polynomial equations without specifying what sorts of solutions we mean. Classical algebraic geometry is concerned with describing the complex solutions to systems of polynomial equations, but in studying number theory, we are more interested in finding solutions whose coordinates are in non-algebraically closed fields such as $\mathbb{Q}$, or even in rings such as $\mathbb{Z}$. That being the case, it makes sense to look at curves given by polynomial equations with rational or integer coefficients.

We call a curve C *rational* if it is the set of zeros of a polynomial having rational coefficients.[1] Note that the solutions of the equation $F(X, Y, Z) = 0$ and the equation $cF(X, Y, Z) = 0$ are the same for any non-zero c. This allows us to clear the denominators of the coefficients, so a rational curve is in fact the set of zeros of a polynomial with integer coefficients. All of the examples given above are rational curves, since their equations have integer coefficients.

Let C be a projective curve that is rational, say C is given by an equation $F(X, Y, Z) = 0$ for a homogeneous polynomial F having rational coefficients. The *set of rational points on* C, which we denote by $C(\mathbb{Q})$, is the set of points of C having rational coordinates,

$$C(\mathbb{Q}) = \{[a, b, c] \in \mathbb{P}^2 : F(a, b, c) = 0 \text{ and } a, b, c \in \mathbb{Q}\}.$$

Note that if $P = [a, b, c]$ is in $C(\mathbb{Q})$, it is not necessary that a, b, c themselves be rational, since a point P has many different homogeneous coordinates. All that one can say is that $[a, b, c] \in C$ is a rational point of C if and only if there is a non-zero number t so that ta, tb, and tc are all in $\mathbb{Q}$.

Similarly, if C_0 is an affine curve that is rational, say $C_0 : f(x, y) = 0$, then the set of rational points on C_0, denoted $C_0(\mathbb{Q})$, consists of all $(r, s) \in C_0$ with $r, s \in \mathbb{Q}$. It is easy to see that if C_0 is the affine piece of a projective curve C, then $C(\mathbb{Q})$ consists of $C_0(\mathbb{Q})$, together with those points at infinity that happen to be rational. Some of the most famous theorems in number theory involve the set of rational points $C(\mathbb{Q})$ on certain curves. For example, the N'th Fermat curve C_N is the projective curve

$$C_N : X^N + Y^N = Z^N,$$

and Wiles' theorem (Fermat's last theorem) says that $C_N(\mathbb{Q})$ consists of only those points with one of X, Y, or Z equal to zero.

The theory of Diophantine equations also deals with integer solutions of polynomial equations. Let C_0 be an affine curve that is rational, say given by an equation $f(x, y) = 0$. We define the *set of integer points of* C_0, which we denote $C_0(\mathbb{Z})$, to be the set of points of C_0 having integer coordinates,

$$C_0(\mathbb{Z}) = \{(r, s) \in \mathbb{A}^2 : f(r, s) = 0 \text{ and } r, s \in \mathbb{Z}\}.$$

---

[1]We must warn the reader than this terminology is non-standard. In the usual language of algebraic geometry, a curve is called rational if it is birationally isomorphic to the projective line $\mathbb{P}^1$, and a curve given by polynomials with rational coefficients is said to be *defined over* $\mathbb{Q}$.

Why do we only talk about integer points on affine curves and not on projective curves? The answer is that for a projective curve, the notions of integer point and rational point coincide. Here we might say that a point $[a, b, c] \in \mathbb{P}^2$ is an integer point if its coordinates are integers. But if $P \in \mathbb{P}^2$ is any point that is given by homogeneous coordinates $P = [a, b, c]$ that are rational, then we can find an integer t to clear the denominators of a, b, c, and so $P = [ta, tb, tc]$ also has homogeneous coordinates that are integers. So for a projective curve C we would have $C(\mathbb{Q}) = C(\mathbb{Z})$.

It is also possible to look at polynomial equations and their solutions in rings and fields other than $\mathbb{Z}$ or $\mathbb{Q}$ or $\mathbb{R}$ or $\mathbb{C}$. For example, one might look at polynomials with coefficients in the finite field $\mathbb{F}_p$ with p elements and ask for solutions whose coordinates are also in the field $\mathbb{F}_p$. You may worry about your geometric intuitions in situations like this. How can one visualize points and curves and directions in $\mathbb{A}^2$ when the points of $\mathbb{A}^2$ are pairs (x, y) with $x, y \in \mathbb{F}_p$? There are two answers to this question. The first and most reassuring is that you can continue to think of the usual Euclidean plane, i.e., $\mathbb{R}^2$, and most of your geometric intuitions concerning points and curves will still be true when you switch to coordinates in $\mathbb{F}_p$. The second and more practical answer is that the affine and projective planes and affine and projective curves are defined algebraically in terms of ordered pairs (r, s) or homogeneous triples $[a, b, c]$ without any reference to geometry. So in proving things one can work algebraically using coordinates, without worrying at all about geometric intuitions. We might summarize this general philosophy as:

$$\boxed{\textit{Think Geometrically, Prove Algebraically}}$$

One of the fundamental questions answered by the differential calculus is that of finding the tangent line to a curve. If $C : f(x, y) = 0$ is an affine curve, then implicit differentiation gives the relation

$$\frac{\partial f}{\partial x} + \frac{\partial f}{\partial y} \frac{dy}{dx} = 0.$$

So if $P = (r, s)$ is a point on C, the tangent line to C at P is given by the equation

$$\frac{\partial f}{\partial x}(r, s)(x - r) + \frac{\partial f}{\partial y}(r, s)(y - s) = 0.$$

This is the answer provided by elementary calculus. But we clearly have a problem if both partial derivatives are 0. For example, this happens for each of the curves

$$C_1 : y^2 = x^3 + x^2 \quad \text{and} \quad C_2 : y^2 = x^3$$

at the point $P = (0,0)$. If we sketch these curves, we see that they look at bit strange at P; see Figures 1.13 and 1.15 in Section 1.3. The curve C_1 crosses over itself at P, so it has two distinct tangent directions there. The curve C_2, on the other hand, has a cusp at P, which means that it comes to a sharp point at P. We say that P is a *singular point* of the curve $C : f(x,y) = 0$ if

$$\frac{\partial f}{\partial x}(P) = \frac{\partial f}{\partial y}(P) = 0.$$

We call P a *non-singular point* if it is not singular, i.e., if at least one of the partial derivatives does not vanish, and we say that C is a *non-singular curve* (or a *smooth curve*) if every point of C is non-singular. If $P = (r,s)$ is a non-singular point of C, then we define the *tangent line to C at P* to be the line

$$\frac{\partial f}{\partial x}(r,s)(x - r) + \frac{\partial f}{\partial y}(r,s)(y - s) = 0,$$

as discussed above.

For a projective curve $C : F(X,Y,Z) = 0$ described by a homogeneous polynomial, we make similar definitions. More precisely, if $P = [a,b,c]$ is a point on C with $c \neq 0$, then we go to the affine part of C and check whether the point

$$P_0 = \left(\frac{a}{c}, \frac{b}{c}\right) \quad \text{is singular on the affine curve} \quad C_0 : F(x,y,1) = 0.$$

And if $c = 0$, then we can dehomogenize in some other way. For example, if $a \neq 0$, then we check whether the point

$$P_0 = \left(\frac{b}{a}, \frac{c}{a}\right) \quad \text{is singular on the affine curve} \quad C_0 : F(1,y,z) = 0.$$

We say that C is non-singular (or smooth) if all of its points, including the points at infinity, are non-singular. If P is a non-singular point of C, we define the tangent line to C at P by dehomogenizing, finding the tangent line to the affine part of C at P, and then homogenizing the equation of the tangent line to get a line in $\mathbb{P}^2$. (An alternative method to check for singularities and find tangent lines on projective curves is described in Exercise A.5.)

When one is faced with a complicated equation, it is natural to try to make a change of variables in order to simplify it. Probably the first significant example of this that you have seen is the process of completing the square to solve a quadratic equation. Thus to solve $Ax^2 + Bx + C = 0$, we multiply by $4A$ and rewrite the equation as

$$(2Ax + B)^2 + 4AC - B^2 = 0.$$

This suggest the substitution $x' = 2Ax + B$, and then we can solve

$$x'^2 + 4AC - B^2 = 0 \quad \text{to get} \quad x' = \pm\sqrt{B^2 - 4AC}.$$

The crucial final step uses the fact that our substitution is invertible, so we can solve for x in terms of x' to obtain the usual quadratic formula

$$x = \frac{-B + x'}{2A} = \frac{-B \pm \sqrt{B^2 - 4AC}}{2A}.$$

More generally, suppose that we are given a projective curve of degree d, say defined by an equation $C : F(X, Y, Z) = 0$. In order to change coordinates on $\mathbb{P}^2$, we make a substitution

$$\begin{aligned} X &= m_{11}X' + m_{12}Y' + m_{13}Z', \\ Y &= m_{21}X' + m_{22}Y' + m_{23}Z', \\ Z &= m_{31}X' + m_{32}Y' + m_{33}Z'. \end{aligned} \qquad (*)$$

Then we get a new curve C' given by the equation $F'(X', Y', Z') = 0$, where F' is the polynomial

$$\begin{aligned} F'(X', Y', Z') = F(&m_{11}X' + m_{12}Y' + m_{13}Z', \\ &m_{21}X' + m_{22}Y' + m_{23}Z', m_{31}X' + m_{32}Y' + m_{33}Z') \end{aligned}$$

The change of coordinates $(*)$ gives a map from C' to C, that is, given a point $[a', b', c'] \in C'$, we substitute $X' = a$, $Y' = b$, and $Z' = c$ into $(*)$ to get a point $[a, b, c] \in C$. Further, this map $C' \to C$ has an inverse provided that the matrix $M = (m_{ij})_{1 \le i, j \le 3}$ is invertible. More precisely, if $M^{-1} = N = (n_{ij})$, then the change of coordinates

$$\begin{aligned} X' &= n_{11}X + n_{12}Y + n_{13}Z, \\ Y' &= n_{21}X + n_{22}Y + n_{23}Z, \\ Z' &= n_{31}X + n_{32}Y + n_{33}Z, \end{aligned}$$

maps C to C' We call a change of coordinates on $\mathbb{P}^2$ given by an invertible 3×3 matrix a *projective transformation*. Note that if the matrix has rational coefficients, then the corresponding projective transformation gives a one-to-one correspondence between $C(\mathbb{Q})$ and $C'(\mathbb{Q})$. So the number theoretic problem of finding the rational points on the curve C is equivalent to the problem of finding the rational points on the curve C'.

A.3   Intersections of Projective Curves

Recall that our geometric construction of the projective plane was based on the desire that every pair of distinct lines should intersect in exactly one point. In this section we are going to discuss the intersection of curves of higher degree.

How many intersection points should two curves have? Let's begin with a thought experiment, and then we'll consider some examples and see to what extent our intuition is correct. Let C_1 be an affine curve of degree d_1 and let C_2 be an affine curve of degree d_2. Thus C_1 and C_2 are given by polynomials

$$C_1 : f_1(x,y) = 0 \qquad \text{with } \deg(f_1) = d_1,$$
$$C_2 : f_2(x,y) = 0 \qquad \text{with } \deg(f_2) = d_2.$$

The points in the intersection $C_1 \cap C_2$ are solutions to the simultaneous equations

$$f_1(x,y) = f_2(x,y) = 0.$$

Suppose now that we consider f_1 as a polynomial in the variable y whose coefficients are polynomials in x. Then $f_1(x,y) = 0$, being a polynomial of degree d_1 in y, should in principle have d_1 roots $y_1, \ldots, y_d$. Now we substitute each of these roots into the second equation $f_2(x,y)$ to find d_1 equations for x, namely

$$f_2(x,y_1) = 0, \quad f_2(x,y_2) = 0, \quad \ldots \quad f_2(x,y_{d_1}) = 0.$$

Each of these equations is a polynomial in x of degree d_2, so in principle each equation should yield d_2 values for x. Altogether we appear to get $d_1 d_2$ pairs (x,y) that satisfy $f_1(x,y) = f_2(x,y) = 0$, which seems to indicate that we should have $\#(C_1 \cap C_2) = d_1 d_2$. For example, a curve of degree 2 and a curve of degree 4 should intersect in 8 points, as illustrated in Figure A.3. This assertion, that curves of degree d_1 and d_2 intersect in $d_1 d_2$ points, is indeed true provided that it is interpreted properly. However, matters are considerably more complicated than they appear at first glance, as will be clear from the following examples. [Can you find all of the ways in which our plausibility argument fails to be a valid proof? For example, the "roots" $y_1, \ldots, y_d$ really depend on x, so we should write $f_2\bigl(x, y_i(x)\bigr) = 0$, and then it is not at all clear how many roots we should expect.]

Curves of degree one are lines, and curves of degree two are called *conics* (short for conic sections). We already know that two lines in $\mathbb{P}^2$ intersect in

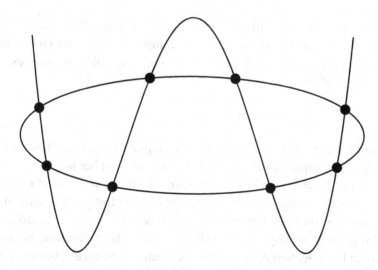

Figure A.3: Curves of degree two and degree four intersect in eight points

a unique point, so the next simplest case is the intersection of a line and a conic. Our discussion above leads us to expect two intersection points, so we look at some examples to see what really happens. The (affine) line and conic

$$C_1 : x + y + 1 = 0 \quad \text{and} \quad C_2 : x^2 + y^2 = 1$$

intersect in the two points $(-1, 0)$ and $(0, -1)$, as is easily seen by substituting $y = -x - 1$ into the equation for C_2 and solving the resulting quadratic equation for x; see Figure A.4(a). Similarly,

$$C_1 : x + y = 0 \quad \text{and} \quad C_2 : x^2 + y^2 = 1$$

intersect in the two points $(\frac{1}{2}\sqrt{2}, -\frac{1}{2}\sqrt{2})$ and $(-\frac{1}{2}\sqrt{2}, \frac{1}{2}\sqrt{2})$. Note that we have to allow real coordinates for the intersection points, even though C_1 and C_2 are rational curves; see Figure A.4(b).

What about the intersection of the line and conic

$$C_1 : x + y + 2 = 0 \quad \text{and} \quad C_2 : x^2 + y^2 = 1?$$

They do not intersect at all in the usual Euclidean plane $\mathbb{R}^2$, as illustrated in Figure A.4(c), but if we allow complex numbers then we again find two intersection points,

$$\left(-1 + \frac{\sqrt{2}}{2}i, -1 - \frac{\sqrt{2}}{2}i\right) \quad \text{and} \quad \left(-1 - \frac{\sqrt{2}}{2}i, -1 + \frac{\sqrt{2}}{2}i\right).$$

Of course, it is reasonable to allow complex coordinates, since even for polynomials of one variable we need to use complex numbers to ensure that a polynomial of degree d actually has d roots counted with multiplicities.

Next we look at

$$C_1 : x + 1 = 0 \quad \text{and} \quad C_2 : x^2 - y = 0.$$

These curves appear to intersect in the single point $(-1, 1)$ as shown in Figure A.4(d), but appearances can be deceiving. Remember that even for two lines, we may need to also look at the points at infinity in $\mathbb{P}^2$. In our case, the line C_1 is in the vertical direction, and the tangent lines to the parabola C_2 approach the vertical direction, so geometrically C_1 and C_2 should have a common point at infinity corresponding to the vertical direction. Following our maxim from Section A.2, we now check this assertion algebraically. First we homogenize the equations for C_1 and C_2 to get the corresponding projective curves

$$\tilde{C}_1 : X + Z = 0 \quad \text{and} \quad \tilde{C}_2 : X^2 - YZ = 0.$$

Then $\tilde{C}_1 \cap \tilde{C}_2$ consists of the two points $[-1, 1, 1]$ and $[0, 1, 0]$, as may be seen by substituting $X = -Z$ into the equation for $\tilde{C}_2$. So we get the expected two points provided that we work with projective curves.

All of this looks very good, but the next example illustrates another problem that may occur. Consider the intersection of the line and conic

$$C_1 : x + y = 2 \quad \text{and} \quad C_2 : x^2 + y^2 = 2;$$

see Figure A.4(e). Then $C_1 \cap C_2$ consists of the single point $(1, 1)$, and even if we go to projective curves

$$\tilde{C}_1 : X + Y = 2Z \quad \text{and} \quad \tilde{C}_2 : X^2 + Y^2 = 2Z^2,$$

we still find the single intersection point $[1, 1, 1]$. What is wrong?

Geometrically we immediately see the problem, namely the line C_1 is tangent to the circle C_2 at the point $(1, 1)$, so in some sense that point should count double. We can also see this algebraically. If we substitute the relation $y = 2 - x$ from C_1 into the equation for C_2 and simplify, we get the equation $2x^2 - 4x + 2 = 0$, or equivalently $2(x - 1)^2 = 0$. So we do have a quadratic equation to solve for x, and normally we would expect to find two distinct roots, but in this case we happen to find one root repeated twice. This makes sense, since even a degree d polynomial of one variable can only be

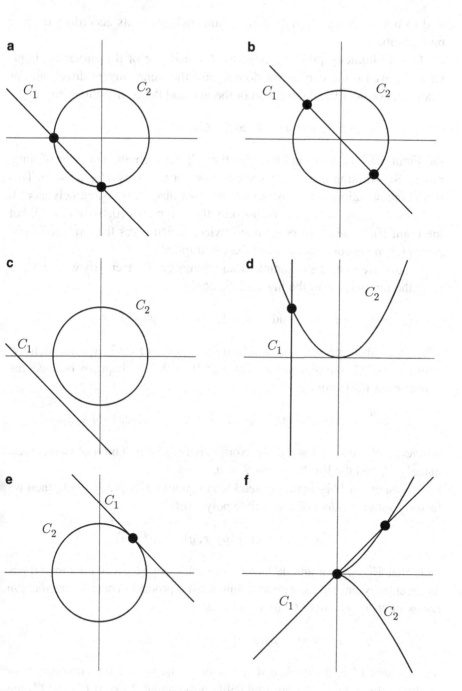

Figure A.4: Some of the ways in which curves may intersect

said to have d complex roots if we count multiple roots according to their multiplicities.

This multiplicity problem may also occur if one of the curves is singular at P, even if the two curves do not have the same tangent direction. For example, consider the intersection of the line and the degree three curve

$$C_1 : x - y = 0 \quad \text{and} \quad C_2 : x^3 - y^2 = 0;$$

see Figure A.4(f). Our intuition says that $C_1 \cap C_2$ should consider of three points. Substituting $y = x$ into the equation for C_2 gives $x^3 - x^2 = 0$. This is a cubic equation for x, but it has only two distinct roots, namely $x = 0$ and $x = 1$. Thus $C_1 \cap C_2$ contains only the two points $(0,0)$ and $(1,1)$, but the point $(0,0)$ needs to be counted twice, which gives the expected three points when we count points with their multiplicity.

Finally, we look at an example where things go spectacularly wrong. Consider the intersection of the line and the conic

$$C_1 : x + y + 1 = 0 \quad \text{and} \quad C_2 : 2x^2 + xy - y^2 + 4x + y + 2 = 0.$$

When we substitute $y = -x - 1$ into the equation for C_2, we find that everything cancels out and we are left with $0 = 0$. This happens because the equation for C_2 factors as

$$2x^2 + xy - y^2 + 4x + y + 2 = (x + y + 1)(2x - y + 2),$$

so every point on C_1 lies on C_2. Notice that C_2 is the union of two curves, namely C_1 and the line $2x - y + 2 = 0$.

In general, if C is a curve given by an equation $C : f(x, y) = 0$, then we factor f into a product of irreducible polynomials

$$f(x, y) = p_1(x, y)p_2(x, y) \cdots p_n(x, y).$$

Note that $\mathbb{C}[x, y]$ is a unique factorization domain, so every polynomial has an essentially unique factorization into such a product. Then the *irreducible components of the curve* C are the curves

$$p_1(x, y) = 0, \quad p_2(x, y) = 0, \quad \cdots \quad p_n(x, y) = 0.$$

We say that C is *irreducible* if it has only one irreducible component, or equivalently, if $f(x, y)$ is an irreducible polynomial. Next, if C_1 and C_2 are two curves, we say that C_1 and C_2 *have no common components* if their irreducible components are distinct. It is not hard to prove that $C_1 \cap C_2$ consists of a finite set of points if and only if C_1 and C_2 have no common components. Finally, if we work instead with projective curves C, C_1, C_2, then we make the same definitions using factorizations into products of irreducible homogeneous polynomials in $\mathbb{C}[X, Y, Z]$.

We now consider the general case of projective curves C_1 and C_2, which we assume to have no common components. The intersection $C_1 \cap C_2$ is then a finite set of points with complex coordinates. To each point $P \in \mathbb{P}^2$ we assign a *multiplicity* or *intersection index* $I(C_1 \cap C_2, P)$. This is a non-negative integer reflecting the extent to which C_1 and C_2 are tangent to one another at P or are not smooth at P. We give a formal definition in Section A.4, but one can get a good feeling for the intersection index from the following properties:

(i) If $P \notin C_1 \cap C_2$, then $I(C_1 \cap C_2, P) = 0$.

(ii) If $P \in C_1 \cap C_2$, if P is a non-singular point of C_1 and C_2, and if C_1 and C_2 have different tangent directions at P, then $I(C_1 \cap C_2, P) = 1$. In this case, one says that C_1 and C_2 intersect *transversally at P*.

(iii) If $P \in C_1 \cap C_2$ and if C_1 and C_2 do not intersect transversally at P, then $I(C_1 \cap C_2, P) \geq 2$.

With these preliminaries, we are now ready to formally state the theorem that justifies the plausibility argument that we gave at the beginning of this section.

Theorem A.1 (Bezout's Theorem). *Let C_1 and C_2 be projective curves with no common components. Then*

$$\sum_{P \in C_1 \cap C_2} I(C_1 \cap C_2, P) = (\deg C_1)(\deg C_2),$$

where the sum is over all points of $C_1 \cap C_2$ having complex coordinates. In particular, if C_1 and C_2 are smooth curves with only transversal intersections, then $\#(C_1 \cap C_2) = (\deg C_1)(\deg C_2)$, and in all cases there is an inequality

$$\#(C_1 \cap C_2) \leq (\deg C_1)(\deg C_2).$$

Proof. We give the proof of Bezout's theorem in Section A.4 $\qquad\square$

It would be hard to overestimate the importance of Bezout's theorem in the study of projective geometry. We should stress how amazing a theorem it is. The projective plane was constructed so as to ensure that any two lines, i.e., curves of degree one, intersect in exactly one point, so one could say that the projective plane is formed by taking the affine plane and adding just enough points to make Bezout's theorem true for curves of degree one. It then turns out that the projective plane has enough points to make Bezout's theorem true for all projective curves!

Sometimes Bezout's theorem is used to determine if two curves are the same, or at least have a common component. For example, if C_1 and C_2 are conics, and if C_1 and C_2 have five points in common, then Bezout's theorem tells us that they have a common component. Since the degree of a component can be no larger than the degree of the curve, it follows that there is some line L contained in both C_1 and C_2, or else $C_1 = C_2$. Thus there is only one conic going through any five given points as long as no three of them are collinear. This is analogous to the fact that there is a unique line going through two given points. More generally, one see from Bezout's theorem that if C_1 and C_2 are irreducible curves of degree d with $d^2 + 1$ points in common, then $C_1 = C_2$. Note, however, that for $d \geq 3$, there is in general no curve of degree d going through $d^2 + 1$ preassigned points. This is because the number $d^2 + 1$ of conditions to be met is greater than the number $(d+1)(d+2)/2$ of unknown coefficients of a homogeneous polynomial of degree d.

We now want to consider a slightly more complicated situation. Suppose that C_1 and C_2 are two cubic curves of degree 3, which intersect in 9 distinct points $P_1, \ldots, P_9$. Suppose further that D is another cubic curve that happens to go through the first 8 points $P_1, \ldots, P_8$. We claim that D also goes through the ninth point P_9. To see why this is true, we consider the collection of all cubic curves in $\mathbb{P}^2$, which we denote by $\mathcal{C}^{(3)}$. An element $C \in \mathcal{C}^{(3)}$ is given by a homogeneous equation

$$C : aX^3 + bX^2Y + cXY^2 + dY^3 + eX^2Z + fXYZ$$
$$+ gY^2Z + hXZ^2 + iYZ^2 + jZ^3 = 0,$$

so C is determined by the ten coefficients $a, b, \ldots, j$. Of course, if we multiply the equation for C by any non-zero constant, then we get the same curve, so really C is determined by the homogeneous 10-tuple $[a, b, \ldots, j]$. Conversely, if two 10-tuples give the same curve, then they differ by multiplication by a constant. In other words, the set of cubic curve $\mathcal{C}^{(3)}$ is in a very natural way isomorphic to the projective space $\mathbb{P}^9$.

Suppose that we are given a point $P \in \mathbb{P}^2$ and ask for all cubic curves that go through P. This describes a certain subset of $\mathcal{C}^{(3)} \cong \mathbb{P}^9$, and it is easy to see what this subset is. If P has homogeneous coordinates $P = [X_0.Y_0, Z_0]$, then substituting P into the equation for C shows that C contains P if and only if the 10-tuple $[a, b, \ldots, j]$ satisfies the homogeneous linear equation

$$(X_0^3)a + (X_0^2 Y_0)b + (X_0 Y_0^2)c + (Y_0^3)d + (X_0^2 Z_0)e + (X_0 Z_0^2)f$$
$$+ (Y_0^2 Z_0)g + (Y_0 Z_0^2)h + (Z_0^3)i + (X_0 Y_0 Z_0)j = 0.$$

N.B., this is a linear equation in the 10 variables $a, b, \ldots, j$. In other words, for a given point $P \in \mathbb{P}^2$, the set of cubic curves $C \in \mathcal{C}^{(3)}$ that contain P corresponds to the zeros of a homogeneous linear equation in $\mathbb{P}^9$.

Similarly, if we fix two points $P, Q \in \mathbb{P}^2$, then the set of cubic curves $C \in \mathcal{C}^{(3)}$ containing both P and Q is given by the common solutions of two linear equations in $\mathbb{P}^9$, where one linear equation is specified by P and the other by Q. Continuing in this fashion, we find that for a collection of n points $P_1, \ldots, P_n \in \mathbb{P}^2$, there is a one-to-one correspondence between the two sets

$$\{C \in \mathcal{C}^{(3)} : P_1, \ldots, P_n \in C\} \quad \text{and} \quad \left\{ \begin{array}{c} \text{simultaneous solutions of a} \\ \text{certain system of } n \text{ homo-} \\ \text{geneous linear equations in } \mathbb{P}^9 \end{array} \right\}.$$

For example, suppose that we take $n = 9$. The solutions to a system of 9 homogeneous linear equations in 10 variables generally consists of the multiples of a single solution. In other words, if v_0 is a non-zero solution, then every solution will have the form λv_0 for some constant λ. Now let

$$C_1 : F_1(X, Y, Z) = 0 \quad \text{and} \quad C_2 : F_2(X, Y, Z) = 0$$

be cubic curves in $\mathbb{P}^2$, each going through the given nine points. The coefficients of F_1 and F_2 are then 10-tuples that are solutions to the given system of linear equations, so we conclude that $F_1 = \lambda F_2$, and hence that $C_1 = C_2$. Thus we find that, in general, there is exactly one cubic curve in $\mathbb{P}^2$ that passes through nine given points. Note, however, that for special sets of nine points it is possible to have a one parameter family of cubic curves going through them.

That is the situation in our original problem, to which we now return. Namely, we take two cubic curves C_1 and C_2 in $\mathbb{P}^2$ that intersect in nine distinct points $P_1, \ldots, P_9$. Let C_1 and C_2 be given by the equations

$$C_1 : F_1(X, Y, Z) = 0 \quad \text{and} \quad C_2 : F_2(X, Y, Z) = 0.$$

We consider the set of all cubic curves $C \in \mathcal{C}^{(3)}$ that pass through the first eight points $P_1, \ldots, P_8$. This set corresponds to the simultaneous solutions of eight homogeneous linear equations in ten variables. The set of solutions of this system consists of all linear combinations of two linearly independent 10-tuples. In other words, if v_1 and v_2 are independent solutions, then every solution has the form $\lambda_1 v_1 + \lambda_2 v_2$ for some constants λ_1 and λ_2.[2]

But we already know two cubic curves passing through the eight points $P_1, \ldots, P_8$, namely C_1 and C_2. The coefficients of their equations F_1 and F_2 thus give two 10-tuples solving the system of eight homogeneous linear equations, so they span the complete solution set. This means that if D is any other cubic curve in $\mathbb{P}^2$ that contains the eight points $P_1, \ldots, P_8$, then the equation for D has the form

$$D : \lambda_1 F_1(X, Y, Z) + \lambda_2 F_2(X, Y, Z) = 0 \quad \text{for some constants } \lambda_1, \lambda_2.$$

But the ninth point P_9 is on both C_1 and C_2, so $F_1(P_9) = F_2(P_9) = 0$. It follows from the equation for D that D also contains the point P_9, which is exactly what we have been trying to demonstrate.

More generally, the following theorem is true.

Theorem A.2 (Cayley–Bacharach Theorem). *Let C_1 and C_2 be curves in $\mathbb{P}^2$ without common components of respective degrees d_1 and d_2, and suppose that C_1 and C_2 intersect in $d_1 d_2$ distinct points. Let D be a curve in $\mathbb{P}^2$ of degree $d_1 + d_2 - 3$. If D passes through all but one of the points of $C_1 \cap C_2$, then D must also pass through the remaining point.*

It is not actually necessary that C_1 and C_2 intersect in distinct points. For example, if $P \in C_1 \cap C_2$ is a point of multiplicity two, say because C_1 and C_2 have the same tangent direction at P, then one needs to require that D also has the same tangent direction at P. The most general result is somewhat difficult to state, so we content ourselves with the following version.

Theorem A.3 (Cubic Cayley–Bacharach Theorem). *Let C_1 and C_2 be cubic curves in $\mathbb{P}^2$ without common components, and assume that C_1 is smooth. Suppose that D is another cubic curve that contains eight of the intersection points of $C_1 \cap C_2$ counting multiplicities. This means that if $C_1 \cap C_2 = \{P_1, \ldots, P_r\}$, then*

$$I(C_1 \cap D, P_i) \geq I(C_1 \cap C_2, P_i) \quad \text{for } 1 \leq i < r,$$

---

[2]In principle, the set of solutions might have dimension greater than two. We leave it as a (challenging) exercise for you to check that because the eight points $P_1, \ldots, P_8$ are distinct, the corresponding linear equations are independent; see Exercise A.17.

and

$$I(C_1 \cap D, P_r) \geq I(C_1 \cap C_2, P_r) - 1.$$

Then D goes through the ninth point of $C_1 \cap C_2$, which in terms of multiplicities means that

$$I(C_1 \cap D, P_r) \geq I(C_1 \cap C_2, P_r).$$

We conclude this section of the appendix by applying the Cayley–Bacharach theorem to prove a beautiful geometric result of Pascal. Let C be a smooth conic, for example, a hyperbola, a parabola, or an ellipse. Choose any six points lying on the conic, say labeled consecutively $P_1, P_2, \ldots, P_6$, and play connect-the-dots to draw a hexagon. Now take the lines through opposite sides of the hexagon and extend them to find the intersection points as illustrated in Figure A.5, say

$$\overleftrightarrow{P_1 P_2} \cap \overleftrightarrow{P_4 P_5} = \{Q_1\}, \quad \overleftrightarrow{P_2 P_3} \cap \overleftrightarrow{P_5 P_6} = \{Q_2\}, \quad \overleftrightarrow{P_3 P_4} \cap \overleftrightarrow{P_6 P_1} = \{Q_3\}.$$

Theorem A.4 (Pascal's Theorem). *The three points Q_1, Q_2, Q_3 described above lie on a line.*

To prove Pascal's theorem, we consider the two cubic curves

$$C_1 = \overleftrightarrow{P_1 P_2} \cup \overleftrightarrow{P_3 P_4} \cup \overleftrightarrow{P_5 P_6} \quad \text{and} \quad C_2 = \overleftrightarrow{P_2 P_3} \cup \overleftrightarrow{P_4 P_5} \cup \overleftrightarrow{P_6 P_1}.$$

Why do we call C_1 and C_2 cubic curves? The answer is that if we choose an equation for the line $\overleftrightarrow{P_j P_j}$, say

$$\alpha_{ij} X + \beta_{ij} Y + \gamma_{ij} Z = 0,$$

then C_1 is given by the homogeneous cubic equation

$$(\alpha_{12} X + \beta_{12} Y + \gamma_{12} Z)(\alpha_{34} X + \beta_{34} Y + \gamma_{34} Z)(\alpha_{56} X + \beta_{56} Y + \gamma_{56} Z) = 0,$$

and similarly for C_2.

Notice that all nine of the points

$$P_1, P_2, P_3, P_4, P_5, P_6, Q_1, Q_2, Q_3 \quad \text{are on both } C_1 \text{ and } C_2.$$

This sets us up to use the Cayley–Bacharach theorem. We take D to be the cubic curve that is the union of our original conic C with the line through Q_1 and Q_2,

$$D = C \cup \overleftrightarrow{Q_1 Q_2}.$$

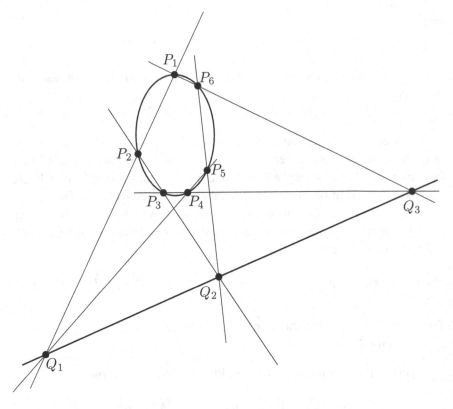

Figure A.5: Pascal's theorem

Clearly D contains the eight points $P_1, P_2, P_3, P_4, P_5, P_6, Q_1, Q_2$. The Cayley–Bacharach theorem then tells us that D contains the ninth point in $C_1 \cap C_2$, namely Q_3. Now Q_3 does not lie on C, since otherwise the line $\overleftrightarrow{P_6 P_1}$ would intersect the conic in the three points P_6, P_1, Q_3, contradicting Bezout's theorem. Therefore Q_3 must be on the line $\overleftrightarrow{Q_1 Q_2}$. In other words, the points Q_1, Q_2, and Q_3 are collinear, which completes the proof of Pascal's theorem.

A.4   Intersection Multiplicities and a Proof of Bezout's Theorem

We give the proof of Bezout's theorem in the form of a long exercise with hints. It is quite elementary. For the first weak inequality, which is all that is needed in many important applications of the theorem, we use only linear algebra and the notion of dimension of a vector space. After that, we need the concepts of commutative ring, ideal, and quotient ring, and the fact that unique factorization holds in polynomial rings, but that is about all.

Let C_1 and C_2 be curves in $\mathbb{P}^2$ of respective degrees n_1 and n_2, without common components. Until the last step of the proof we assume that the line at infinity is not a component of either curve, and we work with affine coordinates x and y. Let

$$C_1 : f_1(x, y) = 0 \quad \text{and} \quad C_2 : f_2(x, y) = 0$$

be the equations for the two curves in the affine plane $\mathbb{A}^2$. The assumptions we have made mean that the polynomials f_1 and f_2 have no common factor and are of degree n_1 and n_2, respectively.

The proof is pure algebra, although the geometric ideas behind it should be apparent, and it works over any algebraically closed field k. The reader is welcome to take $k = \mathbb{C}$, but k could also be an algebraic closure of a finite field $\mathbb{F}_p$, for example. We also note that in this section, $\dim V$ means the dimension of V as a k-vector space.

Let $R = k[x, y]$ be a polynomial ring in two variables, and let $(f_1, f_2) = f_1 R + f_2 R$ be the ideal of R generated by the polynomials f_1 and f_2. The steps in the proof of Bezout's theorem are as follows:

(1) We prove the following two inequalities which, on eliminating the middle term, show that the number of intersection points of C_1 and C_2 in $\mathbb{A}^2$ is at most $n_1 n_2$:

$$\#(C_1 \cap C_2 \cap \mathbb{A}^2) \overset{(A)}{\leq} \dim\left(R/(f_1, f_2)\right) \overset{(B)}{\leq} n_1 n_2.$$

(2) We show that (B) is an equality if C_1 and C_2 do not meet at infinity.

(3) We strengthen (A) to get

$$\sum_{P \in C_1 \cap C_2 \cap \mathbb{A}^2} I(C_1 \cap C_2, P) \overset{(A^+)}{\leq} \dim\left(R/(f_1, f_2)\right),$$

where $I(C_1 \cap C_2, P)$ is a suitably defined *intersection multiplicity* of C_1 and C_2 at P.

(4) We show that (A^+) is in fact an equality.

The fact that k is algebraically closed is not needed for the proofs of the inequalities in (1) and (3), but it is essential for verifying the equalities in (2) and (4). Taken together, (2) and (4) give Bezout's theorem in the case that C_1 and C_2 do not meet at infinity. To get it in general, there is one more step.

(5) We show that the definition of intersection multiplicity does not change when we make a projective transformation, and that there is a line L in $\mathbb{P}^2$ not meeting any intersection point. Changing coordinates so that the line L is the line at infinity, we then get Bezout in general.

    To round out the argument, we include one more segment:

(6) We prove some basic properties satisfied by the intersection multiplicity $I(C_1 \cap C_2, P)$ and show that it depends only on the initial part of the Taylor expansions of f_1 and f_2 at P.

    Now we sketch the proof as a series of exercises with hints, breaking each of the segments (1)–(5) into smaller steps.

(1.1) Let $P_1, P_2, \ldots, P_m$ be m different points in the (x, y)-plane. Show that for each i there is a polynomial $h_i = h_i(x, y)$ such that $h_i(P_i) = 1$ and $h_i(P_j) = 0$ for $j \neq i$. (*Idea.* Construct h_i as a product of linear polynomials, using the fact that for each $j \neq i$ there is a line through P_j not meeting P_i.)

(1.2) Suppose that the m points P_i from (1.1) lie in $C_1 \cap C_2$. Prove that the polynomials h_i are linearly independent modulo (f_1, f_2), and consequently that

$$m \leq \dim\left(R/(f_1, f_2)\right).$$

This proves inequality (A). (*Idea.* Consider a possible dependence

$$c_1 h_1 + c_2 h_2 + \cdots + c_m h_m = g_1 f_1 + g_2 f_2 \in (f_1, f_2)$$

with $c_i \in k$. Substitute P_i into the equation to show that every $c_i = 0$.)

    This takes care of inequality (A). To prove (B), for each integer $d \geq 0$ we define:

$$\phi(d) = \frac{1}{2}(d+1)(d+2) = \frac{1}{2}d^2 + \frac{3}{2}d + 1,$$
$$R_d = (\text{vector space of polynomial } f(x, y) \text{ of degree } \leq d),$$
$$W_d = R_{d-n_1} f_1 + R_{d-n_2} f_2.$$

Thus W_d is the k-vector space of polynomials of the form

$$f = g_1 f_1 + g_2 f_2 \quad \text{with } \deg g_i \leq d - n_i \text{ for } i = 1, 2.$$

Notice that $W_d = 0$ if $d < \max\{n_1, n_2\}$, and in any case, $W_d \subset (f_1, f_2)$.

(1.3) Show that $\dim R_d = \phi(d)$. (*Idea.* One way to see this is to note that

$$\phi(d) - \phi(d-1) = \text{(number of monomials } x^i y^j \text{ of degree } d) = d + 1$$

and use induction on d.)

(1.4) For $d \geq n_1 + n_2$, show that

$$R_{d-n_1} f_1 \cap R_{d-n_2} f_2 = R_{d-n_1-n_2} f_1 f_2.$$

Here is where we use the hypothesis that f_1 and f_2 have no common factor.

(1.5) Prove that for $d \geq n_1 + n_2$,

$$\dim R_d - \dim W_d = \phi(d) - \phi(d-n_1) - \phi(d-n_2) - \phi(d-n_1-n_2) = n_1 n_2.$$

(*Idea.* If f is a non-zero polynomial, then $g \mapsto fg$ defines an isomorphism $R_{d-j} \xrightarrow{\sim} R_{d-j} f$, and hence $\dim R_{d-j} f = \phi(d-j)$. Now use the lemma from linear algebra which says that

$$\dim(U + V) = \dim(U) + \dim(V) - \dim(U \cap V)$$

for subspaces U and V of a finite dimensional vector space.)

(1.6) Prove inequality (B) by showing that if $g_1, g_2, \ldots, g_{n_1 n_2 + 1}$ are elements of R, then they are linearly dependent modulo (f_1, f_2). (*Idea.* Take d so large that the g_j are in R_d and so (1.5) holds. Then use (1.5) to show that there is a non-trivial linear combination $g = \sum c_j g_j$ such that $g \in W_d \subset (f_1, f_2)$.)

This finishes segment (1). For segment (2), we begin by recalling how one computes the intersections of an affine curve $f(x, y) = 0$ with the line at infinity.

(2.1) For each non-zero polynomial $f = f(x, y)$, let f^* denote the homogeneous part of f of highest degree. In other words, if

$$f = \sum_{i,j} c_{ij} x^i y^j \quad \text{has degree } n, \text{ then} \quad f^* = \sum_{i+j=n} c_{ij} x^i y^j.$$

Because k is algebraically closed, we can factor f^* into linear factors,

$$f^*(x, y) = \prod_{i=1}^{n} (a_i x + b_i y) \quad \text{with } a_i, b_i \in k \text{ and } n = \deg f = \deg f^*.$$

Show that the points at infinity on the curve $f(x, y) = 0$ are the points with homogeneous coordinates

$$[X, Y, Z] = [b_i, -a_i, 0].$$

(*Idea.* Put $x = X/Z$, $y = Y/Z$, etc.)

An example should make this clearer. Consider the polynomials

$$f(x, y) = x^4 - x^2 y^2 + 3x^3 + xy^2 + 2y^3 + 2y^2 + 8x + 3,$$
$$f^*(x, y) = x^4 - x^2 y^2 = x^2(x + y)(x - y),$$

each of which has degree 4. The quartic curve $f(x, y) = 0$ thus meets the line at infinity in the points $[0, 1, 0]$, $[1, -1, 0]$, and $[1, 1, 0]$. The fact that x^2 divides $f^*(x, y)$ means that the curve is tangent to the line at infinity at the point $[0, 1, 0]$.

The remaining steps in segment 2 are as follows:

(2.2) If C_1 and C_2 do not meet at infinity, show that f_1^* and f_2^* have no common factor.

(2.3) If f_1^* and f_2^* have no common factor, show that $(f_1, f_2) \cap R_d = W_d$ for all $d \geq n_1 + n_2$.

(2.4) If $(f_1, f_2) \cap R_d = W_d$ and $d \geq n_1 + n_2$, show that

$$\dim\left(R/(f_1, f_2)\right) \geq n_1 n_2.$$

(*Idea.* (2.2) is an easy consequence of (2.1). To do (2.3), we suppose that $f \in (f_1, f_2) \cap R_d$ is written in the form $f = g_1 f_1 + g_2 f_2$ with g_1 and g_2 of smallest possible degree. If $\deg g_1 > d - n_1$, then looking at the terms of highest degree shows that $g_1^* f_1^* + g_2^* f_2^* = 0$. Then use the fact that f_1^* and f_2^* are relatively prime to show that there is an h such that

$$\deg(g_1 + hf_2) < \deg(g_1) \quad \text{and} \quad \deg(g_2 + hf_1) < \deg(g_2).$$

Deduce that $\deg g_i \leq d - n$, and hence that $f \in W_d$. For (2.4), note that by (1.5) there are $n_1 n_2$ elements in R_d that are linearly independent modulo W_d, and that if $(f_1, f_2) \cap R_d = W_d$, then they are linearly independent as elements of R modulo (f_1, f_2). Hence $\dim R/(f_1, f_2) \geq n_1 n_2$.)

To define intersection multiplicity, we introduce the important notion of the *local ring* $\mathcal{O}_P$ of a point $P \in \mathbb{A}^2$. Let $K = k(x, y)$ be the fraction field of $R = k[x, y]$, that is, K is the field of rational functions of x

and y. For a point $P = (a, b)$ in the (x, y)-plane and a rational function $\phi = f(x, y)/g(x, y) \in K$, we say that ϕ *is defined at* P if $g(a, b) \neq 0$, and then we put

$$\phi(P) = \frac{f(a, b)}{g(a, b)} = \frac{f(P)}{g(P)}.$$

For a given point P, we define the *local ring of* P to be the set

$$\mathcal{O}_P = \{\phi \in K : \phi \text{ is defined at } P\}.$$

We leave the following basic properties of $\mathcal{O}_P$ as exercises. First, $\mathcal{O}_P$ is a subring of K, and the evaluation map

$$\mathcal{O}_P \longrightarrow k, \quad \phi \longrightarrow \phi(P),$$

is a ring homomorphism of $\mathcal{O}_P$ onto k that is the identity on k. Let

$$\mathcal{M}_P = \{\phi \in \mathcal{O}_P : \phi(P) = 0\}$$

be the kernel of the evaluation homomorphism. Then $\mathcal{O}_P$ is equal to the direct sum $\mathcal{O}_P = k + \mathcal{M}_P$ and $\mathcal{O}_P/\mathcal{M}_P \cong k$. An element $\phi \in \mathcal{O}_P$ has an inverse in $\mathcal{O}_P$ if and only if $\phi \notin \mathcal{M}_P$. Every ideal of $\mathcal{O}_P$, other than $\mathcal{O}_P$ itself, is contained in $\mathcal{M}_P$, so $\mathcal{M}_P$ is the unique maximal ideal of $\mathcal{O}_P$. (A ring having a unique maximal ideal is called a *local ring*. We used another local ring $R_p \subset \mathbb{Q}$ in Section 2.4; see also Exercise 2.7.)

Now let $(f_1, f_2)_P = \mathcal{O}_P f_1 + \mathcal{O}_P f_2$ denote the ideal in $\mathcal{O}_P$ generated by f_1 and f_2. Our definition of *intersection multiplicity of* C_1 *and* C_2 *at* P, also called the *intersection index*, is

$$I(C_1 \cap C_2, P) = \dim\left(\mathcal{O}_P/(f_1, f_2)_P\right).$$

We are now ready to do segment (3), which means taking inequality (A) and strengthening it to inequality (A^+).

(3.1) Show that

$$\dim\left(\mathcal{O}_P/(f_1, f_2)_P\right) \leq \dim\left(R/(f_1, f_2)\right).$$

Deduce from inequality (B) that the intersection multiplicity $I(C_1 \cap C_2, P)$ is finite. (*Idea.* Note that any finite set of elements in $\mathcal{O}_P$ can be written over a common denominator. Show that if $g_1/h, g_2/h, \ldots, g_r/h$ are elements of $\mathcal{O}_P$ that are linearly independent modulo $(f_1, f_2)_P$, then $g_1, g_2, \ldots, g_r$ are elements of R that are linearly independent modulo (f_1, f_2).)

(3.2) Show that $\mathcal{O}_P = R + (f_1, f_2)_P$. (*Idea.* By (3.1), we may suppose that the elements g_i/h span $\mathcal{O}_P$ modulo $(f_1, f_2)_P$, and because $h^{-1} \in \mathcal{O}_P$, it follows that the polynomials g_i span $\mathcal{O}_P$ modulo $(f_1, f_2)_P$.)

(3.3) Show that if $P \notin C_1 \cap C_2$, then $I(C_1 \cap C_2, P) = 0$. Show that if $P \in C_1 \cap C_2$, then

$$(f_1, f_2)_P \subset \mathcal{M}_P \quad \text{and} \quad I(C_1 \cap C_2, P) = 1 + \dim\left(\mathcal{M}_P / (f_1, f_2)_P\right).$$

Conclude that if $P \in C_1 \cap C_2$, then $I(C_1 \cap C_2, P) \geq 1$, with equality if and only if $(f_1, f_2)_P = \mathcal{M}_P$.

(3.4) Suppose that $P \in C_1 \cap C_2$. Let r satisfy $r \geq \dim(\mathcal{O}_P / (f_1, f_2)_P)$. Show that $\mathcal{M}_P^r \subset (f_1, f_2)_P$. (*Idea.* We are to prove that, given any collection of r elements $t_1, t_2, \ldots, t_r$ in $\mathcal{M}_P$, their product $t_1 t_2 \cdots t_r$ is in $(f_1, f_2)_P$. Define a sequence of ideals J_i in $\mathcal{O}_P$ by

$$J_i = t_1 t_2 \cdots t_i \mathcal{O}_P + (f_1, f_2)_P \quad \text{for } 1 \leq i \leq r, \text{ and} \quad J_{r+1} = (f_1, f_2)_P.$$

Then

$$\mathcal{M}_P \supset J_1 \supset J_2 \supset \cdots \supset J_r \supset J_{r+1} = (f_1, f_2)_P.$$

Since $r \geq \dim(\mathcal{O}_P / (f_1, f_2)_P)$, it follows that $J_i = J_{i+1}$ for some i with $1 \leq i \leq r$. If $i = r$, then $t_1 t_2 \cdots t_r \in (f_1, f_2)_P$ and we are done. If $i < r$, then we have

$$t_1 t_2 \cdots t_i = t_1 t_2 \cdots t_{i+1} \phi + \psi \quad \text{for some } \phi \in \mathcal{O}_P \text{ and } \psi \in (f_1, f_2)_P,$$

so $t_1 t_2 \cdots t_i (1 - t_{i+1}\phi) = \psi \in (f_1, f_2)_P$. But $(1 - t_{i+1}\phi)(P) = 1$, so we have $(1 - t_{i+1}\phi)^{-1} \in \mathcal{O}_P$. Hence

$$t_1 t_2 \cdots t_r = \psi t_{i+1} \cdots t_r (1 - t_{i+1}\phi)^{-1} \in (f_1, f_2)_P$$

as claimed.)

(3.5) Let $P \in C_1 \cap C_2 \cap \mathbb{A}^2$, and let $\phi \in \mathcal{O}_P$. Show that there exists a polynomial $g \in R$ such that

$$g \equiv \phi \pmod{(f_1, f_2)_P}$$

and

$$g \equiv 0 \pmod{(f_1, f_2)_Q} \quad \text{for all } Q \neq P \text{ with } Q \in C_1 \cap C_2 \cap \mathbb{A}^2.$$

(*Idea.* The inequalities (A) and (B) that we already proved show that only a finite number of points are involved here, in fact, at most $n_1 n_2$ points. Hence, by (1.1), there is a polynomial $h = h(x, y) \in R$ such that $h(P) = 1$ and $h(Q) = 0$ for all $Q \neq P$ with $Q \in C_1 \cap C_2 \cap \mathbb{A}^2$. This means that $h^{-1} \in \mathcal{O}_P$ and $h \in M_Q$ for each of the other points Q. For integers $r \geq 1$ we have $h^{-r} \in \mathcal{O}_P$, and if r is sufficiently large, then (3.4) tells us that $h^r \in (f_1, f_2)_Q$ for the other points Q. By (3.2) there is a polynomial $f \in R$ such that $f \equiv \phi h^{-r} \pmod{(f_1, f_2)_P}$. Then $g = f h^r$ solves the problem.)

(3.6) Show that the natural map

$$R \longrightarrow \prod_{P \in C_1 \cap C_2 \cap \mathbb{A}^2} \mathcal{O}_P/(f_1, f_2)_P,$$

$$f \longmapsto (\ldots, f \bmod (f_1, f_2)_P, \ldots)_{P \in C_1 \cap C_2 \cap \mathbb{A}^2}, \qquad (*)$$

is surjective, and conclude that the inequality (A^+) holds. (*Idea.* Let J be the kernel of the map $(*)$. Then $(f_1, f_2) \subset J$, so $\dim\big(R/(f_1, f_2)\big) \geq \dim(R/J)$. The surjectivity of the map follows easily from (3.5) and implies that

$$\dim R/J = (\text{dimension of the target space})$$

$$= \sum_P \dim\Big(\mathcal{O}_P/(f_1, f_2)_P\Big)$$

$$= \sum_P I(C_1 \cap C_2, P).)$$

To prove that (A^+) is an equality is now seen to be the same as showing that the kernel J of the map $(*)$ is equal to (f_1, f_2). So we must show that $J \subset (f_1, f_2)$, the other inclusion being obvious. Let $f \in J$. Our strategy for showing that $f \in (f_1, f_2)$ is to consider the set

$$L = \{g \in R : gf \in (f_1, f_2)\}$$

and to prove that $1 \in L$.

(4.1) Show that L is an ideal in R and that $(f_1, f_2) \subset L \subset R$.

(4.2) Show that L has the following property:

For every $P \in \mathbb{A}^2$ there is a polynomial $g \in L$ such that $g(P) \neq 0$. $(**)$

In fact, property $(**)$ alone implies that $1 \in L$ by the famous Nullstellensatz of Hilbert. But we don't need the Nullstellensatz in full generality, because we have an additional piece of information about L, namely that $(f_1, f_2) \subset L$, and hence $\dim(R/L)$ is finite. Using this, and assuming that $1 \notin L$ in order to prove a contradiction, verify the following assertion.

(4.3) There is an $a \in k$ such that $1 \notin L + R(x - a)$. (*Idea.* The powers of x cannot all be linearly independent modulo L, so there are constants $c_i \in k$ and an integer n such that $x^n + c_1 x^{n-1} + \cdots + c_n \in L$. Since k is algebraically closed, we can write this as $(x - a_1)(x - a_2) \cdots (x - a_n) \in L$ with suitable $a_i \in k$. Show that if $1 \in L + R(x - a_i)$ for all $i = 1, \ldots, n$, then we get a contradiction to the assumption that $1 \notin L$.)

(4.4) There is a $b \in k$ such that $1 \notin L + R(x - a) + R(y - b)$. (*Idea.* Replace L by $L + R(x - a)$ and x by y and repeat the argument of (4.3).)

(4.5) Let $P = (a, b)$ and show that $g(P) = 0$ for all $g \in L$. This contradicts (4.2) and shows that $1 \in L$. (*Idea.* Write

$$g(x, y) = g\big(a + (x - a), b + (y - b)\big)$$
$$= g(a, b) + g_1(x, y)(x - a) + g_2(x, y)(y - b)$$

and conclude that $g(a, b) \in L$.)

Our next job is to describe K, $\mathcal{O}_P$, $\mathcal{M}_P$, and $(f_1, f_2)_P$ in terms of homogeneous coordinates, so that they make sense also for points P at infinity. This will allow us to check that they are invariant under arbitrary projective coordinate change in $\mathbb{P}^2$. To see what to do we put as usual $x = X/Z$ and $y = Y/Z$, and we view $R = k[x, y] = k[X/Z, Y/Z]$ as a subring of the field $k(X, Y, Z)$ of rational functions of X, Y, Z. Then $K = k(x, y)$ becomes identified with the set of all rational functions $\Phi = F/G$ of X, Y, Z that are *homogeneous of degree* 0 in the sense that F and G are homogeneous polynomials of the same degree. Indeed, for $\phi \in K$, we have

$$\phi(x, y) = \frac{f(x, y)}{g(x, y)} = \frac{Z^n f(X/Z, Y/Z)}{Z^n g(X/Z, Y/Z)} = \frac{F(X, Y, Z)}{G(X, Y, Z)} = \Phi(X, Y, Z),$$

say, where F and G are homogeneous of the same degree

$$n = \max\{\deg f, \deg g\}.$$

On the other hand, if $\Phi = F/G$ is a quotient of forms of the same degree, then $\Phi(tX, tY, tZ) = \Phi(X, Y, Z)$, and

$$\Phi(X, Y, Z) = \Phi(x, y, 1) = \frac{F(x, y, 1)}{G(x, y, 1)} \in K.$$

If $P = [A, B, C]$ is a point in $\mathbb{P}^2$ and $\Phi = F/G \in K$, then we say that Φ is *defined at* P if $G(A, B, C) \neq 0$, i.e., if P is not on the curve $G(X, Y, Z) = 0$. If Φ is defined at P, we put $\Phi(P) = F(A, B, C)/G(A, B, C)$, where this ratio is independent of the homogeneous coordinate triple for P. Clearly we should put

$$\mathcal{O}_P = \{\Phi \in K : \Phi \text{ is defined at } P\},$$
$$\mathcal{M}_P = \{\Phi \in \mathcal{O}_P : \Phi(P) = 0\}.$$

We leave it to the conscientious reader to check the following assertion.

(5.1) If $P = (a, b) = [a, b, 1] \in \mathbb{A}^2$, then these definitions of $\mathcal{O}_P$, of $\Phi(P)$ for $\Phi \in \mathcal{O}_P$, and of $\mathcal{M}_P$ coincide with our earlier definitions.

Now let $C_1 : F_1 = 0$ and $C_2 : F_2 = 0$ be two curves in $\mathbb{P}^2$ without any common components. Let $f_1(x, y) = F_1(x, y, 1)$ and $f_2(x, y) = F_2(x, y, 1)$ be the polynomials defining their affine parts. Define

$$(F_1, F_2)_P = \{F/G \in \mathcal{O}_P : F \text{ is of the form } F = H_1 F_1 + H_2 F_2\}.$$

(Do you see why we cannot just say that $(F_1, F_2)_P$ is the ideal in $\mathcal{O}_P$ generated by F_1 and F_2?)

(5.2) Check that if $P \in \mathbb{A}^2$, then $(F_1, F_2)_P = (f_1, f_2)_P$ is the ideal in $\mathcal{O}_P$ generated by f_1 and f_2.

Of course, we now define the intersection multiplicity of C_1 and C_2 at a point $P \in \mathbb{P}^2$ by

$$I(C_1 \cap C_2, P) = \dim\left(\mathcal{O}_P/(F_1, F_2)_P\right).$$

We know from (5.2) that this coincides with our earlier definition for $P \in \mathbb{A}^2$.

(5.3) Check that the definitions of $\mathcal{O}_P$ and $(F_1, F_2)_P$, and hence also of the intersection multiplicity $I(C_1 \cap C_2, P)$, are independent of our choice of homogeneous coordinates in $\mathbb{P}^2$, i.e., they are invariant under a linear change of the coordinates X, Y, Z.

To finally complete our proof of Bezout's theorem, we must show that there is a line L in $\mathbb{P}^2$ that does not meet $C_1 \cap C_2$. Then we can take a new coordinate system in which L is the line at infinity, and thereby reduce to the case already proved. To show that L exists we use the following:

(5.4) Prove that given any finite set S of points in $\mathbb{P}^2$, there is a line L not meeting S. (*Idea.* Use that an algebraically closed field k is not finite.)

Finally, the next result allows us to apply (5.4).

(5.5) Prove that $C_1 \cap C_2$ is finite. (*Idea.* Use the fact that for every line L that is not a component of either C_1 or C_2, we know, by putting L at infinity and using part (1) of this proof, that $C_1 \cap C_2$ contains a finite number of points not on L.)

This completes our proof of Bezout's Theorem in all its gory detail. To study more closely the properties of the intersection multiplicity $I(C_1 \cap C_2, P)$ at one point P, we may without loss of generality choose coordinates so that $P = (0,0) = [0,0,1]$ is the origin in the affine plane, and we can work with affine coordinates x, y. Let $R = k[x, y]$ as before, and let

$$\mathcal{M} = \{ f = f(x,y) \in R : f(P) = f(0,0) = 0 \}.$$

(6.1) Prove that $\mathcal{M} = (x, y) = Rx + Ry$ and that $\mathcal{M}_P = \mathcal{O}_P x + \mathcal{O}_P y$.

It follows that for each $n \geq 1$, $\mathcal{M}^n$ is the ideal in R generated by the monomials $x^n, x^{n-1}y, \ldots, xy^{n-1}, y^n$. Hence every polynomial $f \in R$ can be written uniquely as a polynomial of degree at most n plus a remainder polynomial $r \in \mathcal{M}^{n+1}$. Thus

$$f(x,y) = c_{00} + c_{10}x + c_{01}y + \cdots + c_{ij}x^i y^j + \cdots$$
$$+ c_{n0}x^n + c_{n-1,1}x^{n-1}y + \cdots + c_{0n}y^n + r. \quad (*)$$

(6.2) Prove that every $\phi = f/g \in \mathcal{O}_P$ can be written uniquely in the form $(*)$ with $c_{ij} \in k$ and $r \in \mathcal{M}_P^{n+1}$. In other words, the inclusion $R \subset \mathcal{O}_P$ induces an isomorphism $R/\mathcal{M}^{n+1} \cong \mathcal{O}_P/\mathcal{M}_P^{n+1}$ for every $n \geq 0$. (*Idea*. We must show that $\mathcal{O}_P = R + \mathcal{M}_P^{n+1}$ and that $R \cap \mathcal{M}_P^{n+1} = \mathcal{M}^{n+1}$. For the first, show that every $\phi \in \mathcal{O}_P$ can be written in the form $\phi = f/(1 - h)$ with $f \in R$ and $h \in \mathcal{M}$. Hence

$$\phi = \frac{f}{1 - h} = f \cdot (1 + h + \cdots + h^n) + \frac{fh^{n+1}}{1 - h} \in R + \mathcal{M}_P^{n+1}.$$

The second reduces to showing that if $gf \in \mathcal{M}^n$ and $g(P) \neq 0$, then $f \in \mathcal{M}^n$. This can be done by considering the terms of lowest degree in g and f and gf.)

Now we can already compute some intersection indices to see if our definitions give answers that are geometrically reasonable. As a matter of notation, we introduce the symbol

$$I(f_1, f_2) = \dim\left(\mathcal{O}_P/(f_1, f_2)_P\right)$$

for the intersection multiplicity of two curves $f_1 = 0$ and $f_2 = 0$ at the origin.

(6.3) Check that the curve $y = x^n$ and the x-axis intersect with multiplicity n at the origin, i.e., show that $I(y - x^n, y) = n$. (*Idea*. Note first that the ideals

$(y - x^n, y)$ and (x^n, y) are equal, and that this ideal contains $\mathcal{M}^n$. Then, using what we know from (6.2) about $\mathcal{O}_P/\mathcal{M}_P^n$, show that $1, x, \ldots, x^{n-1}$ is a basis for the vector space $\mathcal{O}_P/(x^n, y)\mathcal{O}_P$.

(6.4) (Nakayama's Lemma) Suppose that J is an ideal of $\mathcal{O}_P$ contained in a finitely generated ideal $\Phi = (\phi_1, \phi_2, \ldots, \phi_m)\mathcal{O}_P$. Suppose some elements of J generate Φ modulo $\mathcal{M}_P\Phi$, i.e., $\Phi = J + \mathcal{M}_P\Phi$. Then $J = \Phi$. (*Idea.* The case $\Phi = (\phi_1, \phi_2)\mathcal{O}_P$ is all that we need. To prove that case, write

$$\phi_1 = j_1 + \alpha\phi_1 + \beta\phi_2 \quad \text{and} \quad \phi_2 = j_2 + \gamma\phi_1 + \delta\phi_2,$$

with $j_1, j_2 \in J$ and $\alpha, \beta, \gamma, \delta \in \mathcal{M}_P$. Then use the fact that the determinant of the matrix $\begin{pmatrix} 1-\alpha & \beta \\ \gamma & 1-\delta \end{pmatrix}$ is non-zero in order to express the ϕ's in terms of the j's.)

(6.5) Suppose that

$$f_1 = ax + by + (\text{higher terms}) \quad \text{and} \quad f_2 = cx + dy + (\text{higher terms}),$$

where "higher terms" means elements of $\mathcal{M}^2$. Show that the following are equivalent.

(i) The curves $f_1 = 0$ and $f_2 = 0$ meet transversally at the origin, i.e., are smooth with distinct tangent directions there.

(ii) The determinant $ad - bc$ is not equal to zero.

(iii) $(f_1, f_2)_P = \mathcal{M}_P$, i.e., $I(f_1, f_2) = 1$.

(*Idea.* (i) $\Longleftrightarrow$ (ii) follows directly from the definitions. One way to do (ii) $\Longrightarrow$ (iii) is to use (6.4) with $\phi_1 = x$, $\phi_2 = y$, and $J = (f_1, f_2)_P$. To do (iii) $\Longrightarrow$ (ii), note that if $ad - bc = 0$, then

$$\dim\left(\frac{(f_1, f_2)_P + \mathcal{M}_P^2}{\mathcal{M}_P^2}\right) \leq 1,$$

whereas, by (6.2), $\dim(\mathcal{M}_P/\mathcal{M}_P^2) = 2$.)

(6.6) Let $f(x, y) \in R$. Show that $I\big(f(x, y), y\big) = m$, where x^m is the highest power of x dividing $f(x, 0)$. (*Idea.* Use the fact that the ideal $\big(f(x, y), y\big)$ is the same as the ideal $\big(f(x, 0), y\big)$. Then argue as in (6.3).)

(6.7) Let $C : F(X, Y, Z) = 0$ be a curve in $\mathbb{P}^2$ that does not contain the line $L_\infty : Z = 0$. Show that for each point $Q \in [a, b, 0] \in L_\infty$, we have $I(C \cap L_\infty, Q) = m$, where $(bX - aY)^m$ is the highest power of $(bZ - aY)$ dividing $F(X, Y, 0)$. (*Idea.* Make a suitable coordinate change to reduce to (6.6).)

A.5   Reduction Modulo p

Let $\mathbb{P}^2(\mathbb{Q})$ denote the set of rational points in $\mathbb{P}^2$. We say that a homogeneous coordinate triple $[A, B, C]$ is *normalized* if A, B, C are integers with no common factors. Each point $P \in \mathbb{P}^1(\mathbb{Q})$ has a normalized coordinate triple that is unique up to sign. To obtain it we start with any triple of rational coordinates, multiply through by a common denominator, and then divide the resulting triple of integers by their greatest common divisor. For example,

$$\left[\frac{4}{5}, -\frac{2}{3}, 2 \right] = [12, -10, 30] = [6, -5, 15].$$

The other normalized coordinate triple for this point is $[-6, 5, -15]$.

Let p be a fixed prime number, and for each integer $m \in \mathbb{Z}$, let $\tilde{m} \in \mathbb{F}_p = \mathbb{Z}/p\mathbb{Z}$ denote its residue modulo p. If $[l, m, n]$ is a normalized coordinate triple for a point $P \in \mathbb{P}^2(\mathbb{Q})$, then the triple $[\tilde{l}, \tilde{m}, \tilde{n}]$ defines a point $\tilde{P}$ in $\mathbb{P}^2(\mathbb{F}_p)$, since at least one of the three numbers l, m, and n is not divisible by p. Since P determines the triple $[l, m, n]$ up to sign, the point $\tilde{P}$ depends only on P, not on the choice of coordinates for P. Thus $P \mapsto \tilde{P}$ gives a well-defined map

$$\mathbb{P}^2(\mathbb{Q}) \longrightarrow \mathbb{P}^2(\mathbb{F}_p),$$

called for obvious reasons the *reduction mod p map*. Note that reduction mod p does not map $\mathbb{A}^2(\mathbb{Q})$ to $\mathbb{A}^2(\mathbb{F}_p)$. For example,

$$P = \left(\frac{1}{p}, 0 \right) = \left[\frac{1}{p}, 0, 1 \right] = [1, 0, p] \longmapsto \widetilde{[1, 0, p]} = [1, 0, 0] \notin \mathbb{A}^2(\mathbb{F}_p).$$

In fact, if $P = (a, b) = [a, b, 1] \in \mathbb{A}^2(\mathbb{Q})$, then its reduction $\tilde{P}$ is in $\mathbb{A}^2(\mathbb{F}_p)$ if and only if the rational numbers a and b are p-integral, i.e., have denominators that are prime to p.

Let $C : F(X, Y, Z) = 0$ be a rational curve in $\mathbb{P}^2$. By rational we mean as usual that the coefficients of F are rational numbers. Clearing the denominators of the coefficients and then dividing by the greatest common divisor of their numerators, we may suppose that the coefficients of F are integers with greatest common divisor one. Call such an F *normalized*. Then $\tilde{F}$, the polynomial that we obtain by reducing the coefficients of F modulo p, is non-zero and defines a curve $\tilde{C}$ in characteristic p. If $[l, m, n]$ is a normalized coordinate triple and if $F(l, m, n) = 0$, then $F(\tilde{l}, \tilde{m}, \tilde{n}) = 0$, because $x \to \tilde{x}$ is a homomorphism. In other words, if P is a rational point on C, then $\tilde{P}$ is a point on $\tilde{C}$, so reduction mod p takes $C(\mathbb{Q})$ and maps it into $C(\mathbb{F}_p)$.

If C_1 and C_2 are curves, it follows that

$$\left(\widetilde{C_1(\mathbb{Q}) \cap C_2(\mathbb{Q})}\right) \subset \tilde{C}_1(\mathbb{F}_p) \cap \tilde{C}_2(\mathbb{F}_p).$$

Is there some sense in which $\widetilde{(C_1 \cap C_2)} = \tilde{C}_1 \cap \tilde{C}_2$ if we count multiplicities? After all, the degrees of the reduced curves $\tilde{C}_i$ are the same as those of the C_i, so by Bezout's theorem the intersection before and after reduction has the same number of points if we count multiplicities. But Bezout's theorem requires that the ground field be algebraically closed, and we don't have the machinery to extend our reduction mod p map to that case. However, if we assume that all of the complex intersection points are rational, then everything is okay. We treat only the special case in which one of the curves is a line. This case suffices for the application to elliptic curves that we are after, and it is easy to prove.

Proposition A.5. *Suppose that C is a rational curve and L is a rational line in $\mathbb{P}^2$. Suppose that all of the complex intersection points of C and L are rational. Let $C \cap L = \{P_1, P_2, \ldots, P_d\}$, where $d = \deg(C)$ and each P_i is repeated in the list as many times as its multiplicity. Assume that $\tilde{L}$ is not a component of $\tilde{C}$. Then $\tilde{C} \cap \tilde{L} = \{\tilde{P}_1, \tilde{P}_2, \ldots, \tilde{P}_d\}$ with the correct multiplicities.*

Proof. Suppose first that L is the line at infinity $Z = 0$. Let $F(X, Y, Z) = 0$ be a normalized equation for C. The assumption that $\tilde{L}$ is not a component of $\tilde{C}$ means that $\tilde{F}(X, Y, 0) \neq 0$, i.e., some coefficient of $F(X, Y, 0)$ is not divisible by p. For each intersection point P_i, let $P_i = [l_i, m_i, 0]$ in normalized coordinates. Then

$$F(X, Y, 0) = c \prod_{i=1}^{d} (m_i X - l_i Y) \tag{$*$}$$

for some constant c. This is true because the intersection points of a curve $F = 0$ with the line $Z = 0$ correspond, with the correct multiplicities, to the linear factors of $F(X, Y, 0)$. Since each of the linear polynomials on the right of $(*)$ is normalized and since some coefficient of F is not divisible by p, we see that c must be an integer that is not divisible by p. Therefore we can reduce $(*)$ modulo p to obtain

$$\tilde{F}(X, Y, 0) = \tilde{c} \prod_{i=1}^{d} (\tilde{m}_i X - \tilde{l}_i Y), \tag{$\tilde{*}$}$$

which shows that $\tilde{C} \cap \tilde{L} = \{\tilde{P}_1, \tilde{P}_2, \ldots, \tilde{P}_d\}$ as claimed.

What if the line L is not the line $Z = 0$. Then we just make a linear change of coordinates

$$\begin{pmatrix} X' \\ Y' \\ Z' \end{pmatrix} = \begin{pmatrix} n_{11} & n_{12} & n_{13} \\ n_{21} & n_{22} & n_{23} \\ n_{31} & n_{32} & n_{33} \end{pmatrix} \begin{pmatrix} X \\ Y \\ Z \end{pmatrix}$$

so that L is the line $Z' = 0$ in the new coordinate system.

Is that all there is to it? No, we must be careful to make sure that our change of coordinates is compatible with reduction modulo p. This is not true for general changes with $n_{ij} \in \mathbb{Q}$. However, if we change using a matrix (n_{ij}) with integer entries and determinant 1, then the inverse matrix (m_{ij}) will have integer entries, and the reduced matrices $(\tilde{n}_{ij})$ and $(\tilde{m}_{ij})$ are inverses giving the corresponding coordinate change in characteristic p. And clearly if we change coordinates with (n_{ij}) and reduce mod p, the result will be the same as if we first reduce mod p and then change coordinates with $(\tilde{n}_{ij})$.

Thus, to complete our proof we must show that for every rational line in $\mathbb{P}^2$ there is an "integral" coordinate change such that in the new coordinates, the line L is the line at infinity. To do this, we let

$$L : aX + bY + cZ = 0$$

be a normalized equation for the line L and use the following result.

Lemma A.6. *Let (a, b, c) be a triple of integers satisfying $\gcd(a, b, c) = 1$. Then there exists a 3×3 matrix with integer coefficients, determinant 1, and bottom line (a, b, c).*

Proof. Let $d = \gcd(b, c)$, choose integers r and s such that $rc - sb = d$, and note for later use that r and s are necessarily relatively prime. Now $\gcd(a, d) = 1$, so we can choose t and u such that $td + ua = 1$. Finally, since $\gcd(r, s) = 1$, we can choose v and w such that $vs - wr = u$. Then the matrix

$$\begin{pmatrix} t & v & w \\ 0 & r & s \\ a & b & c \end{pmatrix}$$

has the desired properties. $\square$

Using Lemma A.6 completes the proof of Proposition A.5. $\square$

Finally, we apply Proposition A.5 to show that the reduction mod p map respects the group law on a cubic curve.

Corollary A.7. *Let C be a non-singular rational cubic curve in $\mathbb{P}^2$ and let $\mathcal{O}$ be a rational point on C, which we take as the origin for the group law on C. Suppose that $\tilde{C}$ is non-singular and take $\tilde{\mathcal{O}}$ as the origin for the group law on $\tilde{C}$. Then the reduction mod p map $P \mapsto \tilde{P}$ is a group homomorphism $C(\mathbb{Q}) \to \tilde{C}(\mathbb{F}_p)$.*

Proof. Let $P, Q \in C(\mathbb{Q})$, and let $R = P + Q$. This means that there are lines L_1 and L_2 and a rational point $S \in C(\mathbb{Q})$ such that, in the notation of Proposition A.5,

$$C \cap L_1 = \{P, Q, S\} \quad \text{and} \quad C \cap P_2 = \{S, \mathcal{O}, R\}.$$

Putting tildes on everything, which is allowed by the proposition, we conclude that $\tilde{P} + \tilde{Q} = \tilde{R}$. $\qquad\square$

Exercises

A.1. Let $\mathbb{P}^2$ be the set of homogeneous triples $[a, b, c]$ as usual, and recall that with this definition a line in $\mathbb{P}^2$ is defined to be the set of solutions of an equation of the form

$$\alpha X + \beta Y + \gamma Z = 0$$

for some numbers α, β, γ not all zero.
  (a) Prove directly from this definition that any two distinct points in $\mathbb{P}^2$ are contained in a unique line.
  (b) Similarly, prove that any two distinct lines in $\mathbb{P}^2$ intersect in a unique point.

A.2. Let K be a field, for example K might be the rational numbers or the real numbers or a finite field. Define a relation $\sim$ on $(n + 1)$-tuples $[a_0, a_1, \ldots, a_n]$ of elements of K by the following rule:

$$[a_0, a_1, \ldots, a_n] \sim [a'_0, a'_1, \ldots, a'_n] \quad \text{if there is a non-zero } t \in K$$
$$\text{so that } a_0 = ta'_0, \, a_1 = ta'_1, \ldots, a_n = ta'_n.$$

(a) Prove that $\sim$ is an equivalence relation. That is, prove that for any $(n + 1)$-tuples $a = [a_0, a_1, \ldots, a_n]$, $b = [b_0, b_1, \ldots, b_n]$, and $c = [c_0, c_1, \ldots, c_n]$, the relation $\sim$ satisfies the following three conditions:
  (i) $a \sim a$ (Reflexive)
  (ii) $a \sim b \Longrightarrow b \sim a$ (Symmetric)
  (iii) $a \sim b$ and $b \sim c \Longrightarrow a \sim c$ (Transitive)
(b) Which of these properties (i), (ii), (iii) fails to be true if K is replaced by a ring R that is not a field? (There are several answers to this question, depending on what the ring R looks like.)

A.3. We saw in Section A.1 that the directions in the affine plane $\mathbb{A}^2$ correspond to the points of the projective line $\mathbb{P}^1$. In other words, $\mathbb{P}^1$ can be described as the set of lines is $\mathbb{A}^2$ going through the origin.

(a) Prove similarly that $\mathbb{P}^2$ can be described as the set of lines in $\mathbb{A}^3$ going through the origin.

(b) Let $\Pi \subset \mathbb{A}^3$ be a plane in $\mathbb{A}^3$ that goes through the origin, and let S_Π be the collection of lines in $\mathbb{A}^3$ going through the origin and contained in Π. From (a), S_Π defines a subset L_Π of $\mathbb{P}^2$. Prove that L_Π is a line in $\mathbb{P}^2$, and conversely that every line in $\mathbb{P}^2$ can be constructed in this way.

(c) Generalize (a) by showing the $\mathbb{P}^n$ can be described as the set of lines in $\mathbb{A}^{n+1}$ going through the origin.

A.4. Let $F(X, Y, Z) \in \mathbb{C}[X, Y, Z]$ be a homogeneous polynomial of degree d.

(a) Prove that the three partial derivatives of F are homogeneous polynomials of degree $d - 1$.

(b) Prove that

$$X\frac{\partial F}{\partial X} + Y\frac{\partial F}{\partial Y} + Z\frac{\partial F}{\partial Z} = d \cdot F(X, Y, Z).$$

(*Hint.* Differentiate $F(tX, tY, tZ) = t^d F(X, Y, Z)$ with respect to t.)

A.5. Let $C : F(X, Y, Z) = 0$ be a projective curve given by a homogeneous polynomial $F \in \mathbb{C}[X, Y, Z]$, and let $P \in \mathbb{P}^2$ be a point.

(a) Prove that P is a singular point of C if and only if

$$\frac{\partial F}{\partial X}(P) = \frac{\partial F}{\partial Y}(P) = \frac{\partial F}{\partial Z}(P) = 0.$$

(b) If P is a non-singular point of C, prove that the tangent line to C at P is given by the equation

$$\frac{\partial F}{\partial X}(P)X + \frac{\partial F}{\partial Y}(P)Y + \frac{\partial F}{\partial Z}(P)Z = 0.$$

A.6. Let C be the projective curve given by the equation

$$C : Y^2 Z - X^3 - Z^3 = 0.$$

(a) Show that C has only one point at infinity, namely the point $[0, 1, 0]$ corresponding to the vertical direction $x = 0$.

(b) Let $C_0 : y^2 - x^3 - 1 = 0$ be the affine part of C, and let (r_i, s_i) be a sequence of points on C_0 with $r_i \to \infty$. Let L_i be the tangent line to C_0 at the point (r_i, s_i). Prove that as $i \to \infty$, the slopes of the lines L_i approach infinity, i.e., they approach the slope of the line $x = 0$.

A.7. Let $f(x, y)$ be a polynomial.

(a) Expand $f(tx, ty)$ as a polynomial in t whose coefficients are polynomials in x and y. Prove that the degree of $f(tx, ty)$, considered as a polynomial in the variable t, is equal to the degree of the polynomial $f(x, y)$.

(b) Prove that the homogenization $F(X, Y, Z)$ of $f(x, y)$ is given by

$$F(X, Y, Z) = Z^d f\left(\frac{X}{Z}, \frac{Y}{Z}\right), \quad \text{where } d = \deg(f).$$

A.8. For each of the given affine curves C_0, find a projective curve C whose affine part is C_0. Then find all of the points at infinity on the projective curve C.

(a) $C_0 : 3x - 7y + 5 = 0$.
(b) $C_0 : x^2 + xy - 2y^2 + x - 5y + 7 = 0$.
(c) $C_0 : x^3 + x^2 y - 3xy^2 - 3y^3 + 2x^2 - 2 + 5 = 0$.

A.9. For each of the following curves C and points P, either find the tangent line to C at P or else verify that C is singular at P

(a) $C : y^2 = x^3 - x$, $\qquad\qquad P = (1, 0)$.
(b) $C : X^2 + Y^2 = Z^2$, $\qquad\quad P = [3, 4, 5]$.
(c) $C : x^2 + y^4 + 2xy + 2x + 2y + 1 = 0$, $P = (-1, 0)$.
(d) $C : X^3 + Y^3 + Z^3 = XYZ$, $\quad P = [1, -1, 0]$.

A.10. (a) Prove that a projective transformation of $\mathbb{P}^2$ sends lines to lines.
(b) More generally, prove that a projective transformation of $\mathbb{P}^2$ sends curves of degree d to curves of degree d.

A.11. Let P, P_1, P_2, P_3 be points in $\mathbb{P}^2$, and let L be a line in $\mathbb{P}^2$.
(a) If P_1, P_2, and P_3 do not lie on a line, prove that there is a projective transformation of $\mathbb{P}^2$ so that

$$P_1 \longmapsto [0, 0, 1], \quad P_2 \longmapsto [0, 1, 0], \quad P_3 \longmapsto [1, 0, 0].$$

(b) If no three of P_1, P_2, P_3, and P lie on a line, prove that there is a unique projective transformation as in (a) that also sends P to $[1, 1, 1]$.
(c) Prove that there is a projective transformation of $\mathbb{P}^2$ so that L is sent to the line $Z' = 0$.
(d) More generally, if P does not lie on L, prove that there is a projective transformation of $\mathbb{P}^2$ so that L is sent to the line $Z' = 0$ and P is sent to the point $[0, 0, 1]$.

A.12. For each of the pairs of curves C_1, C_2, find all of the points in the intersection $C_1 \cap C_2$. Be sure to include points with complex coordinates and points at infinity.

(a) $C_1 : x - y = 0$, $\qquad\quad C_2 : x^2 - y = 0$.
(b) $C_1 : x - y - 1 = 0$, $\qquad C_2 : x^2 - y^2 + 2 = 0$.
(c) $C_1 : x - y - 1 = 0$, $\qquad C_2 : x^2 - 2y^2 - 5 = 0$.
(d) $C_1 : x - 2 = 0$, $\qquad\quad\; C_2 : y^2 - x^3 + 2x = 0$.

A.13. For each of the pairs of curves C_1, C_2, compute the intersection index $I(C_1 \cap C_2, P)$ at the indicated point P. Also sketch the curves and the point in $\mathbb{R}^2$.

(a) $C_1 : x - y = 0$, $C_2 : x^2 - y = 0$, $P = (0,0)$.
(b) $C_1 : y = 0$, $C_2 : x^2 - y = 0$, $P = (0,0)$.
(c) $C_1 : x - y = 0$, $C_2 : x^3 - y^2 = 0$, $P = (0,0)$.
(d) $C_1 : x^2 - y = 0$, $C_2 : x^3 - y = 0$, $P = (0,0)$.
(e) $C_1 : x + y = 2$, $C_2 : x^2 + y^2 = 2$, $P = (1,1)$.

A.14. Let $\mathcal{C}^{(d)}$ be the collection of curves of degree d in $\mathbb{P}^2$.
(a) Show that $\mathcal{C}^{(d)}$ is naturally isomorphic to the projective space $\mathbb{P}^N$ for a certain value of N, and find N explicitly in terms of d.
(b) In Section A.3 we gave a plausibility argument for why the Cayley–Bacharach theorem is true for curves of degree d. Give a similar argument for general curves C_1, C_2, and D of degrees d_1, d_2, and $d_1 + d_2 - 3$, respectively.

A.15. Let $P \in \mathbb{A}^2$. In this exercise we ask you to verify various properties of $\mathcal{O}_P$, the local ring at P, as defined in Section A.4.
(a) Prove that $\mathcal{O}_P$ is a subring of $K = k(x, y)$.
(b) Prove that the map $\phi \mapsto \phi(P)$ is a homomorphism of $\mathcal{O}_P$ onto k. Let $\mathcal{M}_P$ be the kernel of this homomorphism.
(c) Prove that $\mathcal{O}_P$ equals the direct sum $k + \mathcal{M}_P$.
(d) Prove that $\phi \in \mathcal{O}_P$ is a unit if and only if $\phi \notin \mathcal{M}_P$.
(e) Let $I \subset \mathcal{O}_P$ be an ideal of $\mathcal{O}_P$. Prove that either $I = \mathcal{O}_P$, or else $I \subset \mathcal{M}_P$. Deduce that $\mathcal{M}_P$ is the unique maximal ideal of $\mathcal{O}_P$.

A.16. Let P_1, P_2, P_3, P_4, P_5 be five distinct points in $\mathbb{P}^2$.
(a) Show that there exists a conic C, i.e., a curve of degree two, passing through the five points.
(b) Show that C is unique if and only if no four of the five points lie on a line.
(c) Show that C is irreducible if and only if no three of the five points lie on a line.

A.17. In this exercise we guide you in proving the cubic Cayley–Bacharach theorem in the case that the eight points are distinct. Let $C_1 : F_1 = 0$ and $C_2 : F_2 = 0$ be cubic curves in $\mathbb{P}^2$ without common component which have eight distinct points $P_1, P_2, \ldots, P_8$ in common. Suppose that $C_3 : F_3 = 0$ is a third cubic curve passing through these same eight points. Prove that C_3 is on the "line of cubics" joining C_1 and C_2, i.e., prove that there are constants λ_1 and λ_2 such that

$$F_3 = \lambda_1 F_1 + \lambda_2 F_2.$$

In order to prove this result, assume that no such λ_1, λ_2 exist and derive a contradiction as follows:
(i) Show that F_1, F_2, and F_3 are linearly independent.
(ii) Let P' and P'' be any two points in $\mathbb{P}^2$ different from each other and different from the P_i. Show that there is a cubic curve passing through all ten points $P_1, \ldots, P_8, P', P''$. (*Hint.* Show that there exist constants $\lambda_1, \lambda_2, \lambda_3$ such that $F = \lambda_1 F_1 + \lambda_2 F_2 + \lambda_3 F_3$ is not identically zero and such that the curve $F = 0$ does the job.)

(iii) Show that no four of the eight points P_i are collinear, and no seven of them lie on a conic. (*Hint.* Use the fact that C_1 and C_2 have no common component.)

(iv) Use the previous exercise to observe that there is a unique conic Q going through any five of the eight points $P_1, \ldots, P_8$.

(v) Show that no three of the eight points P_i are collinear. (*Hint.* If three are on a line L, let Q be the unique conic going through the other five, choose P' on L and P'' not on L. Then use (ii) to get a cubic which has L as a component, so is of the form $C = L \cup Q'$ for some conic Q'. This contradicts the fact that Q is unique.)

(vi) To get the final contradiction, let Q be the conic through the five points $P_1, P_2, \ldots, P_5$. By (iii), at least one (in fact two) of the remaining three points is not on Q. Call it P_6, and let L be the line joining P_7 to P_8. Choose P' and P'' on L so that again the cubic C through the ten points has L as a component. Show that this gives a contradiction.

A.18. Show that if C_1 and C_2 are both singular at the point P, then their intersection index satisfies $I(C_1 \cap C_2, P) \geq 3$.

A.19. Consider the affine curve $C : y^4 - xy - x^3 = 0$. Show that at the origin $(x, y) = (0, 0)$, the curve C meets the y-axis four times, the x-axis three times, and every other line through the origin twice.

A.20. Show that the separation of real conics into hyperbolas, parabolas, and ellipses is an affine business and has no meaning projectively, by giving an example of a quadratic homogeneous polynomial $F(X, Y, Z)$ with real coefficients such that:

$F(x, y, 1) = 0$ is a hyperbola in the real (x, y)-plane,

$F(x, 1, z) = 0$ is a parabola in the real (x, z)-plane,

$F(1, y, z) = 0$ is an ellipse in the real (y, z)-plane.

Appendix B

Transformation to Weierstrass Form

We illustrate the transformation of a cubic equation to Weierstrass form, using the procedure described in Section 1.3, for the curve

$$C : X^3 + 2Y^3 + 4Z^3 - 7XYZ = 0 \quad \text{and the point} \quad \mathcal{O} = [1, 1, 1].$$

Before starting, we observe that in general, the tangent line in $\mathbb{P}^2$ to a curve described by a homogeneous equation

$$F(X, Y, Z) = 0$$

at the point $P_0 = [X_0, Y_0, Z_0] \in \mathbb{P}^2$ is given by the homogeneous linear equation

$$\frac{\partial F}{\partial X}(P_0)X + \frac{\partial F}{\partial Y}(P_0)Y + \frac{\partial F}{\partial Z}(P_0)Z = 0.$$

Looking at Figure 1.10, we see that a good first step is to move the point $\mathcal{O}$ to the point $[1, 0, 0]$, so we make the substitution

$$X_1 = X, \quad Y_1 = Y - X, \quad Z_1 = Z - X.$$

This transforms the equation for C into

$$C : X_1^2 Y_1 + 6X_1 Y_1^2 + 2Y_1^3 + 5X_1^2 Z_1 - 7X_1 Y_1 Z_1 + 12X_1 Z_1^2 + 4Z_1^3 = 0.$$

The tangent line to C at $\mathcal{O} = [1, 0, 0]$ is $Y_1 - 5Z_1 = 0$, and according to Figure 1.10, we want this tangent line to be the line $Z = 0$. So we make the substitution

© Springer International Publishing Switzerland 2015
J.H. Silverman, J.T. Tate, *Rational Points on Elliptic Curves*,
Undergraduate Texts in Mathematics, DOI 10.1007/978-3-319-18588-0

$$X_2 = X_1, \quad Y_2 = Y_1, \quad Z_2 = Y_1 - 5Z_1,$$

which gives the equation

$$C : 635X_2Y_2^2 + 254Y_2^3 - 125X_2^2Z_2 + 55X_2Y_2Z_2 - 12Y_2^2Z_2$$
$$+ 60X_2Z_2^2 + 12Y_2Z_2^2 - 4Z_2^3 = 0.$$

The tangent line at $\mathcal{O} = [1,0,0]$ is now the line $Z_2 = 0$. To find the other intersection point of this line with C, we substitute $Z_2 = 0$ into the equation for C. This leads to $127Y_2^2(5X_2 + 2Y_2) = 0$, and thus the third intersection point is

$$\mathcal{O} * \mathcal{O} = [2, -5, 0].$$

Again looking at Figure 1.10, we move this point to $[0,1,0]$ by making the substitution

$$X_3 = 5X_2 + 2Y_2, \quad Y_3 = Y_2, \quad Z_3 = Z_2,$$

which gives

$$C : 127X_3Y_3^2 - 5X_3^2Z_3 + 31X_3Y_3Z_3 - 54Y_3^2Z_3 + 12X_3Z_3^2$$
$$- 12Y_3Z_3^2 - 4Z_3^3 = 0.$$

The tangent line to C at the point $[0,1,0]$ is now easily computed; it turns out to be $127X_3 - 54Z_3 = 0$. A final look at Figure 1.10 shows that this line should be moved to $X = 0$, so we make the substitution (note that we want the line $Z = 0$ and the point $[1,0,0]$ to stay where they are)

$$X_4 = 127X_3 - 54Z_3, \quad Y_4 = Y_3, \quad Z_4 = Z_3.$$

This transforms C into

$$C : 16129X_4Y_4^2 - 5X_4^2Z_4 + 3937X_4Y_4Z_4 + 984X_4Z_4^2$$
$$+ 19050Y_4Z_4^2 + 32000Z_4^3 = 0.$$

Don't despair, we're almost done. We dehomogenize using $x_5 = X_4/Z_4$ and $y_5 = Y_4/Z_4$ to get

$$C : 3200 + 984x_5 - 5x_5^2 + 19050y_5 + 3937x_5y_5 + 16129x_5y_5^2 = 0.$$

Next we multiply by x_5 and let $x_6 = x_5$ and $y_6 = x_5y_5$, which gives

$$C : 3200x_6 + 984x_6^2 - 5x_6^3 + 19050y_6 + 3937x_6y_6 + 16129y_6^2 = 0.$$

To make the coefficient of x_6^3 equal to 1 and the coefficient of y_6^2 equal to 4, we set $x_7 = 20x_6$ and $y_7 = 2540y_6 = 4 \cdot 5 \cdot 127y_6$ and obtain

$$C : 256000x_7 + 3936x_7^2 - x_7^3 + 12000y_7 + 124x_7y_7 + 4y_7^2 = 0.$$

Finally, we complete the square in y_7 by setting

$$x = x_7 \quad \text{and} \quad y = 2y_7 + 31x_7 + 3000,$$

which puts C into Weierstrass form,[1]

$$C : y^2 = x^3 - 2975x^2 - 70000x + 9000000.$$

Tracing through all of the substitutions, we find that the transformation taking the original equation

$$C : X^3 + 2Y^3 + 4Z^3 - 7XYZ = 0$$

to the Weierstrass equation is given by the formulas

$$x = \frac{100(33X + 40Y + 54Z)}{4X + Y - 5Z},$$
$$y = \frac{-63500(6X^2 - 7XY - 18Y^2 + 21XZ - 14YZ + 12Z^2)}{(4X + Y - 5Z)^2}.$$

---

[1]The further substitution $(x, y) = (25x_0, 125y_0)$ gives an equation with smaller integer coefficients, $y_0^2 = x_0^3 - 119x_0^2 - 112x_0 + 576$.

List of Notation

$P * Q$	third intersection of $\overleftrightarrow{PQ}$ and cubic curve,	9
$\mathcal{O}$	specified base point on cubic curve,	11
$+$	addition on a cubic curve,	12
$-P$	inverse of point on cubic curve,	13
$C(\mathbb{Q})$	the rational points on a cubic curve,	39
$C(\mathbb{R})$	the real points on a cubic curve,	39
$C(\mathbb{C})$	the complex points on a cubic curve,	39
g_2, g_3	Eisenstein series, coefficients of Weierstrass equation,	41
$\wp$	Weierstrass $\wp$-function,	41
$\operatorname{ord}(x)$	the p-adic order of a rational number x,	48
$C(p^\nu)$	points on C with specified p-divisibility,	49
R, R_p	ring of integers localized at p,	50
$\operatorname{Disc}(f)$	discriminant of the polynomial f,	60
$H(P)$	height of a point on a cubic curve,	66
$h(P)$	height of a point on a cubic curve,	66
Γ	notation for $C(\mathbb{Q})$,	80
$\overline{C}$	curve with degree two map $C \to \overline{C}$,	81
$\overline{\overline{C}}$	curve with degree two map $\overline{C} \to \overline{\overline{C}}$,	81
$\phi(\Gamma)$	the image of Γ by ϕ,	89
$\mathbb{Q}^*$	the multiplicative group of non-zero rational numbers,	91
$\mathbb{Q}^{*2}$	the group of squares of elements of $\mathbb{Q}^*$,	91
α	a homomorphism $\Gamma \to \mathbb{Q}^*/\mathbb{Q}^{*2}$,	91
$\mathbb{Z}$	the group of integers,	95
$\mathbb{Z}_m$	the cyclic group $\mathbb{Z}/m\mathbb{Z}$,	95
$\Gamma[2]$	subgroup of Γ of points of order dividing two,	96
C_{ns}	non-singular points on a singular cubic curve,	107
$\mathbb{F}_q$	finite field with q elements,	117
$C(\mathbb{F}_p)$	the points on the curve C with coordinates in $\mathbb{F}_p$,	118
R, S, T	cosets of cubes in $\mathbb{F}_p^*$,	123
$[XYZ]$	number of solutions of $x + y + z = 0$,	123
ζ	a p'th root of unity,	125
θ_p	angle whose cosine is $(M_p - p - 1)/2\sqrt{p}$,	132

© Springer International Publishing Switzerland 2015

315

J.H. Silverman, J.T. Tate, *Rational Points on Elliptic Curves*,
Undergraduate Texts in Mathematics, DOI 10.1007/978-3-319-18588-0

References

[1] A. Baker, Contributions to the theory of Diophantine equations. II. The Diophantine equation $y^2 = x^3 + k$. Philos. Trans. R. Soc. Lond. Ser. A **263**, 193–208 (1967/1968)

[2] A. Baker, *Transcendental Number Theory*. Cambridge Mathematical Library, 2nd edn. (Cambridge University Press, Cambridge, 1990)

[3] T. Barnet-Lamb, D. Geraghty, M. Harris, R. Taylor, A family of Calabi-Yau varieties and potential automorphy II. Publ. Res. Inst. Math. Sci. **47**(1), 29–98 (2011)

[4] G. Billing, K. Mahler, On exceptional points on cubic curves. J. Lond. Math. Soc. **15**, 32–43 (1940)

[5] B.J. Birch, How the number of points of an elliptic curve over a fixed prime field varies. J. Lond. Math. Soc. **43**, 57–60 (1968)

[6] A. Bremner, J.W.S. Cassels, On the equation $Y^2 = X(X^2 + p)$. Math. Comput. **42**(165), 257–264 (1984)

[7] C. Breuil, B. Conrad, F. Diamond, R. Taylor, On the modularity of elliptic curves over **Q**: wild 3-adic exercises. J. Am. Math. Soc. **14**(4), 843–939 (electronic) (2001)

[8] E. Brieskorn, H. Knörrer, *Plane Algebraic Curves*. Modern Birkhäuser Classics (Birkhäuser/Springer Basel AG, Basel, 1986). Translated from the German original by John Stillwell, [2012] reprint of the 1986 edition

[9] J.S. Chahal, *Topics in Number Theory*. The University Series in Mathematics (Plenum Press, New York, 1988)

© Springer International Publishing Switzerland 2015   317
J.H. Silverman, J.T. Tate, *Rational Points on Elliptic Curves*,
Undergraduate Texts in Mathematics, DOI 10.1007/978-3-319-18588-0

[10] G. Chenevier, M. Harris, Construction of automorphic Galois representations, II. Camb. J. Math. **1**(1), 53–73 (2013)

[11] L. Clozel, M. Harris, J.-P. Labesse, B.-C. Ngô (eds.) *On the Stabilization of the Trace Formula*. Stabilization of the Trace Formula, Shimura Varieties, and Arithmetic Applications, vol. 1 (International Press, Somerville, 2011)

[12] L. Clozel, M. Harris, R. Taylor, Automorphy for some l-adic lifts of automorphic mod l Galois representations. Publ. Math. Inst. Hautes Études Sci. **108**, 1–181 (2008). With Appendix A, summarizing unpublished work of Russ Mann, and Appendix B by Marie-France Vignéras

[13] P. Deligne, La conjecture de Weil. I. Inst. Hautes Études Sci. Publ. Math. **43**, 273–307 (1974)

[14] D.S. Dummit, R.M. Foote, *Abstract Algebra*, 3rd edn. (Wiley, Hoboken, 2004)

[15] G. Faltings, Diophantine approximation on abelian varieties. Ann. Math. (2) **133**(3), 549–576 (1991)

[16] W. Fulton, *Algebraic Curves*. Advanced Book Classics (Addison-Wesley, Advanced Book Program, Redwood City, 1989). An introduction to algebraic geometry, Notes written with the collaboration of Richard Weiss, Reprint of 1969 original

[17] P. Griffiths, J. Harris, *Principles of Algebraic Geometry*. Wiley Classics Library (Wiley, New York, 1994). Reprint of the 1978 original

[18] J. Harris, *Algebraic Geometry*. Graduate Texts in Mathematics, vol. 133 (Springer, New York, 1992). A first course

[19] M. Harris, N. Shepherd-Barron, R. Taylor, A family of Calabi-Yau varieties and potential automorphy. Ann. Math. (2) **171**(2), 779–813 (2010)

[20] R. Hartshorne, *Algebraic Geometry*. Graduate Texts in Mathematics, vol. 52 (Springer, New York/Heidelberg, 1977)

[21] H. Hasse, Beweis des Analogons der Riemannschen Vermutung für die Artinschen und F.K. Schmidtschen Kongruenzzetafunktionen in gewissen elliptischen Fällen. Nachr. Ges. Wiss. Göttingen, Math.-Phys. K. 253–262 (1933)

[22] D.R. Heath-Brown, S.J. Patterson, The distribution of Kummer sums at prime arguments. J. Reine Angew. Math. **310**, 111–130 (1979)

[23] I.N. Herstein, *Topics in Algebra*, 2nd edn. (Xerox College Publishing, Lexington/Toronto, 1975)

[24] J. Hoffstein, J. Pipher, J.H. Silverman, *An Introduction to Mathematical Cryptography*. Undergraduate Texts in Mathematics, 2nd edn. (Springer, New York, 2014)

[25] D. Husemöller, *Elliptic Curves*. Graduate Texts in Mathematics, vol. 111, 2nd edn. (Springer, New York, 2004). With appendices by Otto Forster, Ruth Lawrence and Stefan Theisen

[26] N. Jacobson, *Basic Algebra. I, II* (W. H. Freeman and Company, New York, 1985/1989)

[27] N. Koblitz, *A Course in Number Theory and Cryptography*. Graduate Texts in Mathematics, vol. 114, 2nd edn. (Springer, New York, 1994)

[28] E. Kummer, De residuis cubicis disquisitiones nonnullae analyticae. J. Reine Angew. Math. **32**, 341–359 (1846)

[29] S. Lattès, Sur l'iteration des substitutions rationelles et les fonctions de Poincaré. Comptes Rendus Acad. Sci. Paris **166**, 26–28 (1918)

[30] H.W. Lenstra Jr., Factoring integers with elliptic curves. Ann. Math. (2) **126**(3), 649–673 (1987)

[31] J.M. Luck, P. Moussa, M. Waldschmidt (eds.), *Number Theory and Physics*. Springer Proceedings in Physics, vol. 47 (Springer, Berlin, 1990)

[32] B. Mazur, Modular curves and the Eisenstein ideal. Inst. Hautes Études Sci. Publ. Math. **47**, 33–186 (1978/1977)

[33] B. Mazur, Rational isogenies of prime degree (with an appendix by D. Goldfeld). Invent. Math. **44**(2), 129–162 (1978)

[34] J. Milnor, On Lattès maps. Dynamics on the Riemann sphere, 9–43, Eur. Math. Soc., Zürich (2006)

[35] L.J. Mordell, On the rational solutions of the indeterminate equations of the third and fourth degrees, Math. Proc. Cambridge Philos. Soc. **21**, 179–192 (1922)

[36] J.M. Pollard, Theorems on factorization and primality testing. Proc. Camb. Philos. Soc. **76**, 521–528 (1974)

[37] M. Reid, *Undergraduate Algebraic Geometry*. London Mathematical Society Student Texts, vol. 12 (Cambridge University Press, Cambridge, 1988)

[38] K.A. Ribet, On modular representations of $Gal(\overline{Q}/Q)$ arising from modular forms. Invent. Math. **100**(2), 431–476 (1990)

[39] W.M. Schmidt, Simultaneous approximation to algebraic numbers by rationals. Acta Math. **125**, 189–201 (1970)

[40] W.M. Schmidt, *Diophantine Approximation*. Lecture Notes in Mathematics, vol. 785 (Springer, Berlin, 1980)

[41] J.-P. Serre, *Abelian l-Adic Representations and Elliptic Curves*. McGill University Lecture Notes Written with the Collaboration of Willem Kuyk and John Labute (W. A. Benjamin, New York/Amsterdam, 1968)

[42] J.-P. Serre, Propriétés galoisiennes des points d'ordre fini des courbes elliptiques. Invent. Math. **15**(4), 259–331 (1972)

[43] J.-P. Serre, *Linear Representations of Finite Groups*. Graduate Texts in Mathematics, vol. 42 (Springer, New York/Heidelberg, 1977). Translated from the second French edition by Leonard L. Scott

[44] S.W. Shin, Galois representations arising from some compact Shimura varieties. Ann. Math. (2) **173**(3), 1645–1741 (2011)

[45] C.L. Siegel, The integer solutions of the equation $y^2 = ax^n + bx^{n-1} + \cdots + k$. J. Lond. Math. Soc. (2) **1**, 66–68 (1926)

[46] C.L. Siegel, Über einige Anwendungen diophantischer Approximationen (1929), in *Collected Works* (Springer, 1966), pp. 209–266

[47] J.H. Silverman, Integer points and the rank of Thue elliptic curves. Invent. Math. **66**(3), 395–404 (1982)

[48] J.H. Silverman, *Advanced Topics in the Arithmetic of Elliptic Curves*. Graduate Texts in Mathematics, vol. 151 (Springer, New York, 1994)

[49] J.H. Silverman, *The Arithmetic of Elliptic Curves*. Graduate Texts in Mathematics, vol. 106, 2nd edn. (Springer, Dordrecht, 2009)

[50] T. Skolem, *Diophantische Gleichung* (Springer, Berlin, 1938)

[51] J. Tate, A review of non-Archimedean elliptic functions, in *Elliptic Curves, Modular Forms, & Fermat's Last Theorem, Hong Kong, 1993.* Series Number Theory, vol. I (International Press, Cambridge, 1995), pp. 162–184

[52] R. Taylor, Automorphy for some l-adic lifts of automorphic mod l Galois representations. II. Publ. Math. Inst. Hautes Études Sci. **108**, 183–239 (2008)

[53] R. Taylor, A. Wiles, Ring-theoretic properties of certain Hecke algebras. Ann. Math. (2) **141**(3), 553–572 (1995)

[54] A. Thue, Über annäherungswerte Algebraischer Zahlen. J. Reine Angew. Math. **135**, 284–305 (1909)

[55] P. Vojta, *Diophantine Approximations and Value Distribution Theory.* Lecture Notes in Mathematics, vol. 1239 (Springer, Berlin, 1987)

[56] P. Vojta, Siegel's theorem in the compact case. Ann. Math. (2) **133**(3), 509–548 (1991)

[57] R.J. Walker, *Algebraic Curves* (Springer, New York, 1978). Reprint of the 1950 edition

[58] A. Weil, *Sur les courbes algébriques et les variétés qui s'en déduisent.* Actualités Sci. Ind., no. 1041 = Publ. Inst. Math. Univ. Strasbourg **7** (1945). Hermann et Cie., Paris, 1948

[59] A. Weil, Numbers of solutions of equations in finite fields. Bull. Am. Math. Soc. **55**, 497–508 (1949)

[60] A. Wiles, Modular elliptic curves and Fermat's last theorem. Ann. Math. (2) **141**(3), 443–551 (1995)

Index